Mathematical Biology

Mathematical Biology

Edited by Fiona Palmer

SYRAWOOD
PUBLISHING HOUSE

New York

Published by Syrawood Publishing House,
750 Third Avenue, 9th Floor,
New York, NY 10017, USA
www.syrawoodpublishinghouse.com

Mathematical Biology
Edited by Fiona Palmer

International Standard Book Number: 978-1-64740-099-6 (Hardback)

Cataloging-in-Publication Data

Mathematical biology / edited by Fiona Palmer.
     p. cm.
Includes bibliographical references and index.
ISBN 978-1-64740-099-6
1. Biology--Mathematical models. 2. Biomathematics. I. Palmer, Fiona.
QH323.5 .M38 2022
570.151 95--dc23

TABLE OF CONTENTS

PREFACE

The branch of biology which employs mathematical models and abstractions of the living organisms is known as mathematical biology. These models and abstractions are employed for the purpose of investigating the principles which govern the development, behavior and structure of the systems. This discipline uses techniques and tools of applied mathematics for the modeling and mathematical representation of biological processes. Mathematical models are primarily used for describing systems in a quantitative manner. This facilitates the simulation of their behavior which in turn helps in predicting their properties, which might not have been visible otherwise. Some of the various areas of research under this discipline are abstract relational biology, algebraic biology, complex systems biology and evolutionary biology. While understanding the long-term perspectives of the topics, the book makes an effort in highlighting their impact as a modern tool for the growth of the discipline. The various studies that are constantly contributing towards advancing technologies and evolution of mathematical biology are examined in detail. This book will serve as a reference to a broad spectrum of readers.

This book is a result of research of several months to collate the most relevant data in the field.

When I was approached with the idea of this book and the proposal to edit it, I was overwhelmed. It gave me an opportunity to reach out to all those who share a common interest with me in this field. I had 3 main parameters for editing this text:

1. Accuracy – The data and information provided in this book should be up-to-date and valuable to the readers.

2. Structure – The data must be presented in a structured format for easy understanding and better grasping of the readers.

3. Universal Approach – This book not only targets students but also experts and innovators in the field, thus my aim was to present topics which are of use to all.

Thus, it took me a couple of months to finish the editing of this book.

I would like to make a special mention of my publisher who considered me worthy of this opportunity and also supported me throughout the editing process. I would also like to thank the editing team at the back-end who extended their help whenever required.

Editor

Descriptor-based Fitting of Lysophosphatidic Acid Receptor 3 Antagonists into a Single Predictive Mathematical Model

Olaposi Idowu Omotuyi, Hiroshi Ueda
Department of Pharmacology and Therapeutic Innovation
University Graduate School of Biomedical Sciences, 852-8521
Nagasaki, Japan
Email: bbis11r104@cc.nagasaki-u.ac.jp

Abstract—Sixty six diverse compounds previously reported as Lysophosphatidic Acid Receptor (LPA$_3$) inhibitors have been used to derive a mathematical model based on partial least square (PLS) clustering of 41 molecular descriptors and pIC$_{50}$ values. The pre- and post- cross-validated correlation coefficient (R^2) is 0.94462 (RMSE=0.21390) and 0.74745 (RMSE=0.49055) respectively. Bivariate contingency analysis tools implemented in MOE was used to prune the descriptors and refit the equations at a descriptor-pIC$_{50}$ correlation coefficient of 0.8 cut-off. A new equation was derived with R^2 and RMSE values estimated at 0.88074 and 0.31388 respectively. Both equations correctly predicted the 95% of the pIC$_{50}$ values of the test dataset. Principal component analysis (PCA) was also used to reduce the dimension and linearly transform the raw data; 8 principal components sufficiently account for more than 98% of the variance of the dataset. The numerical model derived here may be adapted for screening chemical database for LPA$_3$ antagonism.

Keywords-upscaling; LPA$_3$; LPA$_3$ antagonists; Mathematical Model; PCA; Molecular descriptors

I. Introduction

Quantitative structure activity relationship (QSAR) allows statistical analysis of experimental data and building of predictive mathematical models from the dataset. The numerical models built using this approach has been successfully implemented in screening of large database of chemical compounds for hit-compound detection [1]. In the presence of experimental dataset [2], the success of QSAR depends on two key factors: array of descriptors that optimally represent the structural parameters required for molecular interaction or reactions [3] and an appropriate statistical learning and validation algorithms [4]. In practice, physical properties descriptors (1D-descriptor), pharmacophore descriptors (2D-descriptors) and geometrical descriptors (3D-descriptors, often requires prior knowledge of target protein binding-pocket) are the most commonly used descriptor types for QSAR modeling [5,6,7]. We seek to answer a single question here, what combination of molecular predictors would numerically and accurately predict the experimental antagonist activities of LPA$_3$ inhibitors? When answered, the mathematical relationship derived from the descriptors will enable screening of chemical databases for compounds exhibiting LPA$_3$ antagonism required for the treatment of diseased conditions such as ovarian cancer [8] and neuropathic pain [9] with LPA$_3$ etiology.

II. STATISTICAL BASIS OF QSAR MODELING USING PARTIAL LEAST SQUARE METHOD

The QSAR/PLS modeling equations and algorithms have been well described in MOE documentations [10]. Given m molecules of a training dataset, suppose that each of the molecules is described by an n-vector of descriptors $x_i = (x_{i1}, ..., x_{in})$, for one of the molecules denoted as i. Let y_i be a representation of the experimental result (pIC_{50}) for a molecule i. A linear model for y (the experimental result) is given by Eq. (1) [11].

$$y = a_0 + a^T X, \qquad (1)$$

where a_0 is a scalar, and a^T is a n-vector. If each molecule has an importance weight (non-negative) w representing the relative probability that the associated molecule will be encountered, and that the sum of all the weights are designated as W. The mean square error is given as Eq. (2) [12].

$$MSE_{a_0, a} = \frac{1}{w} \sum_{i=1}^{m} [y_i - (y = a_0 + a^T X_i)]^2. \quad (2)$$

Differentiating MSE with respect to the parameters satisfying the normal Eqs (3,4,5,6 &7) solvable by matrix diagonalization:

$$a_0 = y_0 - a^T X_i, \qquad (3)$$

$$y_0 = \frac{1}{w} \sum_{i=1}^{m} [w_i y_i], \qquad (4)$$

$$x_0 = \frac{1}{w} \sum_{i=1}^{m} [w_i x_i], \qquad (5)$$

$$Sa = b = \frac{1}{W} \sum_{i=1}^{m} [w_i y_i (x_i - x_0)], \qquad (6)$$

$$S = \frac{1}{w} \sum_{i=1}^{m} [w_i (x_i - x_0)(x_i - x_0)^T]. \quad (7)$$

Starting from the normal equations above, an estimate of a can be computed if columns of the weight matrix (G_A) (Eq. (8)) is obtained through Gram-Schmidt orthogonalization [13] of the vectors generated by Krylov sequence $b, Sb, S^2 b, ..., S^{A-1} b$ [14]. The Ath PLS coefficient vector is then estimated using Eq. (9).

$$G_A = (g_i, g_2, \ldots, g_A). \qquad (8)$$

$$a = G_A (G_A^T S G_A)^{-1} G_A^T b. \qquad (9)$$

Noting that g_i is the column vectors of length n and A is the degree of the PLS fit; an integer less than or equals n. MOE [10] descriptor calculator was used to generate the numerical representations (a_aro, ASA, ASA_H, a_hyd, SlogP, SlogP_VSA0, SlogP_VSA1, SlogP_VSA2, SlogP_VSA3, SlogP_VSA4, SlogP_VSA5, SlogP_VSA6, SlogP_VSA7, SlogP_VSA8, SlogP_VSA9, SMR_VSA0, SMR_VSA1, SMR_VSA2, SMR_VSA3, SMR_VSA4, SMR_VSA5, SMR_VSA6, SMR_VSA7, a_acc, Kier1, Kier2, Kier3, KierA1, KierA2, KierA3, KierFlex, chi0, chi0v, chi0v_C, chi0_C, chi1, chi1v, chi1v_C, chi1_C, chiral, chiral_u) of the 66 (Supplementary fig. 1) randomly selected LPA$_3$ antagonists retrieved from the European Institute of Bioinformatics dataset (https://www.ebi.ac.uk/chembl/) representing our training dataset (CHEMBL3250). Using the PLS method as described above, Eq. (10) was generated relating the descriptors to the pIC_{50} with a correlation coefficient (R^2) 0.94462 ($RMSE = 0.21390$) (Fig. 1, blue circles and line); when cross validated, R^2 was estimated as 0.74745 ($RMSE = 0.49055$).

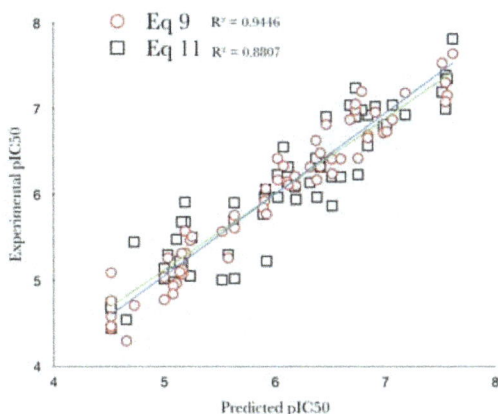

Fig. 1: Scatter plot of the experimental pIC_{50} vs. pIC_{50}-predictions of Eq. (10) (blue) and Eq. (12) (green).

$$pIC_{50} =$$
$$3.57363 - 0.25353 \cdot a\_aro - 0.00361 \cdot ASA$$
$$+ 0.23510 \cdot a\_hyd + 0.05890 \cdot SlogP$$
$$- 0.02287 \cdot SlogP\_VSA0$$
$$+0.00032 \cdot SlogP\_VSA1 + 0.03125 \cdot SlogP\_VSA2$$
$$-0.02059 \cdot SlogP\_VSA3 + 0.02954 \cdot SlogP\_VSA4$$
$$+0.07226 \cdot SlogP\_VSA5 + 0.02879 \cdot SlogP\_VSA6$$
$$+0.04687 \cdot SlogP\_VSA7 + 0.03836 \cdot SlogP\_VSA8$$
$$+0.06880 \cdot SlogP\_VSA9 + 0.04912 \cdot SMR\_VSA0$$
$$+0.02536 \cdot SMR\_VSA1 + 0.08743 \cdot SMR\_VSA2$$
$$+0.00289 \cdot SMR\_VSA3 - 0.01524 \cdot SMR\_VSA4$$
$$+0.04694 \cdot SMR\_VSA5 + 0.09067 \cdot SMR\_VSA6$$
$$- 0.01442 \cdot SMR\_VSA7 + 0.18393 \cdot a\_acc$$
$$- 0.77650 \cdot Kier1 - 0.43968 \cdot Kier2$$
$$- 0.30735 \cdot Kier3 - 0.43752 \cdot KierA1$$
$$- 0.03578 \cdot KierA2 + 0.76916 \cdot KierA3$$
$$- 0.09573 \cdot KierFlex + 0.00332 \cdot chi0$$
$$+ 0.55223 \cdot chi0v + 0.13554 \cdot chi0v\_C$$
$$- 0.16530 \cdot chi0\_C + 0.59498 \cdot chi1$$
$$+ 0.05911 \cdot chi1v - 0.93262 \cdot chi1v\_C$$
$$- 1.22808 \cdot chi1\_C - 0.16986 \cdot chiral$$
$$- 0.56204 \cdot chiral\_u. \qquad (10)$$

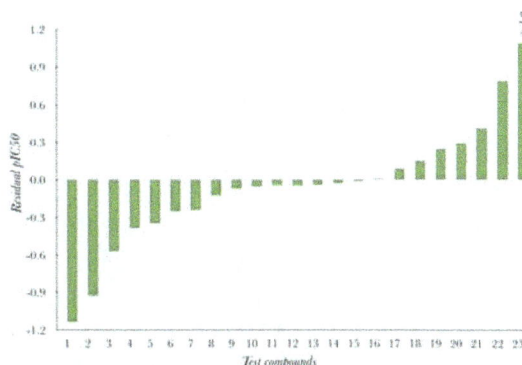

Fig. 2: Bar chart representations of the residual (Experimental pIC_{50}-Predicted pIC_{50} values of the test dataset. Only 1 out of tested compounds (compound 23, see supplementary Fig. 2 for structural details) showed > 1.0 pIC_{50} unit (indication of wrong prediction).

Noting that root mean square error (RMSE) is the square root of MSE function (Eq. (2)) at a given parameter value and the correlation coefficient (R^2) is 1-MSE/YVAR with values raging between 0 and 1 (0= no fit, 1 is perfect fit and YVAR is the sample variance of the y_i values). The predictive suitability of our equation was tested on 23 compounds (Supplementary Fig. 2) with experimentally determined IC_{50} for LPA$_3$ antagonism. If we assume that residual value above 1.0 pIC_{50} unit represents poor fitting. Our data (Fig. 3) suggest that Eq. (10) accurately predicted 22 of the 23 test compounds.

III. DESCRIPTOR CONTINGENCY ANALYSIS

To determine the level of significance of each of the descriptors to the overall equation and we performed contingency analysis. The data presented here provides a window of decision on whether pruning of the descriptor set is required. In MOE [10], QSAR-contingency tool performs a bivariate contingency analysis for each descriptor and the experimental activity value and produces a table of correlation coefficients (Eq. (11)) for each descriptor given that X represents a randomly selected molecular descriptor and Y is a randomly selected activity value for a randomly selected sample m, $Var(X)$ and $Var(Y)$, then the covariance of the random variables X and Y is defined to be $Cov(X, Y) = E(XY) - E(X)E(Y)$ [10, 15].

$$R^2 = \frac{[E(XY) - E(X)E(Y)]^2}{Var(X)Var(Y)} . \qquad (11)$$

Given that the values of R^2 ranges from 0 to 1, and 1 represents a perfectly linear correlation, we therefore proposed that only descriptors R^2 values ≥ 0.8 are useful and that the descriptors outside this range can be pruned. Our data suggest that 31 out of the original 41 descriptors have R^2 values ≥ 0.8 (Fig. 3, Supplementary Table 1). With the exclusion of the descriptors with unsatisfactory coefficient, QSAR is re-calculated using the residual set of descriptors. New numerical relationship was generated (Eq. (12)) with R^2 (0.88074) and $RMSE$ values (0.31388). The scatter plot of the predicted pIC_{50} and the experimental values for the new Eq. (12) is given in Fig. 1 (green circles and line).

$lpIC_{50} =$

$2.23199 - 0.00516xASA - 0.00516xASA\_H$

$- 0.48596xa\_hyd - 0.33917xSlogP$

$-0.05298xSlogP_V SA0 - 0.03967xSlogP_V SA1$

$-0.02243xSlogP_V SA2 + 0.01681xSlogP_V SA7$

$+ 0.02107xSlogP_V SA9$

$-0.00757xSMR_V SA0 - 0.00087xSMR_V SA1$

$- 0.00089xSMR_V SA3$

$-0.01173xSMR_V SA4 + 0.00955xSMR_V SA5$

$- 0.01412xSMR_V SA6$

$- 0.02508xSMR_V SA7 - 0.26771xKier1$

$+ 0.15306xKier20.56650xKier3$

$- 0.30504xKierA2 + 0.98837xKierA3$

$- 0.28849xKierFlex + 0.48535xchi0$

$+ 0.90693xchi0v + 0.10234xchi0v_C$

$+ 0.24407xchi0_C + 0.66154xchi1$

$+ 0.36006xchi1v - 1.03589xchi1v_C$

$- 0.62474xchi1_C - 0.36725xa_aro.$ $\qquad (12)$

When this equation was used for predicting the

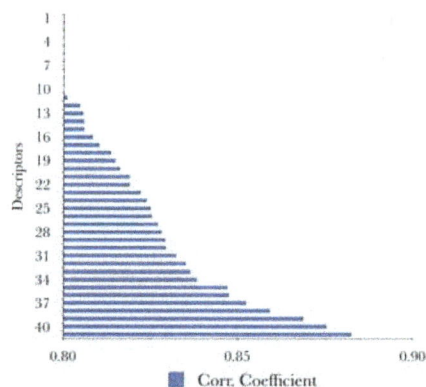

Fig. 3: Bar chart representations of Descriptor-experimental pIC_{50} correlation coefficient. Only 31 out of 41 descriptors lie above 0.8 coefficient cutoff.

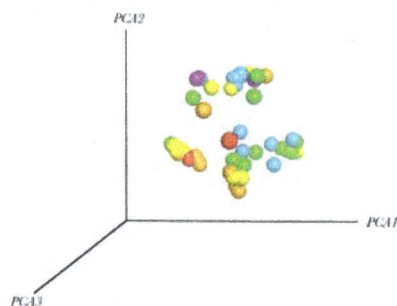

Fig. 4: The $3D$ plot of the first three principal components. Each point represents a compound in the training dataset and each colour represents a distinct cluster of pIC_{50} values.

pIC_{50} values of the test set, only one compound lies above the 1.0 pIC_{50} unit cutoff (data not shown). Thus, Eq. (12) is less bulky and as accurate as Eq. (10) in predicting LPA$_3$ antagonism.

IV. PRINCIPAL COMPONENT ANALYSIS OF EQUATION

We sought to further study the dataset descriptors along the principle components through the reduction of the dimensionality and linear transformation of the raw data [13]. Given the initial 66 training dataset compounds (represented as m) and for one of the compounds say i its descriptors are represented by n-vector of real numbers $x_i =$

$(x_{i1}, ..., x_{in})$, where $n = 1 - 31$, new Eq. (12). Assuming that each molecule i has an associated importance weight w_i, (non-negative, real number) and that the weights is relative probability that the associated molecule x_i will be encountered (adding up to 1); If W denotes the sum of all the weights then, the eigenvalues and eigenvectors for the final data are estimable from the raw data using Eq. (1). If S is a symmetric, semi-definite sample covariance matrix, S can be diagonalized such that $S = Q^T DDQ$ (Q is orthogonal, D is diagonal-sorted in descending order from top left to bottom right) [13, 14].

$$E(x) \approx \overline{x} = x_0 = \frac{1}{w} \sum_{i=1}^{m} [w_i x_i] \qquad (13)$$

$$Cov(x) \approx S = \frac{1}{w} \sum_{i=1}^{m} [w_i x_i x_i^T - \overline{x}\overline{x}^T]. \qquad (14)$$

The effect of the each of the principal components (eigenvectors) on the condition and the variance shows that nine (8) principal components sufficiently accounts for more than 98% of the variance in the dataset [15]. The $3D$-scatter plot of the first three principal components (PCA1, PCA2 and PCA3) with respect to pIC_{50} values is shown in Fig. (4); each point in the plot corresponds to a dataset molecule colored according to clustered pIC_{50} values.

V. Conclusion

Given the good mathematical correlation between the set of descriptors and LPA$_3$ antagonism, it is not unusual to propose that the equation is prejudiced for those set of compounds with highly related descriptor properties and therefore may not be a universal formula for LPA$_3$ antagonist screening. That said, it will however capture the compounds with structural properties found within the dataset accurately and therefore may be piped as into ligand-based screening protocol for more successful hit-compound identification.

Acknowledgment

This work was supported by Platform for Drug Discovery, Informatics, and Structural Life Science from the Ministry of Education, Culture, Sports, Science and Technology, Japan.

Appendix

Supplementary Table 1.0 Showing Correlation coefficient of each Descriptor

S/N	Desciptors	Corr. Coefficient
1	SlogP_VSA6	0.57623
2	chiral_u	0.65734
3	SlogP_VSA4	0.66609
4	SlogP_VSA5	0.6996
5	chiral	0.72218
6	SMR_VSA2	0.76566
7	SlogP_VSA8	0.76621
8	a_acc	0.76922
9	SlogP_VSA3	0.79094
10	KierA1	0.79264
11	a_aro	0.80122
12	SlogP_VSA9	0.80481
13	SlogP_VSA1	0.80675
14	chi0_C	0.806
15	chi1v	0.80603
16	KierFlex	0.80836
17	chi1v_C	0.81041
18	KierA3	0.81376
19	SlogP_VSA2	0.81493
20	SMR_VSA7	0.81623
21	ASA	0.81908
22	ASA_H	0.81908
23	chi0v	0.82223
24	chi0v_C	0.82394
25	SMR_VSA4	0.82512
26	chi1_C	0.82535
27	KierA2	0.82725
28	chi0	0.82827
29	SlogP_VSA7	0.82933
30	SMR_VSA5	0.82941
31	Kier2	0.83257
32	SlogP_VSA0	0.83519
33	Kier1	0.83644
34	SMR_VSA1	0.83639
35	chi1	0.84721
36	SMR_VSA6	0.84762
37	SMR_VSA3	0.8525
38	Kier3	0.85924
39	SlogP	0.86686
40	SMR_VSA0	0.87545
41	a_hyd	0.88264

Scatter plot of the experimental pIC_{50} vs. pIC_{50}-predictions of Eq.(10) (blue) and Eq. (12).

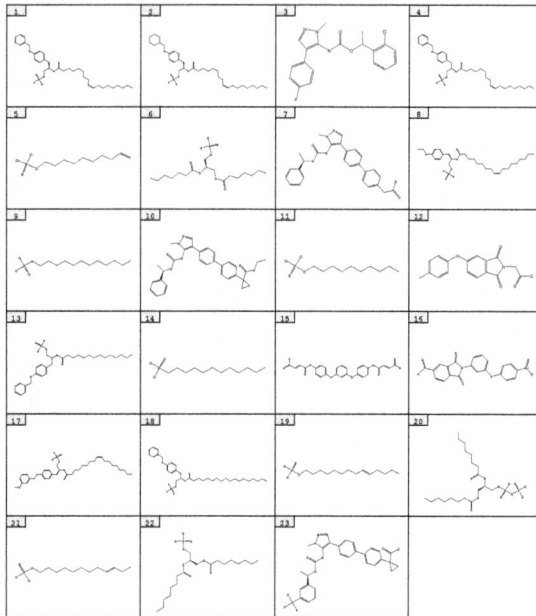

LPA3 INIHIBITORS: TRAINING SET FOR QSAR MODELING

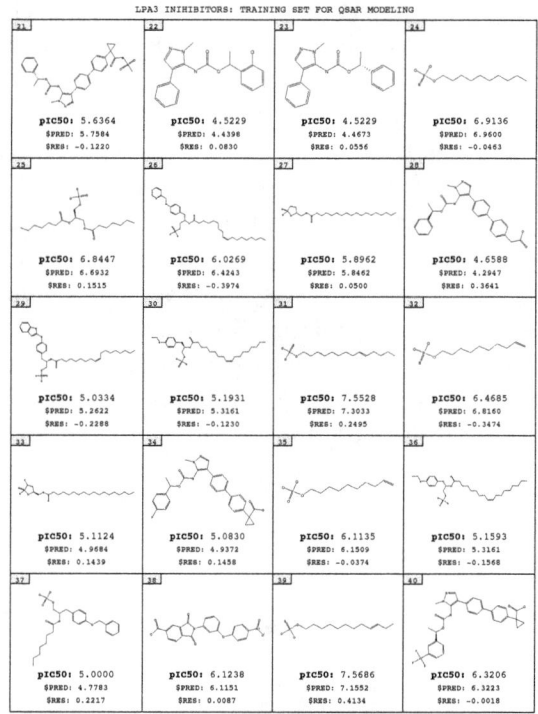

LPA3 INIHIBITORS: TRAINING SET FOR QSAR MODELING

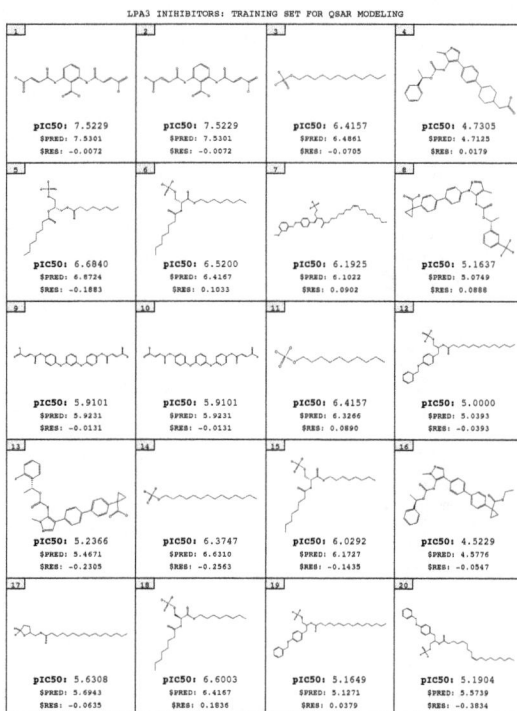

LPA3 INIHIBITORS: TRAINING SET FOR QSAR MODELING

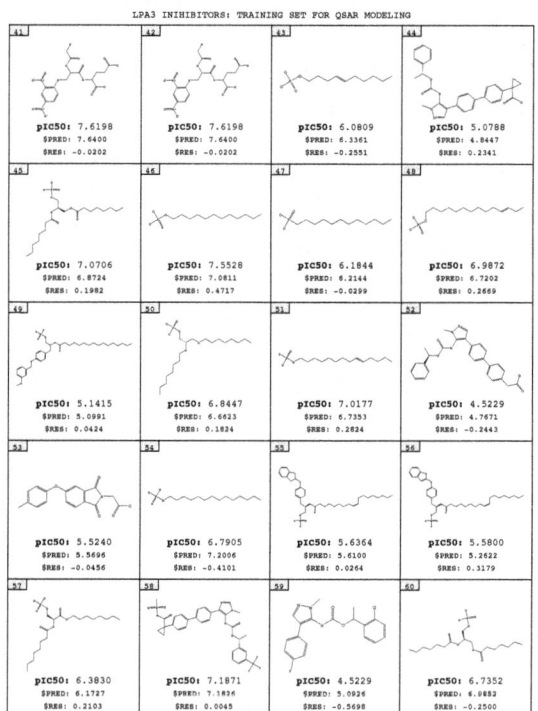

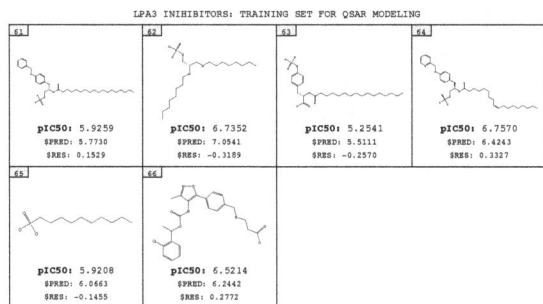

LPA3 INHIBITORS: TRAINING SET FOR QSAR MODELING

REFERENCES

[1] A.M. Helguera, A. Prez-Garrido, A. Gaspar, J. Reis, F. Cagide, D. Vina, M. Cordeiro, F. Borges, Combining QSAR classification models for predictive modeling of human monoamine oxidase inhibitors, Eur J Med Chem. 2013 ;59:75-90.
http://dx.doi.org/10.1016/j.ejmech.2012.10.035

[2] P.P. Roy, J.T. Leonard, K. Roy, Exploring the impact of size of training sets for the development of predictive QSAR models. Chemometrics and Intelligent Laboratory Systems 90 2008 (1): 31-42.

[3] R. Todeschini, V. Consonni. "Molecular Descriptors for Chemoinformatics" (2 volumes), 2009 Wiley-VCH.
http://dx.doi.org/10.1002/9783527628766

[4] T. Scior, J.L. Medina-Franco, QT. Do, K. Martnez-Mayorga, J.A. Yunes-Rojas, P. Bernard. "How to recognize and workaround pitfalls in QSAR studies: a critical review". Curr Med Chem. 2009; 16 (32):4297-313.

[5] B.K. Shoichet, I.D. Kuntz, D.L. Bodian. "Molecular docking using shape descriptors". Journal of Computational Chemistry 13; 2004 (3): 380-397

[6] R.J. Morris, J. Najmanovich, A. Kahraman, J.M. Thornton. "Real spherical harmonic expansion coefficients as 3D shape descriptors for protein binding pocket and ligand comparisons". Bioinformatics 21; 2005 (10): 2347-55.

[7] B.B. Goldman, W.T. Wipke. "QSD quadratic shape descriptors. Molecular docking using quadratic shape descriptors (QSDock)". Proteins 38; 2000 (1): 79-94.

[8] P. Wang, X.H. Wu, W.X. Chen, B.E. Shan, Q. Guo. "Expression of lysophosphatidic acid receptor in human ovarian cancer cell lines 3AO, SKOV3, OVCAR3 and its significance" Di Yi Jun Yi Da Xue Xue Bao. 2005 25(11):1422-4, 1431.

[9] H. Ueda, H. Matsunaga, O.I. Omotuyi, J. Nagai, "Lysophosphatidic acid: chemical signature of neuropathic pain". Biochim Biophys Acta. 2013; 1831(1):61-73.http://dx.doi.org/10.1016/j.bbalip.2012.08.014

[10] Molecular Operating Environment (MOE), 2012.10; Chemical Computing Group Inc., 1010 Sherbooke St. West, Suite 910, Montreal, QC, Canada, H3A 2R7, 2012.

[11] M.J. Wichura, "The coordinate-free approach to linear models". Cambridge Series in Statistical and Probabilistic Mathematics. Cambridge: Cambridge University Press. pp. xiv+199. ISBN 978-0-521-86842-6. 2006. MR 2283455

[12] D. Wackerly, W. Scheaffer. "Mathematical Statistics with Applications" (7 ed.). Belmont, CA, USA: Thomson Higher Education. ISBN 0-49538508-5. 2008

2

Efficient Implicit Runge-Kutta Methods for Fast-Responding Ligand-Gated Neuroreceptor Kinetic Models

Edward T. Dougherty
Department of Mathematics
Rowan University
Glassboro, NJ, USA
Email: doughertye@rowan.edu

Abstract—Neurophysiological models of the brain typically utilize systems of ordinary differential equations to simulate single-cell electrodynamics. To accurately emulate neurological treatments and their physiological effects on neurodegenerative disease, models that incorporate biologically-inspired mechanisms, such as neurotransmitter signalling, are necessary. Additionally, applications that examine populations of neurons, such as multiscale models, can demand solving hundreds of millions of these systems at each simulation time step. Therefore, robust numerical solvers for biologically-inspired neuron models are vital. To address this requirement, we evaluate the numerical accuracy and computational efficiency of three L-stable implicit Runge-Kutta methods when solving kinetic models of the ligand-gated glutamate and γ-aminobutyric acid (GABA) neurotransmitter receptors. Efficient implementations of each numerical method are discussed, and numerous performance metrics including accuracy, simulation time steps, execution speeds, Jacobian calculations, and LU factorizations are evaluated to identify appropriate strategies for solving these models. Comparisons to popular explicit methods are presented and highlight the advantages of the implicit methods. In addition, we show a machine-code compiled implicit Runge-Kutta method implementation that possesses exceptional accuracy and superior computational efficiency.

Keywords-implicit Runge-Kutta; neuroreceptor model; numerical stiffness; ODE simulation

I. INTRODUCTION

Mathematical modeling and computational simulation provide an *in silico* environment for investigating cerebral electrophysiology and neurological therapies including neurostimulation. Traditionally, volume-conduction models have been used to emulate electrical potentials and currents within the head cavity. In particular, these models can reproduced electroencephalograph (EEG) surface potentials [1]–[3], and have been successful in predicting cerebral current density distributions from neurostimulation administrations [1], [4]–[7]. As these models become more refined, their utility in diagnosing, treating, and comprehending neurological disorders greatly increases.

Progress in field of computational neurology has motivated a migration towards models that incorporate cellular-level bioelectromagnetics. For example, bidomain based models have been used to simulate the effects of extracellular electrical current on cellular transmembrane voltage(s) [8]–[13]. In addition, multiscale models have reproduced EEG measurements originating from action potentials [14], [15], and have also demonstrated an ability to simulate the influence of transcranial electrical stimulation on neuronal depolarization [16].

These models typically utilize a system of ordinary differential equations (ODEs) to emulate cellular-level electrophysiology. While the computational expense of simulating a single cell is essentially negligible, this is not the case with large-scale applications that may include hundreds of millions of cells; in multiscale applications, solving this set of ODEs is the computational bottleneck [17]. In these applications, choosing an appropriate numerical solver and using efficient implementation approaches become paramount.

Alterations in neurotransmitter signalling is a hallmark of many neurodegenerative conditions and treatments. Parkinson's disease (PD), for example, which affects approximately one million individuals in the United States alone [18], culminates with pathological glutamate and γ-aminobutyric acid (GABA) binding activity throughout the basal ganglia-thalamocortical network [19], [20]. As a treatment for PD, deep brain stimulation (DBS) electrically stimulates areas of the basal ganglia, such as the subthalamic nucleus (STN) [21], to restore normal glutamate and GABA synaptic concentrations [22]–[24]. Therefore, models that incorporate fundamental neurotransmitter-based signalling provide utility to the neurological research community.

Models of metabotropic and slow-responding ligand-gated receptors, such as the $GABA_B$ and N-methyl-D-aspartate (NMDA) glutamate receptors, can be efficiently solved with explicit Runge-Kutta (ERK) methods [25]. On the contrary, fast-responding ionotropic receptors, such as the α-amino-3-hydroxy-5-methyl-4-isoxazolepropionic acid receptor (AMPAR) and the $GABA_A$ receptor ($GABA_A$R) result in models that are classified as stiff [26], which is an attribute of an ODE system that demands relatively small

step sizes in portions of the numerical solution [27]. For these ODE systems, L-stable implicit Runge-Kutta (IRK) solvers with adaptive time-stepping are ideal given their exceptional stability properties [28].

In this paper, we examine L-stable IRK methods when solving models that represent the AMPA and $GABA_A$ neuroreceptors. Three L-stable IRK methods that are highly effective at solving stiff ODE systems were selected and implemented with custom Matlab [29] programming. Features including adaptive step-sizing, embedded error estimation, error-based step size selection, and simplified Newton iterations are incorporated [30]. Numerical experiments were then used to identify the optimal maximum number of inner Newton iterations for each method. Then, for both the AMPAR and $GABA_A$R models, simulation time step results of each IRK method are compared to commonly used ERK methods. In addition, the numerical accuracy and computational efficiency of each IRK method is compared to one other, as well as the highly-popular fifth order, variable step size Dormand-Prince method. Finally, a C++ based IRK implementation demonstrates exceptionally accurate and expedient performances, showcasing its potential to support large-scale multi-cellular brain simulations.

II. MATERIALS AND METHODS

A. Neuroreceptor models

1) AMPA: Glutamate is the single most abundant neurotransmitter in the human brain [31]. It is produced by glutamatergic neurons, and is classified as excitatory in the sense that it predominately depolarizes post-synaptic neurons towards generating action potentials [32]. Given the large concentration of glutamate in the nervous system, alterations in its production are associated with many neurodegenerative diseases and treatments. In PD patients, for example, stimulating the STN with DBS causes a cascade of cellular effects within the basal-ganglia thalamocortical pathway through its afferent and efferent projections, including increased glutamate secretion to the globus pallidus external (GPe), globus pallidus internal (GPi), and substantia nigra pars reticulata (SNr) [23].

Ligand-gated AMPA receptors for glutamate are permeable to sodium and potassium, have a reversal potential of 0 mV, and possess fast channel opening rates. Therefore, these receptors produce fast excitatory post-synaptic currents [33]. Figure 1a displays the Markov kinetic binding model for the ligand-gated AMPAR that was utilized in this paper [34]. In this network, there is the unbound AMPAR form C_0, singly and doubly bound receptor forms C_1 and C_2, which can lead to desensitized states D_1 and D_2, respectively, and the open receptor form O [35]. In addition, variable T represents neurotransmitter concentration. Mass action kinetics gives the following system of ODEs for the AMPA neuroreceptor model:

$$\frac{dC_0}{dt} = -k_b C_0 T + C_1 k_{u1}, \tag{1a}$$

$$\frac{dC_1}{dt} = k_b C_0 T + k_{u2} C_2 + k_{ud} D_1 - k_{u1} C_1 \\ - k_b C_1 T - k_d C_1, \tag{1b}$$

$$\frac{dC_2}{dt} = k_b C_1 T + k_{ud} D_2 + k_c O - k_{u2} C_2 \\ - k_d C_2 - k_o C_2, \tag{1c}$$

$$\frac{dD_1}{dt} = k_d C_1 - k_{ud} D_1, \tag{1d}$$

$$\frac{dD_2}{dt} = k_d C_2 - k_{ud} D_2, \tag{1e}$$

$$\frac{dO}{dt} = C_2 k_o - k_c O, \tag{1f}$$

$$\frac{dT}{dt} = -k_b C_0 T + k_{u1} C_1 - k_b C_1 T + k_{u2} C_2. \tag{1g}$$

State transition rates were assigned as follows: $k_b = 1.3 \times 10^7$, $k_o = 2.7 \times 10^3$, $k_c = 200$, $k_{u1} = 5.9$, $k_{u2} = 8.6 \times 10^4$, $k_d = 900$, and $k_{ud} = 64$, each with units [1/sec]. Initial concentrations of C_1, C_2, D_1, D_2, and O were set to 0 M [33], and initial values for C_0 and T were computed from a nonlinear least squares fit of the model to the whole cell recording data in Destexhe et al. [35].

2) GABA: GABA is the most abundant inhibitory neurotransmitter in the human brain [36]. Like glutamate, GABA concentrations are altered by neurological disease and treatment. In STN DBS, for example, increased glutamate to the GPe increases GABA secretion to the GPi and SNr, resulting in greater GABA neuroreceptor binding in these regions [24].

There are two main categories of GABA neuroreceptors. Metabotropic GABA$_B$ receptors are slow-responding due to the secondary messenger biochemical network cascade necessary for ion channel activation. On the contrary, ligand-gated GABA$_A$ receptors are fast-responding due to their expedient ion channel opening rates. GABA$_A$ receptors are selective to chlorine with a reversal potential of approximately -70 mV. In addition, this receptor has two bound forms that can both trigger channel activation [35].

Figure 1b displays the kinetic binding model for the GABA$_A$ receptor that was utilized in this paper [26]. In this model, there is the unbound receptor form C_0, singly and doubly bound receptor forms C_1 and C_2, slow and fast desensitized states D_s and D_f, and singly open and doubly open receptor forms O_1 and O_2. This model incorporates the minimal forms needed to accurately reproduce GABA$_A$R kinetics [37]. Mass action kinetics gives the following ODE system for the GABA$_A$ neuroreceptor model:

$$\frac{dC_0}{dt} = -2k_b C_0 T + k_u C_1, \tag{2a}$$

$$\frac{dC_1}{dt} = 2k_b C_0 T - k_u C_1 + k_{uDs} D_s - k_{Ds} C_1 \\ + 2k_u C_2 - k_b C_1 T + k_{c1} O_1 - k_{o1} C_1, \tag{2b}$$

$$\frac{dC_2}{dt} = k_b C_1 T - 2k_u C_2 + k_{c2} O_2 - k_{o2} C_2 \\ + k_{uDf} D_f - k_{Df} C_2, \tag{2c}$$

$$\frac{dD_s}{dt} = k_{fs} D_f - k_{sf} D_s T + k_{Ds} C_1 - k_{uDs} D_s, \tag{2d}$$

$$\frac{dD_f}{dt} = k_{sf} D_s T - k_{fs} D_f + k_{Df} C_2 - k_{uDf} D_f, \tag{2e}$$

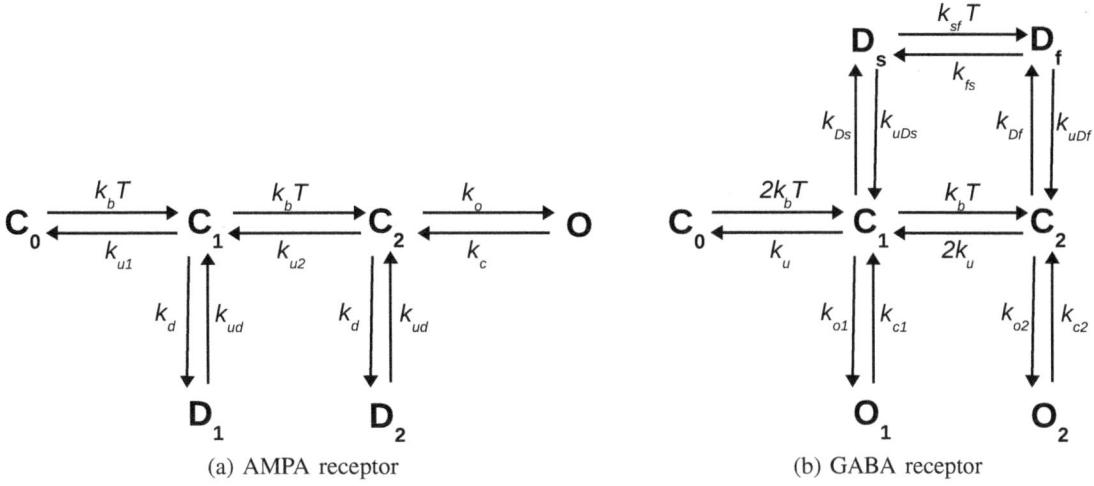

(a) AMPA receptor   (b) GABA receptor

Fig. 1: Kinetic models for ligand-gated neuroreceptors.

$$\frac{dO_1}{dt} = k_{o1}C_1 - k_{c1}O_1, \tag{2f}$$

$$\frac{dO_2}{dt} = k_{o2}C_2 - k_{c2}O_2, \tag{2g}$$

$$\frac{dT}{dt} = k_u C_1 - 2k_b C_0 T + 2k_u C_2 - k_b C_1 T$$
$$+ k_{fs}D_f - k_{sf}D_s T. \tag{2h}$$

Transition rates for the GABA$_A$R ODE system were assigned as follows: $k_b = 5 \times 10^6$, $k_u = 131$, $k_{uDs} = 0.2$, $k_{Ds} = 13$, $k_{c1} = 1100$, $k_{o1} = 200$, $k_{c2} = 142$, $k_{o2} = 2500$, $k_{uDf} = 25$, $k_{Df} = 1250$, $k_{fs} = 0.01$, and $k_{sf} = 2$, each with units [1/sec]. Initial values of C_1, C_2, D_s, D_f, O_1, and O_2 were set 0 M, and C_0 and T were assigned the values 1×10^{-6} M and 4096×10^{-6} M, respectively [26].

B. Stiff ordinary differential equations

The stiffness ratio is defined as

$$L = \frac{\max |Re(\lambda_i)|}{\min |Re(\lambda_i)|},$$

where λ_i is the i_{th} eigenvalue of the local Jacobian matrix [38], given by

$$J_{ij} = \frac{\partial f_i(t, \bar{y})}{\partial y_j}.$$

A general non-linear ODE system is stiff when $L \gg 1$. For each neuroreceptor model, we estimated the eigenvalues numerically; a local Jacobian matrix is computed at each simulation time step using finite differences, and then its eigenvalues are computed using Matlab's `eig` function [39]. For the AMPAR model $L = 1.6 \times 10^{11}$, and for the GABA$_A$R model $L = 3.5 \times 10^{11}$. Thus, both of these systems are classified as stiff.

C. Implicit Runge-Kutta methods

Runge-Kutta methods are a family of numerical integrators that solve ODE systems with trial steps within the time step. These methods can be expressed with the following formulas:

$$\bar{Z}_i = h \sum_{j=1}^{s} a_{ij}\bar{F}(t_n + c_j h, \bar{y}_n + \bar{Z}_j), \quad i = 1, ..., s \tag{3a}$$

$$\bar{y}_{n+1} = \bar{y}_n + h \sum_{j=1}^{s} b_j \bar{F}(t_n + c_j h, \bar{y}_n + \bar{Z}_j), \tag{3b}$$

where $\bar{y}_n$ is the current solution at time t_n, h is the current time step, $[a_{ij}]$ is the Runge-Kutta matrix, $\bar{F}$ is the ODE system, $[c_j]$ represents inter-time trial step nodes, $[b_j]$ is the trial step solution weights, s is the number of stages, and $\bar{y}_{n+1}$ is the numerical solution at time t_{n+1} [28]. A Runge-Kutta method can be fully defined with a Butcher table, i.e. a specific $[a_{ij}]$, $[b_j]$, and $[c_j]$ [40].

L-stable IRK methods are highly effective at solving stiff ODE systems [30]; these methods have no step size constraint to maintain numerical stability and quickly converge [41]. Methods with second and third order accuracy were considered as these orders best match the numerical accuracy of fractional step algorithms typically employed with partial differential equation based multiscale models [16], [39].

The following L-stable IRK methods were selected for examination: SDIRK(2/1) [42], ES-DIRK23A [17], and RadauIIa(3/2) [30], [43]. Each has demonstrated accuracy and computational efficiency when solving extremely stiff ODE systems. In addition, each provide an efficient local error estimator that enables error-based adaptive time-stepping. For simplicity, these solvers will be referred to as SDIRK, ESDIRK, and Radau for the remainder of this paper. Butcher tables for these methods are displayed in Fig. 2.

embedded first order formula for local error estimation. Each trial step, $\bar{Z}_i$, of the SDIRK solver can be solved for sequentially. Specifically, since $a_{12} = 0$ (see Fig. 2a), the first stage of this method can be written as $\bar{Z}_1 = h\left(a_{11}\bar{F}(t_n + c_1 h, \bar{y}_n + \bar{Z}_1)\right)$, and $\bar{Z}_1$ can be solved for first and used directly in the solution of $\bar{Z}_2 = h\left(a_{21}\bar{F}(t_n + c_1 h, \bar{y}_n + \bar{Z}_1) + a_{22}\bar{F}(t_n + c_2 h, \bar{y}_n + \bar{Z}_2)\right)$.

The Radau method has two stages like the SDIRK method (see Fig. 2b), but has third order accuracy with a second order error formula. This method's Runge-Kutta matrix is full, therefore the trial stages are solved as a coupled implicit system:

$$\begin{aligned}
\bar{Z}_1 &= h[a_{11}\bar{F}(t_n + c_1 h, \bar{y}_n + \bar{Z}_1) + \\
&\quad a_{12}\bar{F}(t_n + c_2 h, \bar{y}_n + \bar{Z}_2)], \\
\bar{Z}_2 &= h[a_{21}\bar{F}(t_n + c_1 h, \bar{y}_n + \bar{Z}_1) + \\
&\quad a_{22}\bar{F}(t_n + c_2 h, \bar{y}_n + \bar{Z}_2)].
\end{aligned}$$

Trial steps in the ESDIRK method are solved sequentially like the SDIRK method, after the initial explicit first stage (see Fig. 2c). This method is third order with an embedded second order formula for local error estimation, similar to the Radau solver.

D. Implementation

The three IRK methods were programmed in Matlab using principles specified in [30] and [44]; we refer these resources for a detailed explanation of Runge-Kutta method implementation and in this section provide just a brief overview of key aspects utilized in our implementations.

For each IRK method, Newton's method is used in solving system (3a). Typically, each inner Newton iteration involves computing the local Jacobian matrix and performing an LU factorization. To greatly decrease run-time, at each time step the Jacobian computation and LU factorization are performed just once on the first Newton iteration and retained for all remaining iterations. Execution time is further decreased by retaining the Jacobian in the subsequent time step if the IRK method converges with just one Newton iteration, or $\frac{\|\bar{Z}^{k+1} - \bar{Z}^k\|}{\|\bar{Z}^k - \bar{Z}^{k-1}\|} \leq 10^{-3}$, where k is the number of

$$
\begin{array}{c|cc}
\gamma & \gamma & \\
1 & 1-\gamma & \gamma \\
\hline
b & 1-\gamma & \gamma \\
\hat{b} & 1-\hat{\gamma} & \hat{\gamma}
\end{array}
\qquad
\begin{array}{c|cc}
\frac{1}{3} & \frac{5}{12} & -\frac{1}{12} \\
1 & \frac{3}{4} & \frac{1}{4} \\
\hline
b & \frac{3}{4} & \frac{1}{4} \\
\hat{b} & \frac{3}{4}-\frac{\sqrt{6}}{4} & \frac{1}{4}+\frac{\sqrt{6}}{12}
\end{array}
$$

(a) SDIRK(2/1) (b) RadauIIA(3/2)

$$
\begin{array}{c|cccc}
0 & 0 & & & \\
2\gamma & \gamma & \gamma & & \\
1 & \hat{b}_1 & \hat{b}_2 & \gamma & \\
1 & b_1 & b_2 & b_3 & \gamma \\
\hline
b & \frac{6\gamma-1}{12\gamma} & \frac{-1}{(24\gamma-12)\gamma} & \frac{-6\gamma^2+6\gamma-1}{6\gamma-3} & \gamma \\
\hat{b} & \frac{-4\gamma^2+6\gamma-1}{4\gamma} & \frac{-2\gamma+1}{4\gamma} & \gamma & 0
\end{array}
$$

(c) ESDIRK23A

Fig. 2: Butcher tables for the three implicit Runge-Kutta methods evaluated in this paper. In Fig. 2a, $\gamma = 1 - \frac{\sqrt{2}}{2}$ and $\hat{\gamma} = 2 - \frac{5}{4}\sqrt{2}$, and in Fig. 2c, $\gamma = 0.4358665215$. In each Butcher table, $\hat{b}$ specifies the lower-order trial step solution weights.

The SDIRK method is second order with an

inner iterations for convergence and $\| \cdot \|$ is an error-normalized 2-norm [30], [45].

Efficient starting values for each Newton iteration are produced via a Lagrange interpolation polynomial of degree s [30], [42]. For the Radau method, for example, we use the data points: $q(0) = 0, q(\frac{1}{3}) = \bar{Z}_1$, and $q(1) = \bar{Z}_2$, and obtain the following Lagrange polynomial:

$$q(w) = q(0)\frac{(w - \frac{1}{3})(w - 1)}{(0 - \frac{1}{3})(0 - 1)} +$$
$$q\left(\frac{1}{3}\right)\frac{(w - 0)(w - 1)}{(\frac{1}{3} - 0)(\frac{1}{3} - 1)} +$$
$$q(1)\frac{(w - 0)(w - \frac{1}{3})}{(1 - 0)(1 - \frac{1}{3})}$$
$$= \frac{w(w - 1)}{\frac{-2}{9}}\bar{Z}_1 + \frac{w(w - \frac{1}{3})}{\frac{2}{3}}\bar{Z}_2.$$

Newton iteration starting values are then given by:

$$\bar{Z}_1 = q(1 + wc_1) + \bar{y}_n - \bar{y}_{n+1},$$
$$\bar{Z}_2 = q(1 + wc_2) + \bar{y}_n - \bar{y}_{n+1}, \text{ where } w = \frac{h_{new}}{h_{old}}.$$

For each time step, local error is calculated and used for (i) step acceptance and (ii) subsequent step size prediction. The error at time step t_{n+1} can be computed by $\overline{err} = \hat{y}_{n+1} - \bar{y}_{n+1}$, where

$$\hat{y}_{n+1} = \bar{y}_n + \hat{b}_0 h \bar{F}(t_n, \bar{y}_n) +$$
$$h \sum_{j=1}^{s} \hat{b}_j \bar{F}(t_n + c_j h, \bar{Z}_j + \bar{y}_n). \quad (4)$$

The error calculations in the SDIRK and ES-DIRK methods are suitable for stiff systems [39], [41]. For the Radau method, however, $\hat{y}_{n+1} - \bar{y}_{n+1}$ will become unbounded and is therefore not appropriate for stiff systems [46]. Instead, we use the formula $\overline{err} = (I - h\hat{b}_0 J)^{-1}(\hat{y}_{n+1} - \bar{y}_{n+1})$ which is equivalent to

$$\overline{err} = (I - h\hat{b}_0 J)^{-1}[\hat{b}_0 h \bar{F}(t_n, \bar{y}_n) +$$
$$(\hat{b}_1 - b_1)h\bar{F}(t_n + c_1 h, \bar{Z}_1 + \bar{y}_n) + \quad (5)$$
$$(\hat{b}_2 - b_2)h\bar{F}(t_n + c_2 h, \bar{Z}_2 + \bar{y}_n)],$$

where I is the identity matrix, J is the Jacobian, and $\hat{b}_0 = \frac{\sqrt{6}}{6}$ [46].

We can write $\hat{y}_{n+1} - \bar{y}_{n+1}$ as follows [47]:

$$\hat{y}_{n+1} - \bar{y}_{n+1} = \hat{b}_0 h \bar{F}(t_n, \bar{y}_n) + e_1 \bar{Z}_1 + e_2 \bar{Z}_2. \quad (6)$$

To identify the coefficients e_1 and e_2, we substitute $\bar{Z}_1$ and $\bar{Z}_2$ (3a) into (6):

$$\hat{y}_{n+1} - \bar{y}_{n+1} = \hat{b}_0 h \bar{F}(t_n, \bar{y}_n)$$
$$+ e_1[ha_{11}\bar{F}(t_n + c_1 h, \bar{Z}_1 + \bar{y}_n) +$$
$$ha_{12}\bar{F}(t_n + c_2 h, \bar{Z}_2 + \bar{y}_n)]$$
$$+ e_2[ha_{21}\bar{F}(t_n + c_1 h, \bar{Z}_1 + \bar{y}_n) +$$
$$ha_{22}\bar{F}(t_n + c_2 h, \bar{Z}_2 + \bar{y}_n)].$$

Collecting terms gives:

$$\hat{y}_{n+1} - \bar{y}_{n+1} = \hat{b}_0 h \bar{F}(t_n, \bar{y}_n) +$$
$$(e_1 a_{11} + e_2 a_{21})h\bar{F}(t_n + c_1 h, \bar{Z}_1 + \bar{y}_n) +$$
$$(e_1 a_{12} + e_2 a_{22})h\bar{F}(t_n + c_2 h, \bar{Z}_2 + \bar{y}_n). \quad (7)$$

From (5) and (7), we end up with the following system of equations:

$$\hat{b}_1 - b_1 = e_1 a_{11} + e_2 a_{21},$$
$$\hat{b}_2 - b_2 = e_1 a_{12} + e_2 a_{22}.$$

Using the Radau Butcher table (Fig. 2b) gives $(e_1, e_2) = \hat{b}_0\left(\frac{-9}{2}, \frac{1}{2}\right)$. The error estimation is used to predict step size via the strategy proposed by Gustafsson [45]. Further, step size following a rejected step due to excessive local error, namely $\|\overline{err}\| > 1$, is $\frac{1}{3}h$.

For large-scale simulations, e.g. multiscale applications, hundreds of millions of ODE systems may be solved at each time step. For these computationally intensive simulations, scripting languages such as Matlab are not ideal, and machine-compiled programs are generally necessary to achieve simulation results within reasonable computing time [48]. Due to its superior accuracy in solving both the GABA$_A$R and AMPAR models (see Sec. III), we selected the Radau method and configured a C++ implementation of it. Execution results of this version provide a measure of optimally expected computational performance.

We validated the implementation of each IRK method by comparing their GABA$_A$R simulation results to those presented in Qazi et al. [37], and their AMPAR simulation results to whole cell recording data in Destexhe et al. [35].

E. Simulations

Numerical simulations were performed to assess the robustness of the IRK methods when solving the AMPAR and GABA$_A$R models. Simulations were one second in duration, with rates and initial conditions as specified in Section II-A. Absolute and relative error tolerances were both set to 10^{-8}, and initial step size, h, was set to 10^{-4}. For each IRK method, the optimal number of maximum Newton iterations, k_{max}, was identified by solving the AMPAR and GABA$_A$R models with $k_{max} = 5, 6, ..., 20$. For each value of k_{max}, the mean execution time of five simulations was computed, and the value of k_{max} that produced the lowest mean execution time was selected. Figure 3a displays the k_{max} values selected for each model and method.

For each method, it was observed that a threshold value of k_{max} exists, such that higher values do not result in faster simulations. Therefore, we selected the minimum k_{max} value associated with the fastest execution speed. For example, for the Radau method solving the GABA$_A$R model, simulation times begin to plateau for $k_{max} \geq 10$, and simulation times with $k_{max} \geq 15$ were the same (see Fig. 3b). Therefore, for this model and IRK method, $k_{max} = 15$ was selected.

Figure 3b also shows that faster run times correlate with fewer solution time steps and LU factorizations, until a floor is reached; in the case of the Radau method solving the GABA$_A$R model, this floor is 29 time steps and 30 LU factorizations. To a point, higher values of k_{max} increase the probability of Newton method convergence, resulting in fewer time steps and fewer computationally expensive LU factorizations [30]. For the Radau method solving the GABA$_A$R model, values of $k_{max} \geq 15$ yield the fewest number of simulation times steps in addition to no steps where the Newton iteration fails to converge. Thus, when $k_{max} = 15$, time steps and associated LU factorizations are minimized, yielding the fastest execution speeds.

Model	SDIRK	ESDIRK	Radau
GABA$_A$R	7	10	15
AMPAR	14	12	17

(a) Values of k_{max} selected for each model and method

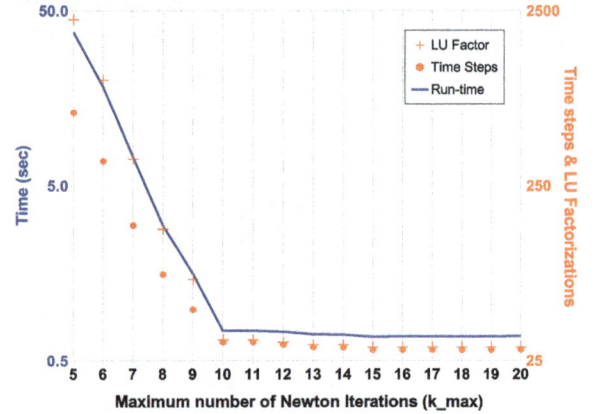

(b) Radau method solving the GABA$_A$R model: run time, time steps, and LU factorizations, for $k_{max} = 5, 6, ..., 20$

Fig. 3: Maximum Newton iteration metrics and results.

To evaluate the advantages that IRK methods have when solving fast-responding neuroreceptor models, we first compare the total number of simulation time steps and simulation step sizes of each IRK method to the following commonly used ERK methods: forward euler (FE), midpoint method (Mid), and 4th order Runge-Kutta (RK4). Next, to compare each IRK method to one another and to the adaptive 5th order Dormand-Prince method (DP5) [49], metrics including local and global error, total simulation time steps, step sizes, execution times, and numbers of Jacobian computations and LU factorizations were evaluated. Absolute and relative error tolerances of the DP5 method were set to 10^{-8}, matching the tolerances of the three implicit methods.

To more comprehensively assess performance differences among the IRK methods, work-precision diagrams using solution run times and scd values, where scd = $-\log_{10}(\|$relative error at $t = 1.0$ sec $\|_\infty)$, were then generated [50]. For the work-precision diagrams, relative error tolerances

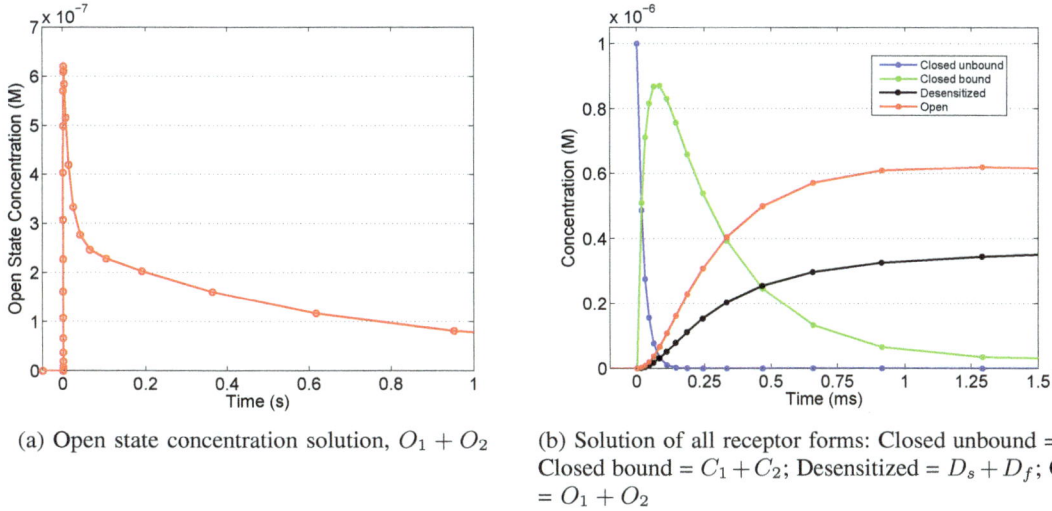

(a) Open state concentration solution, $O_1 + O_2$

(b) Solution of all receptor forms: Closed unbound = C_0; Closed bound = $C_1 + C_2$; Desensitized = $D_s + D_f$; Open = $O_1 + O_2$

Fig. 4: SDIRK method solution of GABA$_A$R model.

were set to $rtol = 10^{-(4+\frac{m}{5})}$, $m = 0, 1, ..., 25$, absolute error tolerance was set to $10^{-4} \cdot rtol$, and initial step size was 10^{-4}. In addition, solution run times presented in these diagrams are the mean of five runs. For all accuracy calculations, solutions with a 5th order adaptive time-stepping L-stable implicit Runge-Kutta method with a maximum step size of 10^{-6} and both absolute and relative tolerances set to 10^{-14} were used as true solutions.

Finally, the execution time of the Radau C++ implementation when solving both neuroreceptor models was assessed. All simulations were run on a Linux machine with an Intel i7 processor with a clock speed of 2.40 GHz.

III. RESULTS AND DISCUSSION

A. GABA$_A$R Model

Figure 4 presents the solution of the GABA$_A$R model with the SDIRK method; ESDIRK and Radau solutions look identical. The sharp transition in the total open state concentration, $O_1(t) + O_2(t)$, at the onset of neurotransmitter stimulus at $t = 0$ displays the necessity for smaller time steps in this region of the solution (Fig. 4a). Upon examining all receptor forms during the first 1.5 ms of the simulation, it is observed that both the unbound closed form, $C_0(t)$, and total bound

closed form, $C_1(t) + C_2(t)$, possess concentration transitions even greater than the open receptor form (Fig. 4b). These results show the stiffness possessed by the GABA$_A$R system.

Table I displays simulation time step metrics for the three IRK methods and the FE, Mid, and RK4 ERK methods. The maximum step size of each explicit method was calculated with the GABA$_A$R model stiffness index and the method's stability region [28], giving the largest step that can be taken while maintaining numerical stability. Then, the number of time steps required for each ERK method was computed by dividing the simulation duration by the maximum step size. The FE and Mid methods both require 2.1 x 10^4 time steps, and the RK4 method requires 1.5 x 10^4, which is lower than the FE and Mid methods due to its larger stability region [30]. On the contrary, each implicit method requires less than 30 simulation time steps. As displayed in Figure 4a, the majority of these time steps for the SDIRK method occur at the beginning of the simulation, within the region of rapid solution transition.

Similarly, the ESDIRK and Radau solvers demand noticeably more time steps at the onset of neurotransmitter stimulation (Fig. 5). Rejected steps, totalling three for the ESDIRK method (Fig.

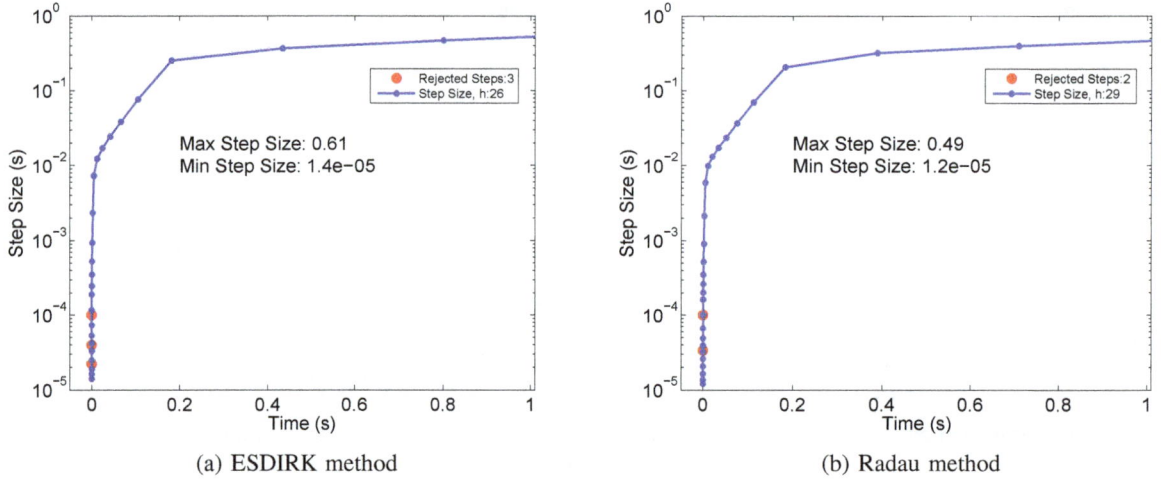

(a) ESDIRK method

(b) Radau method

Fig. 5: Simulation step sizes for the GABA$_A$R model.

TABLE I: Simulation time steps results for the ERK and IRK methods when solving the GABA$_A$R model.

Method (Order)	Max Step Size (s)	Time Steps
FE (1)	4.8×10^{-5}	2.1×10^4
Mid (2)	4.8×10^{-5}	2.1×10^4
RK4 (4)	6.8×10^{-5}	1.5×10^4
SDIRK (2/1)	Adaptive	28
ESDIRK (3/2)	Adaptive	26
Radau (3/2)	Adaptive	29

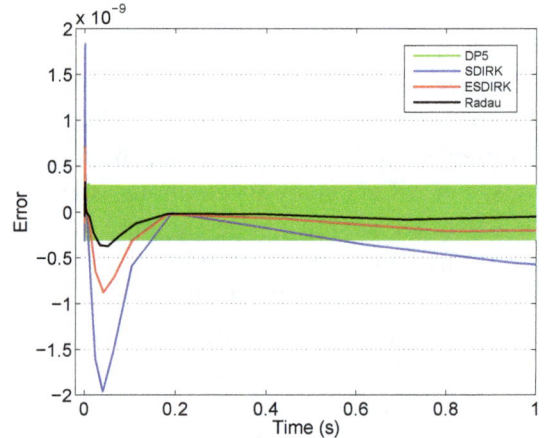

Fig. 6: GABA$_A$R model open state concentration solution error.

5a) and two for the Radau method (Fig. 5b) all occur at time $t = 0$; once the solution in this region has been accurately resolved, no further rejected steps occur. In addition, for all three IRK methods, all Newton iterations converged, which was facilitated by identifying optimal k_{max} values (see Sec. II-E). Further, the *smallest* step sizes of the IRK methods, namely 1.4×10^{-5} for the SDIRK and ESDIRK methods and 1.2×10^{-5} for the Radau method, have the same order of magnitude as the *largest* stable step sizes of the ERK methods.

Next the accuracy and computational efficiency of the IRK methods were compared to one another

and with the DP5 method (Table II). While the DP5 method possesses the lowest maximal local true solution deviation (3.2×10^{-10}), the 2-norm of its global error is one to two orders of magnitude higher than all three IRK methods. These results are explained by the fact that the solution of the DP5 solver oscillates around the true solution (Fig. 6). In addition, the DP5 method requires approximately 50,000 simulation time steps and takes 49.0 seconds to run. In comparison, the

TABLE II: Accuracy and simulation run-time metrics of the DP5 and IRK methods when solving the GABA$_A$R model. Boldface font denotes best results of each column.

Method (Order)	‖ Error ‖$_2$	Max |Error|	Time Steps	Run Time (s)
DP5 (5/4)	252.0×10^{-10}	$\mathbf{3.2 \times 10^{-10}}$	5.0×10^4	49.0
SDIRK (2/1)	45.6×10^{-10}	19.6×10^{-10}	28	**0.21**
ESDIRK (3/2)	18.3×10^{-10}	8.8×10^{-10}	**26**	0.27
Radau (3/2)	$\mathbf{8.7 \times 10^{-10}}$	3.7×10^{-10}	29	0.69

ESDIRK method requires 26 time steps and the SDIRK method executes in 0.21 seconds. DP5 solution accuracy can be improved with either stricter error tolerances or a decreased time step [51], however, these approaches will result in even greater run times.

The Radau method has the greatest execution time of the three IRK methods, at 0.69 seconds. While the number of simulation time steps among the IRK methods are comparable, two factors contribute to the longer run time of the Radau method. First, this solver generally requires a greater number of iterations for Newton's method to converge (Fig. 3a). Second, the Radau method requires 30 Jacobian computations, versus just four for the SDIRK and ESDIRK methods.

Despite its run time disadvantages amongst the IRK methods, the accuracy of the Radau method stands out as superior. It has the lowest global error 2-norm (8.7×10^{-10}), and its maximal deviation from the true solution (-3.7×10^{-10}) is comparable to that of the 5th order DP5 method, the only IRK method examined where this is the case. Further, the Radau method has greater accuracy at every time step than both the SDIRK and ESDIRK methods.

These findings are reinforced by the work-precision diagram for the three IRK methods when solving the GABA$_A$R model (Fig. 7). This diagram highlights the higher precisions attained by the third order methods, and in addition, also confirms the slower execution speeds achieved by the Radau method. However, when comparing graph points of similar relative tolerances, such as the symbols marked in yellow that represent $rtol = 10^{-6}$, the Radau method is consistently more accurate.

B. AMPAR Model

Figure 8 presents solution results of the AMPAR model solved with the Radau method. Like the GABA$_A$R model, the rapid transition in the open state concentration upon neurotransmitter stimulation demands a greater number of time steps (Fig. 8a). Specifically, the first 10% of the simulation (0.1 sec) encompasses approximately 96% of the simulation time steps. Once beyond this initial region, step size eventually increases by seven orders of magnitude (Fig. 8b). Similar to the GABA$_A$R model, both unbound closed and bound closed forms contribute to the system's stiffness.

A noticeable difference, compared to the GABA$_A$R simulation results, is the number of time

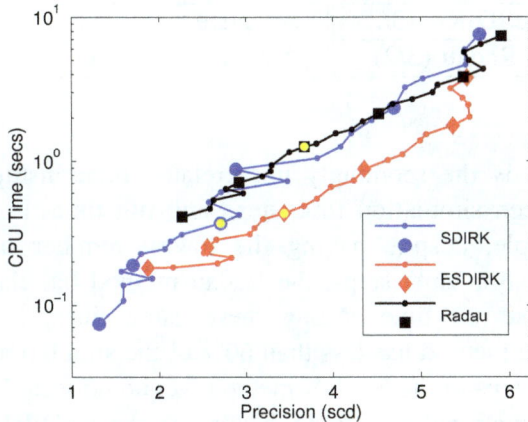

Fig. 7: GABA$_A$R work-precision diagram with solver run time vs. scd for each IRK method. Integer exponential tolerances, i.e. 10^{-4}, 10^{-5}, ..., are presented with enlarged symbols. The symbol for $rtol = 10^{-6}$ is distinguished by the yellow circle.

(a) Open state concentration solution, O

(b) Simulation step sizes

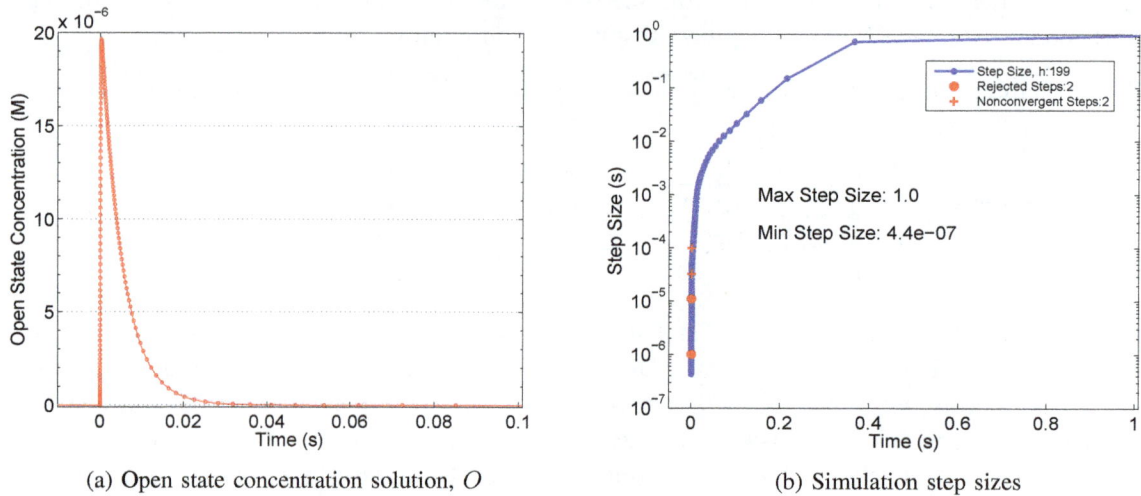

Fig. 8: Radau method solution of the AMPAR model.

steps needed by the implicit methods to solve the AMPAR model. The Radau method, for example, requires 199 time steps (Fig. 8b), a 586% increase from the 29 steps needed to solve the $GABA_AR$ model. Similar increases are observed with the SDIRK and ESDIRK solvers, most notably the 531 steps required by the SDIRK method (Table III). In addition, the smallest step sizes of the IRK methods are two orders of magnitude lower with the AMPAR model (Fig. 8b), due to the stiffness index of the AMPAR system [27]. Despite the elevated simulation time step counts, each IRK method still outperforms the explicit methods (Table III); maximum stable step sizes and simulation time steps for the explicit methods were again computed with their stiffness indices and stability regions [28].

While greater k_{max} values eliminated non-convergent Newton iterations in the $GABA_AR$ model, this is not the case with the AMPAR model. Each IRK method has two instances where Newton's method did not converge. In addition, the SDIRK method has four rejected steps, and the ESDIRK and Radau methods each have two, all occurring at time $t = 0$.

Table IV displays accuracy and execution efficiency results for the IRK methods. An interesting

TABLE III: Simulation time steps results for the ERK and IRK methods when solving the AMPAR model.

Method (Order)	Max Step Size (s)	Time Steps
FE (1)	1.7×10^{-5}	5.9×10^4
Mid (2)	1.7×10^{-5}	5.9×10^4
RK4 (4)	2.4×10^{-5}	4.2×10^4
SDIRK (2/1)	Adaptive	531
ESDIRK (3/2)	Adaptive	211
Radau (3/2)	Adaptive	199

result is the seemingly uncorrelated relationship between simulation time steps and run time. For example, despite having the lowest number of simulation time steps, the Radau method has the longest run time. Along these same lines, the Radau method has less than 50% of the simulation time steps of the SDIRK method, yet no noticeable computational advantage. Moreover, the ESDIRK method has approximately 40% of the SDIRK method's time steps, yet it requires 72% of its run-time.

With a comparable number of rejected and non-convergent steps (Table V), a culprit for this behavior is the number of Jacobian computations

TABLE IV: Accuracy and simulation run-time metrics of the DP5 and IRK methods when solving the AMPAR model. Boldface font denotes best results of each column.

Method (Order)	$\| \text{Error} \|_2$	Max |Error|	Time Steps	Run Time (s)
DP5 (5/4)	3.3×10^{-8}	2.7×10^{-9}	1.1×10^5	32.4
SDIRK (2/1)	3.0×10^{-8}	2.7×10^{-9}	531	1.34
ESDIRK (3/2)	1.7×10^{-8}	2.7×10^{-9}	211	**0.97**
Radau (3/2)	$\mathbf{1.6 \times 10^{-8}}$	2.7×10^{-9}	**199**	1.38

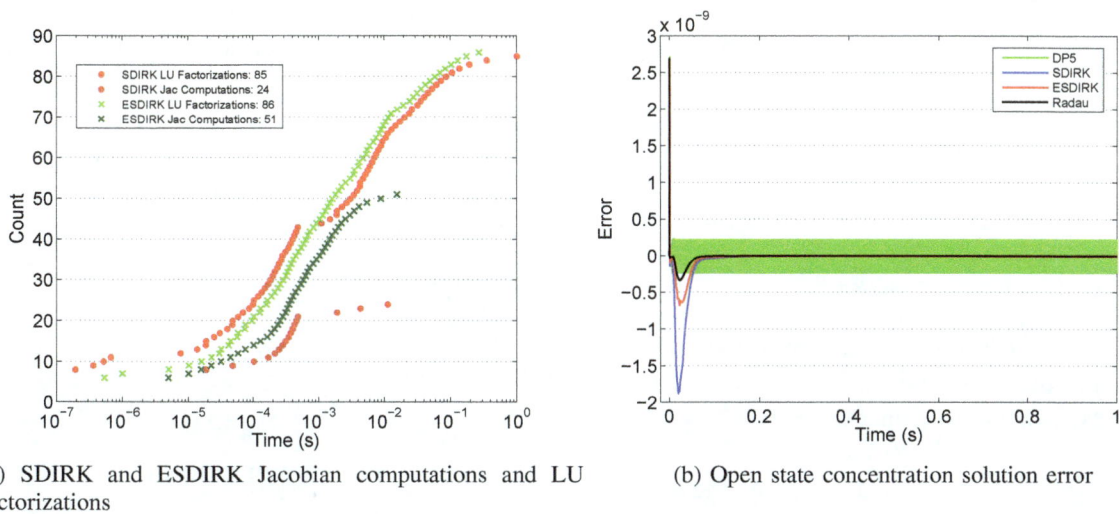

(a) SDIRK and ESDIRK Jacobian computations and LU factorizations

(b) Open state concentration solution error

Fig. 9: Method comparison when solving the AMPAR model.

performed by these solvers. Figure 9a displays the Jacobian computations and LU factorizations of the SDIRK and ESDIRK methods. Each method has a near identical number of LU factorizations, however, the ESDIRK method requires 51 Jacobian computations, which is more than double the 24 performed by the SDIRK method. In addition, the Radau method requires 162 Jacobian computations. Therefore, despite having a lower number of simulation time steps, the computational advantages of the ESDIRK and Radau methods are diminished due to this elevated number of Jacobian computations.

Once again, the accuracy and computational performances of the IRK methods were compared to the DP5 method (Table IV). As observed with the GABA$_A$R model, the DP5 method has inferior execution performance, requiring 1.1×10^5 simula-

TABLE V: Number of rejected and non-convergent steps for each IRK method when solving the AMPAR model.

Model	Rejected	Non-convergent
SDIRK	2	2
ESDIRK	4	2
Radau	2	2

tion time steps and 32.4 seconds for a numerically stable solution, both of which are significantly greater than results attained with the IRK methods. All four methods generate the same maximum local error (2.7×10^{-9}), which occurs at $t = 0$ for all methods. Also, differences among the global errors are relatively smaller with the AMPAR model. The oscillatory nature of the DP5 solution around the true solution (Fig. 9b) contributes to its global error 2-norm (3.3×10^{-8}), which is again larger than those of the three IRK methods. The Radau method once again has the lowest global error 2-norm (1.6×10^{-8}) of all methods inspected.

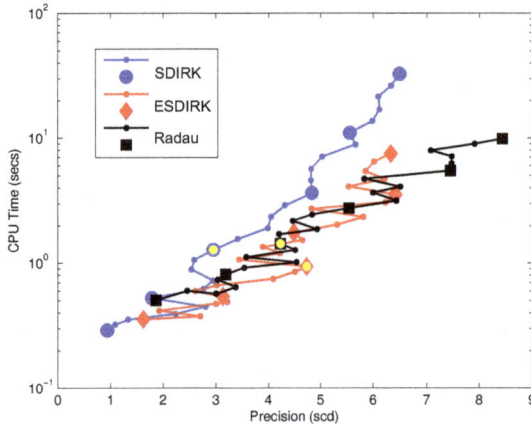

Fig. 10: AMPAR work-precision diagram with solver run time vs. scd for each IRK method. Integer exponential tolerances, i.e. 10^{-4}, 10^{-5}, ..., are presented with enlarged symbols. The symbol for $rtol = 10^{-6}$ is distinguished by the yellow circle.

The work-precision diagram for the AMPAR model (Fig. 10) again confirms the higher precision achieved by the third order ESDIRK and Radau solvers. More noticeable in this graph are the differences in the "slopes" of the curves, where "flatter" curves, i.e. ESDIRK and Radau, have more precision per unit CPU time [30]. For the AMPAR model, the Radau method is slower than the ESDIRK method at all work-precision tolerances examined, yet at relative tolerances greater than 10^{-6}, the Radau method becomes faster than the SDIRK method. Further, the Radau method is generally the most accurate of all three IRK methods.

C. C++ Radau Implementation

The Radau method consistently demonstrates the greatest accuracy of the methods examined, however, its main disadvantage is execution speed. For this reason, we selected the Radau method and

configured a C++ implementation of it. Table VI displays execution times for the previous Radau Matlab implementation, as well as the new C++ version.

As expected, the C++ version is significantly faster. Specifically, the $GABA_AR$ model has a 99.6% decrease in execution time, and the AMPAR model has a 99.7% decrease in execution time. Because the implementation algorithms between the two versions are the same, the C++ version maintains the accuracy of the Matlab prototype.

TABLE VI: Run times (seconds) for the Matlab and C++ Radau method when solving the $GABA_AR$ and AMPAR models.

Implementation	$GABA_AR$	AMPAR
Matlab	0.69	1.38
C++	2.7×10^{-3}	3.5×10^{-3}

IV. CONCLUSIONS

Computational neurology is a valuable contributor in the diagnosis, treatment, and comprehension of neurological disease. To provide maximal utility to the scientific community, computational simulations should incorporate highly-detailed, neurotransmitter-based neuron models. Therefore, large-scale simulations involving populations of neurons will inevitably produce computational challenges. In this paper, we have shown that appropriate numerical solvers with efficient implementation strategies can alleviate computational difficulties.

Commonly used explicit methods are capable in solving a limited number of fast-responding ligand-gated neuroreceptor models. However, we have shown that poor stability properties make them non-ideal for large-scale applications. Rather, by addressing the stiffness possessed by these models, we show that implicit methods are highly advantageous. In particular, we demonstrate that L-stable implicit Runge-Kutta methods offer superior accuracy and run-time efficiency compared to their explicit siblings when solving biologically-based AMPA and $GABA_A$ neuroreceptor models. To accelerate solutions, we utilize a range of strategies including embedded error estimators and simplified Newton iterations. In addition, we show that optimal execution times are achieved when costly Jacobian computations and LU factorizations are minimized.

The third order Radau IRK method demonstrates exceptional local and global accuracy compared to all other explicit and implicit methods examined. In addition, its numerical stability properties yield a relatively low number of simulation time steps and efficient step sizes when solving the AMPA and $GABA_A$ neuroreceptor models. Further, a C++ implementation of the Radau solver displays the computational faculty to enable large-scale multi-cellular simulations. In future work, we plan to continue our investigation of numerical solvers for neurotransmitter-based neuron models by comparing the IRK methods to multi-step methods and exponential integrators.

ACKNOWLEDGMENT

The author is grateful to Professor Jeff Borggaard and Professor James Turner for useful discussions related to this manuscript, and Frank Vogel for assistance with the Radau C++ code.

REFERENCES

[1] A. Datta, X. Zhou, Y. Su, L. C. Parra, and M. Bikson, "Validation of finite element model of transcranial electrical stimulation using scalp potentials: implications for clinical dose," *Journal of Neural Engineering*, vol. 10, no. 3, p. 036018, may 2013. [Online]. Available: http://dx.doi.org/10.1088/1741-2560/10/3/036018

[2] R. Plonsey and D. B. Heppner, "Considerations of quasi-stationarity in electrophysiological systems," *Bulletin of Mathematical Biophysics*, vol. 29, no. 4, pp. 657–664, dec 1967. [Online]. Available: http://dx.doi.org/10.1007/bf02476917

[3] S. Lew, C. Wolters, T. Dierkes, C. Rer, and R. MacLeod, "Accuracy and run-time comparison for different potential approaches and iterative solvers in finite element method based EEG source analysis," *Applied Numerical Mathematics*, vol. 59, no. 8, pp. 1970–1988, aug 2009. [Online]. Available: http://dx.doi.org/10.1016/j.apnum.2009.02.006

[4] T. Neuling, S. Wagner, C. H. Wolters, T. Zaehle, and C. S. Herrmann, "Finite-element model predicts current density distribution for clinical applications of tDCS and tACS," *Front. Psychiatry*, vol. 3, 2012. [Online]. Available: http://dx.doi.org/10.3389/fpsyt.2012.00083

[5] F. Gasca, L. Marshall, S. Binder, A. Schlaefer, U. G. Hofmann, and A. Schweikard, "Finite element simulation of transcranial current stimulation in realistic rat head model," in *2011 5th International IEEE/EMBS Conference on Neural Engineering*. IEEE, apr 2011. [Online]. Available: http://dx.doi.org/10.1109/ner.2011.5910483

[6] P. C. Miranda, M. Lomarev, and M. Hallett, "Modeling the current distribution during transcranial direct current stimulation," *Clinical Neurophysiology*, vol. 117, no. 7, pp. 1623–1629, jul 2006. [Online]. Available: http://dx.doi.org/10.1016/j.clinph.2006.04.009

[7] M. Åstrm, L. U. Zrinzo, S. Tisch, E. Tripoliti, M. I. Hariz, and K. Wårdell, "Method for patient-specific finite element modeling and simulation of deep brain stimulation," *Medical & Biological Engineering & Computing*, vol. 47, no. 1, pp. 21–28, oct 2008. [Online]. Available: http://dx.doi.org/10.1007/s11517-008-0411-2

[8] R. Sadleir, "A Bidomain Model for Neural Tissue," *International Journal of Bioelectromagnetism*, vol. 12, no. 1, pp. 2–6, 2010.

[9] E. Mandonnet and O. Pantz, "The role of electrode direction during axonal bipolar electrical stimulation: a bidomain computational model study," *Acta Neurochirurgica*, vol. 153, no. 12, pp. 2351–2355, sep 2011. [Online]. Available: http://dx.doi.org/10.1007/s00701-011-1151-x

[10] W. Ying and C. S. Henriquez, "Hybrid finite element method for describing the electrical response of biological cells to applied fields," *IEEE Transactions on Biomedical Engineering*, vol. 54, no. 4, pp. 611–620, apr 2007. [Online]. Available: http://dx.doi.org/10.1109/tbme.2006.889172

[11] A. Agudelo-Toro and A. Neef, "Computationally efficient simulation of electrical activity at cell membranes interacting with self-generated and externally imposed electric fields," *Journal of Neural Engineering*, vol. 10, no. 2, p. 026019, mar 2013. [Online]. Available: http://dx.doi.org/10.1088/1741-2560/10/2/026019

[12] K. W. Altman and R. Plonsey, "Development of a model for point source electrical fibre bundle stimulation," *Med. Biol. Eng. Comput.*, vol. 26, no. 5, pp. 466–475, sep 1988. [Online]. Available: http://dx.doi.org/10.1007/bf02441913

[13] R. Szmurlo, J. Starzynski, S. Wincenciak, and A. Rysz, "Numerical model of vagus nerve electrical stimulation," *COMPEL*, vol. 28, no. 1, pp. 211–220, jan 2009. [Online]. Available: http://dx.doi.org/10.1108/03321640910919002

[14] R. Szmurlo, J. Starzynski, B. Sawicki, and S. Wincenciak, "Multiscale finite element model of the electrically active neural tissue," in *EUROCON*

2007 - The International Conference on "Computer as a Tool". IEEE, 2007. [Online]. Available: http://dx.doi.org/10.1109/eurcon.2007.4400409

[15] R. Szmurlo, J. Starzynski, B. Sawicki, S. Wincenciak, and A. Cichocki, "Bidomain formulation for modeling brain activity propagation," in *2006 12th Biennial IEEE Conference on Electromagnetic Field Computation*. IEEE, 2006. [Online]. Available: http://dx.doi.org/10.1109/cefc-06.2006.1633138

[16] E. T. Dougherty, J. C. Turner, and F. Vogel, "Multiscale coupling of transcranial direct current stimulation to neuron electrodynamics: modeling the influence of the transcranial electric field on neuronal depolarization," *Comput Math Methods Med*, vol. 2014, pp. 1–14, 2014. [Online]. Available: http://dx.doi.org/10.1155/2014/360179

[17] J. Sundnes, G. T. Lines, and A. Tveito, "Efficient solution of ordinary differential equations modeling electrical activity in cardiac cells," *Mathematical Biosciences*, vol. 172, no. 2, pp. 55–72, aug 2001. [Online]. Available: http://dx.doi.org/10.1016/s0025-5564(01)00069-4

[18] C. N. V. S. Reports, "Deaths: Preliminary Data for 2011," *NVSS*, vol. 61, no. 6, 2012.

[19] J.-A. Girault and P. Greengard, "The neurobiology of dopamine signaling," *Archives of Neurology*, vol. 61, no. 5, p. 641, may 2004. [Online]. Available: http://dx.doi.org/10.1001/archneur.61.5.641

[20] D. Tarsy, J. L. Vitek, P. A. Starr, and M. S. Okun, Eds., *Deep Brain Stimulation in Neurological and Psychiatric Disorders*. Humana Press, 2008. [Online]. Available: http://dx.doi.org/10.1007/978-1-59745-360-8

[21] S. Miocinovic, S. Somayajula, S. Chitnis, and J. L. Vitek, "History, applications, and mechanisms of deep brain stimulation," *JAMA Neurol*, vol. 70, no. 2, p. 163, feb 2013. [Online]. Available: http://dx.doi.org/10.1001/2013.jamaneurol.45

[22] S. Miocinovic, C. C. McIntyre, M. Savasta, and J. L. Vitek, "Mechanisms of deep brain stimulation," in *Deep Brain Stimulation in Neurological and Psychiatric Disorders*. Humana Press, 2008, pp. 151–177. [Online]. Available: http://dx.doi.org/10.1007/978-1-59745-360-8_8

[23] M. D. Johnson, S. Miocinovic, C. C. McIntyre, and J. L. Vitek, "Mechanisms and targets of deep brain stimulation in movement disorders," *Neurotherapeutics*, vol. 5, no. 2, pp. 294–308, apr 2008. [Online]. Available: http://dx.doi.org/10.1016/j.nurt.2008.01.010

[24] F. Windels, N. Bruet, A. Poupard, C. Feuerstein, A. Bertrand, and M. Savasta, "Influence of the frequency parameter on extracellular glutamate and gamma-aminobutyric acid in substantia nigra and globus pallidus during electrical stimulation of subthalamic nucleus in rats," *Journal of Neuroscience Research*, vol. 72, no. 2, pp. 259–267, apr 2003. [Online]. Available: http://dx.doi.org/10.1002/jnr.10577

[25] E. Suli and D. F. Mayers, *An Introduction to Numerical Analysis*. Cambridge University Press, 2003. [Online]. Available: http://dx.doi.org/10.1017/cbo9780511801181

[26] S. Qazi, M. Caberlin, and N. Nigam, "Mechanism of psychoactive drug action in the brain: Simulation modeling of GABAA receptor interactions at non-equilibrium conditions," *Current Pharmaceutical Design*, vol. 13, no. 14, pp. 1437–1455, may 2007. [Online]. Available: http://dx.doi.org/10.2174/138161207780765972

[27] U. M. Ascher and L. R. Petzold, *Computer Methods for Ordinary Differential Equations and Differential-Algebraic Equations*. SIAM, jan 1998. [Online]. Available: http://dx.doi.org/10.1137/1.9781611971392

[28] E. Hairer, S. P. Nørsett, and G. Wanner, *Solving Ordinary Differential Equations I*. Springer Berlin Heidelberg, 1987. [Online]. Available: http://dx.doi.org/10.1007/978-3-662-12607-3

[29] MATLAB, *version 8.2.0.701 (R2013b)*. Natick, Massachusetts: The MathWorks Inc., 2013.

[30] E. Hairer and G. Wanner, *Solving Ordinary Differential Equations II*. Springer Berlin Heidelberg, 1996. [Online]. Available: http://dx.doi.org/10.1007/978-3-642-05221-7

[31] R. Sapolsky, *Biology and Human Behavior: The Neurological Origins of Individuality, 2nd Edition*. American Psychological Association (APA). [Online]. Available: http://dx.doi.org/10.1037/e526622012-001

[32] B. S. Meldrum, "Glutamate as a neurotransmitter in the brain: review of physiology and pathology," *J. Nutr.*, vol. 130, no. 4S Suppl, pp. 1007S–15S, Apr 2000.

[33] A. Destexhe, Z. F. Mainen, and T. J. Sejnowski, "Synthesis of models for excitable membranes, synaptic transmission and neuromodulation using a common kinetic formalism," *Journal of Computational Neuroscience*, vol. 1, no. 3, pp. 195–230, aug 1994. [Online]. Available: http://dx.doi.org/10.1007/bf00961734

[34] D. K. Patneau and M. L. Mayer, "Kinetic analysis of interactions between kainate and AMPA: Evidence for activation of a single receptor in mouse hippocampal neurons," *Neuron*, vol. 6, no. 5, pp. 785–798, may 1991. [Online]. Available: http://dx.doi.org/10.1016/0896-6273(91)90175-y

[35] A. Destexhe, Z. F. Mainen, and T. J. Sejnowski, "Kinetic models of synaptic transmission," in *Methods in Neuronal Modeling: From Synapse to Networks*, C. Koch and I. Segev, Eds. MIT press, 1998, pp. 1–25.

[36] A. Meir, S. Ginsburg, A. Butkevich, S. G. Kachalsky, I. Kaiserman, R. Ahdut, S. Demirgoren, and R. Rahamimoff, "Ion channels in presynaptic nerve terminals and control of transmitter release," *Physiol. Rev.*, vol. 79, no. 3, pp. 1019–1088, Jul 1999.

[37] S. Qazi, A. Beltukov, and B. A. Trimmer, "Simulation modeling of ligand receptor interactions at non-equilibrium conditions: processing of noisy inputs by ionotropic receptors," *Mathematical Biosciences*, vol. 187, no. 1, pp. 93–110, jan 2004. [Online]. Available: http://dx.doi.org/10.1016/j.mbs.2003.01.001

[38] J. D. Lambert, *Numerical methods for ordinary differential systems : the initial value problem.* Chichester New York: Wiley, 1991.

[39] J. Sundnes, G. T. Lines, X. Cai, F. N. Bjorn, K. A. Mardal, and A. Tveito, *Computing the Electrical Activity in the Heart.* Springer Berlin Heidelberg, 2006. [Online]. Available: http://dx.doi.org/10.1007/3-540-33437-8

[40] A. Iserles, *A First Course in the Numerical Analysis of Differential Equations.* Cambridge University Press, 2008. [Online]. Available: http://dx.doi.org/10.1017/cbo9780511995569

[41] A. Kværnø, "Singly diagonally implicit runge–kutta methods with an explicit first stage," *BIT Numerical Mathematics*, vol. 44, no. 3, pp. 489–502, aug 2004. [Online]. Available: http://dx.doi.org/10.1023/b:bitn.0000046811.70614.38

[42] L. M. Skvortsov, "An efficient scheme for the implementation of implicit runge-kutta methods," *Computational Mathematics and Mathematical Physics*, vol. 48, no. 11, pp. 2007–2017, nov 2008. [Online]. Available: http://dx.doi.org/10.1134/s0965542508110092

[43] L. Brugnano, F. Iavernaro, and C. Magherini, "Efficient implementation of radau collocation methods," *Applied Numerical Mathematics*, vol. 87, pp. 100–113, jan 2015. [Online]. Available: http://dx.doi.org/10.1016/j.apnum.2014.09.003

[44] J. J. de Swart, "A simple ODE solver based on 2-stage radau IIA," *Journal of Computational and Applied Mathematics*, vol. 84, no. 2, pp. 277–280, oct 1997. [Online]. Available: http://dx.doi.org/10.1016/s0377-0427(97)00141-6

[45] K. Gustafsson, "Control-theoretic techniques for stepsize selection in implicit runge-kutta methods," *ACM Trans. Math. Softw.*, vol. 20, no. 4, pp. 496–517, dec 1994. [Online]. Available: http://dx.doi.org/10.1145/198429.198437

[46] J. Wang, J. Rodriguez, and R. Keribar, "Integration of flexible multibody systems using radau IIA algorithms," *J. Comput. Nonlinear Dynam.*, vol. 5, no. 4, p. 041008, 2010. [Online]. Available: http://dx.doi.org/10.1115/1.4001907

[47] N. Guglielmi and E. Hairer, "Implementing radau IIA methods for stiff delay differential equations," *Computing*, vol. 67, no. 1, pp. 1–12, jul 2001. [Online]. Available: http://dx.doi.org/10.1007/s006070170013

[48] H. P. Langtangen, *Computational Partial Differential Equations: Numerical Methods and Diffpack Programming*, ser. Texts in Computational Science and Engineering. Springer Berlin Heidelberg, 2003.

[49] J. Dormand and P. Prince, "A family of embedded runge-kutta formulae," *Journal of Computational and Applied Mathematics*, vol. 6, no. 1, pp. 19–26, mar 1980. [Online]. Available: http://dx.doi.org/10.1016/0771-050x(80)90013-3

[50] F. Mazzia and C. Magherini, "Testset for initial value problem solvers, release 2.4," http://www.dm.unipi.it/testset/testsetivpsolvers/, University of Bari and INdAM, Tech. Rep., 02 2008.

[51] M. Caberlin, *Stiff Ordinary and Delay Differential Equations in Biological Systems*, ser. McGill theses. McGill University, 2002.

3

On the Approximation of the Cut and Step Functions by Logistic and Gompertz Functions

Anton Iliev*[†], Nikolay Kyurkchiev[†], Svetoslav Markov[†]

*Faculty of Mathematics and Informatics
Paisii Hilendarski University of Plovdiv, Plovdiv, Bulgaria
Email: aii@uni-plovdiv.bg
[†]Institute of Mathematics and Informatics
Bulgarian Academy of Sciences, Sofia, Bulgaria
Emails: nkyurk@math.bas.bg, smarkov@bio.bas.bg

Abstract—**We study the uniform approximation of the sigmoid cut function by smooth sigmoid functions such as the logistic and the Gompertz functions. The limiting case of the interval-valued step function is discussed using Hausdorff metric. Various expressions for the error estimates of the corresponding uniform and Hausdorff approximations are obtained. Numerical examples are presented using *CAS MATHEMATICA*.**

Keywords-**cut function; step function; sigmoid function; logistic function; Gompertz function; squashing function; Hausdorff approximation.**

I. Introduction

In this paper we discuss some computational, modelling and approximation issues related to several classes of sigmoid functions. Sigmoid functions find numerous applications in various fields related to life sciences, chemistry, physics, artificial intelligence, etc. In fields such as signal processing, pattern recognition, machine learning, artificial neural networks, sigmoid functions are also known as "activation" and "squashing" functions. In this work we concentrate on several practically important classes of sigmoid functions. Two of them are the cut (or ramp) functions and the step functions. Cut functions are continuous but they are not smooth (differentiable) at the two endpoints of the interval where they increase. Step functions can be viewed as limiting case of cut functions; they are not continuous but they are Hausdorff continuous (H-continuous) [4], [43]. In some applications smooth sigmoid functions are preferred, some authors even require smoothness in the definition of sigmoid functions. Two familiar classes of smooth sigmoid functions are the logistic and the Gompertz functions. There are situations when one needs to pass from nonsmooth sigmoid functions (e. g. cut functions) to smooth sigmoid functions, and vice versa. Such a necessity rises the issue of approximating nonsmooth sigmoid functions by smooth sigmoid functions.

One can encounter similar approximation problems when looking for appropriate models for fitting time course measurement data coming e. g. from cellular growth experiments. Depending on the general view of the data one can decide to use initially a cut function in order to obtain rough initial values for certain parameters, such as the maximum growth rate. Then one can use a more sophisticate model (logistic or Gompertz) to obtain a better fit to the measurement data. The presented results may be used to indicate to what extend and in what sense a model can be improved by another one and how the two models can be compared.

Section 2 contains preliminary definitions and motivations. In Section 3 we study the uniform and Hausdorff approximation of the cut functions by logistic functions. Curiously, the uniform distance between a cut function and the logistic function of best uniform approximation is an absolute constant not depending on the slope of the functions, a result observed in [18]. By contrast, it turns out that the Hausdorff distance (H-distance) depends on the slope and tends to zero when increasing the slope. Showing that the family of logistic functions cannot approximate the cut function arbitrary well, we then consider the limiting case when the cut function tends to the step function (in Hausdorff sense). In this way we obtain an extension of a previous result on the Hausdorff approximation of the step function by logistic functions [4]. In Section 4 we discuss the approximation of the cut function by a family of squashing functions induced by the logistic function. It has been shown in [18] that the latter family approximates uniformly the cut function arbitrary well. We propose a new estimate for the H-distance between the cut function and its best approximating squashing function. Our estimate is then extended to cover the limiting case of the step function. In Section 5 the approximation of the cut function by Gompertz functions is considered using similar techniques as in the previous sections. The application of the logistic and Gompertz functions in life sciences is briefly discussed. Numerical examples are presented throughout the paper using the computer algebra system *MATHEMATICA*.

II. PRELIMINARIES

Sigmoid functions. In this work we consider *sigmoid functions* of a single variable defined on the real line, that is functions s of the form $s : \mathbb{R} \longrightarrow \mathbb{R}$. Sigmoid functions can be defined as bounded monotone non-decreasing functions on $\mathbb{R}$. One usually makes use of normalized sigmoid functions defined as monotone non-decreasing functions $s(t), t \in \mathbb{R}$, such that $\lim s(t)_{t \to -\infty} = 0$ and $\lim s(t)_{t \to \infty} = 1$. In the fields of neural networks and machine learning sigmoid-like functions of many variables are used, familiar under the name *activation functions*. (In some applications the sigmoid functions are normalised so that the lower asymptote is assumed -1: $\lim s(t)_{t \to -\infty} = -1$.)

Cut (ramp) functions. Let $\Delta = [\gamma - \delta, \gamma + \delta]$ be an interval on the real line $\mathbb{R}$ with centre $\gamma \in \mathbb{R}$ and radius $\delta \in \mathbb{R}$. A cut function (on Δ) is defined as follows:

Definition 1. *The cut function $c_{\gamma,\delta}$ on Δ is defined for $t \in \mathbb{R}$ by*

$$c_{\gamma,\delta}(t) = \begin{cases} 0, & \text{if } t < \Delta, \\ \dfrac{t - \gamma + \delta}{2\delta}, & \text{if } t \in \Delta, \\ 1, & \text{if } \Delta < t. \end{cases} \tag{1}$$

Note that the slope of function $c_{\gamma,\delta}(t)$ on the interval Δ is $1/(2\delta)$ (the slope is constant in the whole interval Δ). Two special cases are of interest for our discussion in the sequel.

Special case 1. For $\gamma = 0$ we obtain a cut function on the interval $\Delta = [-\delta, \delta]$:

$$c_{0,\delta}(t) = \begin{cases} 0, & \text{if } t < -\delta, \\ \dfrac{t + \delta}{2\delta}, & \text{if } -\delta \leq t \leq \delta, \\ 1, & \text{if } \delta < t. \end{cases} \tag{2}$$

Special case 2. For $\gamma = \delta$ we obtain the cut function on $\Delta = [0, 2\delta]$:

$$c_{\delta,\delta}(t) = \begin{cases} 0, & \text{if } t < 0, \\ \dfrac{t}{2\delta}, & \text{if } 0 \leq t \leq 2\delta, \\ 1, & \text{if } 2\delta < t. \end{cases} \tag{3}$$

Step functions. The step function (with "jump" at $\gamma \in \mathbb{R}$) can be defined by

$$h_\gamma(t) = c_{\gamma,0}(t) = \begin{cases} 0, & \text{if} \quad t < \gamma, \\ [0,1], & \text{if} \quad t = \gamma, \\ 1, & \text{if} \quad t > \gamma, \end{cases} \quad (4)$$

which is an *interval-valued function* (or just *interval function*) [4], [43]. In the literature various point values, such as $0, 1/2$ or 1, are prescribed to the step function (4) at the point γ; we prefer the interval value $[0,1]$. When the jump is at the origin, that is $\gamma = 0$, then the step function is known as the Heaviside step function; its "interval" formulation is:

$$h_0(t) = c_{0,0}(t) = \begin{cases} 0, & \text{if} \quad t < 0, \\ [0,1], & \text{if} \quad t = 0, \\ 1, & \text{if} \quad t > 0. \end{cases} \quad (5)$$

H-distance. The step function can be perceived as a limiting case of the cut function. Namely, for $\delta \to 0$, the cut function $c_{\delta,\delta}$ tends in "Hausdorff sense" to the step function. Here "Hausdorff sense" means *Hausdorff distance*, briefly *H-distance*. The H-distance $\rho(f,g)$ between two interval functions f, g on $\Omega \subseteq \mathbb{R}$, is the distance between their completed graphs $F(f)$ and $F(g)$ considered as closed subsets of $\Omega \times \mathbb{R}$ [24], [41]. More precisely,

$$\rho(f,g) = \max\{ \sup_{A \in F(f)} \inf_{B \in F(g)} ||A - B||, \quad (6)$$

$$\sup_{B \in F(g)} \inf_{A \in F(f)} ||A - B||\},$$

wherein $||.||$ is any norm in $\mathbb{R}^2$, e. g. the maximum norm $||(t, x)|| = \max |t|, |x|$.

To prove that (3) tends to (5) let h be the H-distance between the step function (5) and the cut function (3) using the maximum norm, that is a square (box) unit ball. By definition (6) h is the side of the smallest unit square, centered at the point $(0, 1)$ touching the graph of the cut function. Hence we have $1 - c_{\delta,\delta}(h) = h$, that is $1 - h/(2\delta) = h$, implying

$$h = \frac{2\delta}{1 + 2\delta} = 2\delta + O(\delta^2).$$

For the sake of simplicity throughout the paper we shall work with some of the special cut functions (2), (3), instead of the more general (arbitrary shifted) cut function (1); these special choices will not lead to any loss of generality concerning the results obtained. Moreover, for all sigmoid functions considered in the sequel we shall define a "basic" sigmoid function such that any member of the corresponding class is obtained by replacing the argument t by $t - \gamma$, that is by shifting the basic function by some $\gamma \in \mathbb{R}$.

Logistic and Gompertz functions: applications to life-sciences. In this work we focus on two familiar smooth sigmoid functions, namely the Gompertz function and the Verhulst logistic function. Both their inventors, B. Gompertz and P.-F. Verhulst, have been motivated by the famous demographic studies of Thomas Malthus.

The Gompertz function was introduced by Benjamin Gompertz [22] for the study of demographic phenomena, more specifically human aging [38], [39], [47]. Gompertz functions find numerous applications in biology, ecology and medicine. A. K. Laird successfully used the Gompertz curve to fit data of growth of tumors [32]; tumors are cellular populations growing in a confined space where the availability of nutrients is limited [1], [2], [15], [19].

A number of experimental scientists apply Gompertz models in bacterial cell growth, more specifically in food control [10], [31], [42], [48], [49], [50]. Gompertz models prove to be useful in animal and agro-sciences as well [8], [21], [27], [48]. The Gompertz model has been applied in modelling aggregation processes [25], [26]; it is a subject of numerous theoretical modelling studies as well [6], [7], [9], [20], [37], [40].

The logistic function was introduced by Pierre François Verhulst [44]–[46], who applied it to human population dynamics. Verhulst derived his logistic equation to describe the mechanism of the self-limiting growth of a biological population. The equation was rediscovered in 1911 by A. G. McKendrick [35] for the bacterial growth in broth and was tested using nonlinear parameter estimation. The logistic function finds applications in an wide range of fields, including biology, ecology, population dynamics, chemistry, demography, economics, geoscience, mathematical psychology, probability, sociology, political science, financial mathematics, statistics, fuzzy set theory, to name a few [12], [13], [11], [14], [18].

Logistic functions are often used in artificial neural networks [5], [16], [17], [23]. Any neural net element computes a linear combination of its input signals, and applies a logistic function to the result; often called "activation" function. Another application of logistic curve is in medicine, where the logistic differential equation is used to model the growth of tumors. This application can be considered an extension of the above-mentioned use in the framework of ecology. In (bio)chemistry the concentration of reactants and products in autocatalytic reactions follow the logistic function.

Other smooth sigmoid functions. The integral (antiderivative) of any smooth, positive, "bump-shaped" or "bell-shaped" function will be sigmoidal. A famous example is the error function, which is the integral (also called the cumulative distribution function) of the Gaussian normal distribution. The logistic function is also used as a base for the derivation of other sigmoid functions, a notable example is the generalized logistic function, also known as Richards curve [37]. Another example is the Dombi-Gera-squashing function introduced and studied in [18] obtained as an antiderivative (indefinite integral) of the difference of two shifted logistic functions.

In what follows we shall be interested in the approximation of the cut function by smooth sigmoid functions, more specifically the Gompertz, the logistic and the Dombi-Gera-squashing function. We shall focus first on the Verhulst logistic function.

III. APPROXIMATION OF THE CUT FUNCTION BY LOGISTIC FUNCTIONS

Definition 2. *Define the logistic (Verhulst) function v on $\mathbb{R}$ as [44]–[46]*

$$v_{\gamma,k}(t) = \frac{1}{1 + e^{-4k(t-\gamma)}}. \tag{7}$$

Note that the logistic function (7) has an inflection at its "centre" $(\gamma, 1/2)$ and its slope at γ is equal to k.

Proposition 1. [18] The function $v_{\gamma,k}(t)$ defined by (7) with $k = 1/(2\delta)$: i) is the logistic function of best uniform one-sided approximation to function $c_{\gamma,\delta}(t)$ in the interval $[\gamma, \infty)$ (as well as in the interval $(-\infty, \gamma]$); ii) approximates the cut function $c_{\gamma,\delta}(t)$ in uniform metric with an error

$$\rho = \rho(c, v) = \frac{1}{1 + e^2} = 0.11920292.... \tag{8}$$

Proof. Consider functions (1) and (7) with same centres $\gamma = \delta$, that is functions $c_{\delta,\delta}$ and $v_{\delta,k}$. In addition chose c and v to have same slopes at their coinciding centres, that is assume $k = 1/(2\delta)$, cf. Figure 1. Then, noticing that the largest uniform distance between the cut and logistic functions is achieved at the endpoints of the underlying interval $[0, 2\delta]$, we have:

$$\rho = v_{\delta,k}(0) - c_{\delta,\delta} = \frac{1}{1 + e^{4k\delta}} = \frac{1}{1 + e^2}. \tag{9}$$

This completes the proof of the proposition.

We note that the uniform distance (9) is an absolute constant that does not depend on the width of the underlying interval Δ, resp. on the slope k. The next proposition shows that this is not the case whenever H-distance is used.

Proposition 2. The function $v(t) = v_{0,k}(t)$ with $k = 1/(2\delta)$ is the logistic function of best Hausdorff one-sided approximation to function $c(t) = c_{0,k}(t)$ in the interval $[0, \infty)$ (resp. in the interval $(-\infty, 0]$). The function $v(t)$, approximates function $c(t)$ in H-distance with an error $h = h(c, v)$ that satisfies the relation:

$$\ln \frac{1-h}{h} = 2 + 4kh. \tag{10}$$

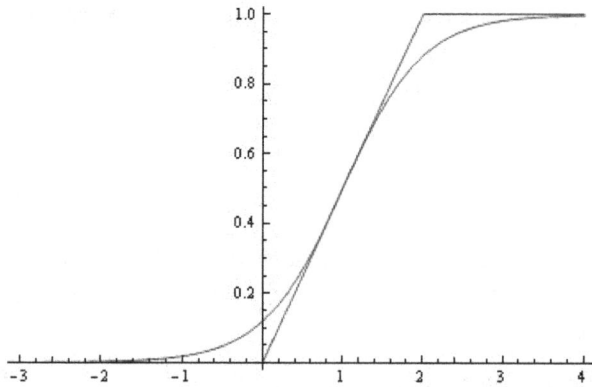

Fig. 1.   The cut and logistic functions for $\gamma = \delta = 1$, $k = 1/2$.

Proof. Using $\delta = 1/(2k)$ we can write $\delta + h = (1 + 2hk)/(2k)$, resp.:

$$v(-\delta - h) = \frac{1}{1 + e^{2(1+2hk)}}.$$

The H-distance h using square unit ball (with a side h) satisfies the relation $v(-\delta - h) = h$, which implies (10). This completes the proof of the proposition.

Relation (10) shows that the H-distance h depends on the slope k, $h = h(k)$. The next result gives additional information on this dependence.

Proposition 3. *For the H-distance $h(k)$ the following holds for $k > 5$:*

$$\frac{1}{4k+1} < h(k) < \frac{\ln(4k+1)}{4k+1}. \qquad (11)$$

Proof. We need to express h in terms of k, using (10). Let us examine the function

$$f(h) = 2 + 4hk - \ln(1-h) - \ln\frac{1}{h}.$$

From

$$f'(h) = 4k + \frac{1}{1-h} + \frac{1}{h} > 0$$

we conclude that function f is strictly monotone increasing. Consider the function

$$g(h) = 2 + h(1 + 4k) - \ln\frac{1}{h}.$$

Then $g(h) - f(h) = h + \ln(1 - h) = O(h^2)$ using the Taylor expansion $\ln(1 - h) = -h + O(h^2)$. Hence $g(h)$ approximates $f(h)$ with $h \to 0$ as $O(h^2)$. In addition $g'(h) = 1 + 4k + 1/h > 0$, hence function g is monotone increasing. Further, for $k \geq 5$

$$g\left(\frac{1}{1+4k}\right) = 3 - \ln(1 + 4k) < 0,$$

$$g\left(\frac{\ln(4k+1)}{4k+1}\right) = 2 + \ln\ln(1 + 4k) > 0.$$

This completes the proof of the proposition.

Relation (11) implies that when the slope k of functions c and v tends to infinity, the h-distance $h(c, v)$ between the two functions tends to zero (differently to the uniform distance $\rho(c, v)$ which remains constant).

The following proposition gives more precise upper and lower bounds for $h(k)$. For brevity denote $K = 4k + 1$.

Proposition 4. *For the H-distance h the following inequalities hold for $k \geq 5$:*

$$\frac{\ln K}{K} - \frac{2 + \ln\ln K}{K\left(1 + \frac{1}{\ln K}\right)} < h(k) < \qquad (12)$$

$$\frac{\ln K}{K} + \frac{2 + \ln\ln K}{K\left(\frac{\ln\ln K}{1 - \ln K} - 1\right)}, K = 4k + 1.$$

Proof. Evidently, the second derivative of $g(h) = 2 + h(1 + 4k) - \ln(1/h)$, namely $g''(h) = -\frac{1}{h^2} < 0$, has a constant sign on $[\frac{1}{K}, \frac{\ln K}{K}]$. The straight line, defined by the points $\left(\frac{1}{K}, g(\frac{1}{K})\right)$ and $\left(\frac{\ln K}{K}, g(\frac{\ln K}{K})\right)$, and the tangent to g at the point $\left(\frac{\ln K}{K}, g(\frac{\ln K}{K})\right)$ cross the abscissa at the points

$$\frac{\ln K}{K} + \frac{2 + \ln\ln K}{K\left(\frac{\ln\ln K}{1 - \ln K} - 1\right)}, \quad \frac{\ln K}{K} - \frac{2 + \ln\ln K}{K\left(1 + \frac{1}{\ln K}\right)},$$

respectively. This completes the proof of the Proposition.

Propositions 2, 3 and 4 extend similar results from [4] stating that the Heaviside interval-valued step function is approximated arbitrary well by logistic functions in Hausdorff metric. The Hausdorff approximation of the Heaviside step function by sigmoid functions is discussed from various computational and modelling aspects in [28], [29], [30].

IV. APPROXIMATION OF THE CUT FUNCTION BY A SQUASHING FUNCTION

The results obtained in Section 3 state that the cut function cannot be approximated arbitrary well by the family of logistic functions. This result justifies the discussion of other families of smooth sigmoid functions having better approximating properties. Such are the squashing functions proposed in [18] further denoted DG-squashing functions.

Definition 3. The DG-squashing function s_Δ on the interval $\Delta = [\gamma - \delta, \gamma + \delta]$ is defined by

$$s_\Delta^{(\beta)}(t) = s_{\gamma,\delta}^{(\beta)}(t) = \frac{1}{2\delta} \ln \left(\frac{1 + e^{\beta(t-\gamma+\delta)}}{1 + e^{\beta(t-\gamma-\delta)}} \right)^{\frac{1}{\beta}}. \tag{13}$$

Note that the squashing function (13) has an inflection at its "centre" γ and its slope at γ is equal to $(2\delta)^{-1}$.

The squashing function (13) with centre $\gamma = \delta$:

$$s_{\delta,\delta}^{(\beta)}(t) = \frac{1}{2\delta} \ln \left(\frac{1 + e^{\beta t}}{1 + e^{\beta(t-2\delta)}} \right)^{\frac{1}{\beta}}, \tag{14}$$

is the function of best uniform approximation to the cut function (3). Indeed, functions $c_{\delta,\delta}$ and $s_{\gamma,\delta}^{(\beta)}$ have same centre $\gamma = \delta$ and equal slopes $1/(2\delta)$ at their coinciding centres. As in the case with the logistic function, one observes that the uniform distance $\rho = \rho(c, s)$ between the cut and squashing function is achieved at the endpoints of the interval Δ, more specifically at the origin. Denoting the width of the interval Δ by $w = 2\delta$ we obtain

$$\rho = s_{\delta,\delta}^{(\beta)}(0) = \frac{1}{w} \ln(\frac{2}{1+e^{\beta(-w)}})^{1/\beta} < \tag{15}$$

$$\frac{\ln 2}{w} \frac{1}{\beta} = \text{const} \frac{1}{\beta}.$$

The estimate (15) has been found by Dombi and Gera [18]. This result shows that any cut

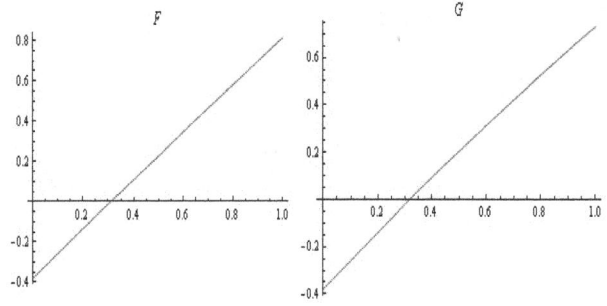

Fig. 2. The functions $F(d)$ and $G(d)$.

function c_Δ can be approximated arbitrary well by squashing functions $s_\Delta^{(\beta)}$ from the class (13). The approximation becomes better with the increase of the value of the parameter β. Thus β affects the quality of the approximation; as we shall see below the practically interesting values of β are integers greater than 4.

In what follows we aim at an analogous result using Hausdorff distance. Let us fix again the centres of the cut and squashing functions to be $\gamma = \delta$ so that the form of the cut function is $c_{\delta,\delta}$, namely (3), whereas the form of the squashing function is $s_{\delta,\delta}^{(\beta)}$ as given by (14). Both functions $c_{\delta,\delta}$ and $s_{\delta,\delta}^{(\beta)}$ have equal slopes $1/w$, $w = 2\delta$, at their centres δ.

Denoting the square-based H-distance between $c_{\delta,\delta}$ and $s_{\delta,\delta}^{(\beta)}$ by $d = d(w; \beta)$, $w = 2\delta$, we have the relation

$$s_{\delta,\delta}^{(\beta)}(w + d) = \frac{1}{w} \ln \left(\frac{1 + e^{\beta(w+d)}}{1 + e^{\beta d}} \right)^{\frac{1}{\beta}} = 1 - d$$

or

$$\ln \frac{1 + e^{\beta(w+d)}}{1 + e^{\beta d}} = \beta w(1 - d). \tag{16}$$

The following proposition gives an upper bound for $d = d(w; \beta)$ as implicitly defined by (16):

Proposition 5. *For the distance d the following holds for $\beta \geq 5$:*

$$d < \ln 2 \frac{\ln(4\beta w + 1)}{4w\beta + 1}. \tag{17}$$

Proof. We examine the function:

$$F(d) = -\beta w(1-d) + \ln(1 + e^{\beta(w+d)}) + \ln \frac{1}{1 + e^{\beta d}}.$$

From $F'(d) > 0$ we conclude that function $F(d)$ is strictly monotone increasing. We define the function

$$G(d) = -\beta w + \ln(1 + e^{\beta w}) +$$

$$d\beta \left(w + \frac{e^{\beta w}}{1 + e^{\beta w}} \right) + \ln \frac{1}{1 + e^{\beta d}}.$$

We examine $G(d) - F(d)$:

$$G(d) - F(d) =$$

$$\ln(1 + e^{\beta w}) + \frac{e^{\beta w}\beta d}{1 + e^{\beta w}} - \ln(1 + e^{\beta(w+d)}).$$

From Taylor expansion

$$\ln(1 + e^{\beta(w+d)}) = \ln(1 + e^{\beta w}) + \frac{e^{\beta w}\beta d}{1 + e^{\beta w}} + O(d^2)$$

we see that function $G(d)$ approximates $F(d)$ with $d \to 0$ as $O(d^2)$ (cf. Fig. 2).

In addition $G(0) < 0$ and $G\left(\ln 2 \frac{\ln(4\beta w + 1)}{4w\beta + 1} \right) > 0$ for $\beta \geq 5$. This completes the proof of the proposition.

Some computational examples using relation (16) and (17) for various β and w are presented in Table 1.

w	β	$d(w;\beta)$ from(16)	$d(w;\beta)$ from(17)
1	30	0.016040	0.027472
5	10	0.012639	0.018288
6	100	0.001068	0.002247
14	5	0.009564	0.013908
50	100	0.000137	0.000343
500	1000	1.38×10^{-6}	5.02×10^{-6}
1000	5000	1.3×10^{-7}	5.8×10^{-7}

TABLE I
BOUNDS FOR $d(w;\beta)$ COMPUTED BY (16) AND (17), RESPECTIVELY

The numerical results are plotted in Fig. 3 (for the case $\beta = 5$, $w = 3$; $d = 0.0398921$) and Fig. 4 (for the case $\beta = 10$, $w = 4$; $d = 0.0154697$).

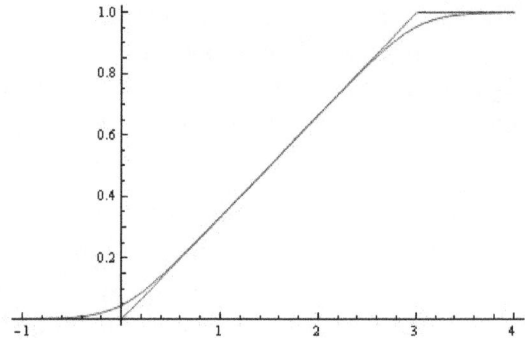

Fig. 3.   Functions $c_{\delta,\delta}$ and $s_{\delta,\delta}^{(\beta)}$ for $\beta = 5$, $w = 3$; $d \leq 0.4$.

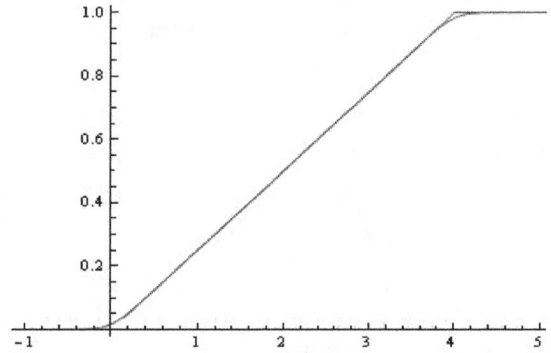

Fig. 4.   Functions $c_{\delta,\delta}$ and $s_{\delta,\delta}^{(\beta)}$ for $\beta = 10$, $w = 4$; $d \leq 0.016$.

V. APPROXIMATION OF THE STEP FUNCTION BY THE GOMPERTZ FUNCTION

In this section we study the Hausdorff approximation of the step function by the Gompertz function and obtain precise upper and lower bounds for the Hausdorff distance. Numerical examples, illustrating our results are given.

Definition 4. The Gompertz function $\sigma_{\alpha,\beta}(t)$ is defined for α, $\beta > 0$ by [22]:

$$\sigma_{\alpha,\beta}(t) = e^{-\alpha e^{-\beta t}}. \tag{18}$$

Special case 3. For $\alpha^* = \ln 2 = 0.69314718...$ we obtain the special Gompertz function:

$$\sigma_{\alpha^*,\beta}(t) = e^{-\alpha^* e^{-\beta t}}, \tag{19}$$

such that $\sigma_{\alpha^*,\beta}(0) = 1/2$.

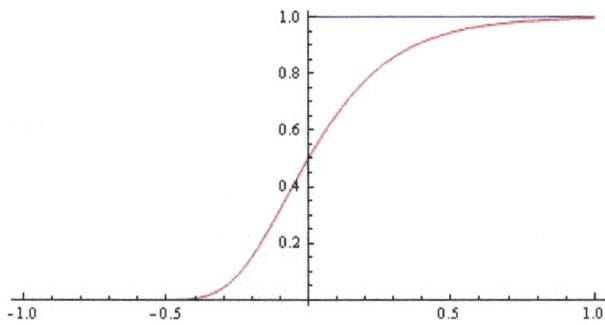

Fig. 5. The Gompertz function with $\alpha = \ln 2$ and $\beta = 5$; H-distance $d = 0.212765$.

We study the Hausdorff approximation of the Heaviside step function $c_0 = h_0(t)$ by Gompertz functions of the form (18) and find an expression for the error of the best approximation.

The H-distance $d = d(\alpha^*, \beta)$ between the Heaviside step function $h_0(t)$ and the Gompertz function (19) satisfies the relation

$$\sigma_{\alpha^*,\beta}(d) = e^{-\alpha^* e^{-\beta d}} = 1 - d,$$

or

$$\ln(1 - d) + \alpha^* e^{-\beta d} = 0. \qquad (20)$$

The following theorem gives upper and lower bounds for $d(\alpha^*, \beta)$. For brevity we denote $\alpha = \alpha^*$ in Theorem 1 and its proof.

Theorem 1. The Hausdorff distance $d = d(\alpha, \beta)$ between the step function h_0 and the Gompertz function (19) can be expressed in terms of the parameter β for any real $\beta \geq 2$ as follows:

$$\frac{2\alpha - 1}{1 + \alpha\beta} < d < \frac{\ln(1 + \alpha\beta)}{1 + \alpha\beta}. \qquad (21)$$

Proof. We need to express d in terms of α and β, using (20). Let us examine the function $F(d) = \ln(1 - d) + \alpha e^{-\beta d}$. From

$$F'(d) = -\frac{1}{1 - d} - \alpha\beta e^{-\beta d} < 0$$

we conclude that the function F is strictly monotone decreasing. Consider function $G(d) = \alpha - (1 + \alpha\beta)d$. From Taylor expansion

$$\alpha - (1 + \alpha\beta)d - \ln(1 - d) - \alpha e^{-\beta d} = O(d^2)$$

we obtain $G(d) - F(d) = \alpha - (1 + \alpha\beta)d - \ln(1 - d) - \alpha e^{-\beta d} = O(d^2)$. Hence $G(d)$ approximates $F(d)$ with $d \to 0$ as $O(d^2)$. In addition $G'(d) = -(1 + \alpha\beta) < 0$. Further, for $\beta \geq 2$,

$$G\left(\frac{2\alpha - 1}{1 + \alpha\beta}\right) = 1 - \alpha > 0,$$

$$G\left(\frac{\ln(1 + \alpha\beta)}{1 + \alpha\beta}\right) = \alpha - \ln(1 + \alpha\beta) < 0.$$

This completes the proof of the theorem.

Some computational examples using relation (20) are presented in Table 2.

β	$d(\alpha^*, \beta)$
2	0.310825
5	0.212765
10	0.147136
50	0.0514763
100	0.0309364
500	0.00873829
1000	0.00494117

TABLE II
BOUNDS FOR $d(\alpha^*, \beta)$ COMPUTED BY (20) FOR VARIOUS β.

The calculation of the value of the H-distance between the Gompertz sigmoid function and the Heaviside step function is given in Appendix 1.

The numerical results are plotted in Fig. 5 (for the case $\alpha^* = \ln 2$, $\beta = 5$, H-distance $d = 0.212765$) and Fig. 6 (for the case $\alpha^* = \ln 2$, $\beta = 20$, H-distance $d = 0.0962215$).

Remark 1. For some comparisons of the Gompertz and logistic equation from both practical and theoretical perspective, see [6], [8], [40]. As can be seen from Figure 6 the graph of the Gompertz function is "skewed", it is not symmetric with respect to the inflection point. In biology, the Gompertz function is commonly used to model growth process where the period of increasing growth is shorter than the period in which growth decreases [8], [33].

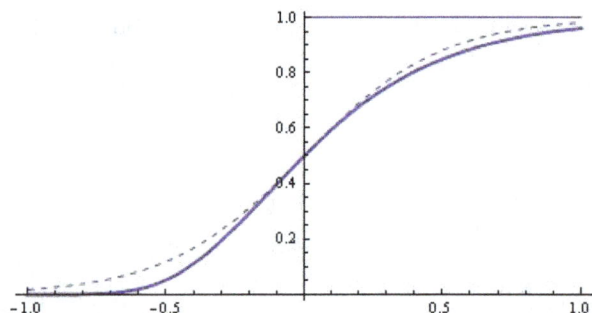

Fig. 6. The logistic (dotted line) and the Gompertz function (dense line) with same point and same rate (at that point).

Remark 2. For $k > 0, \beta > 0$ consider the differential equation

$$y' = ke^{-\beta t}y, \quad \frac{k}{\beta} = \alpha. \tag{22}$$

We have

$$\frac{dy}{dt} = ke^{-\beta t}y; \quad \frac{dy}{y} = ke^{-\beta t}dt$$

$$\ln y = -\frac{k}{\beta}e^{-\beta t} = -\alpha e^{-\beta t}; \quad y = e^{-\alpha e^{-\beta t}}.$$

We see that the solution of differential equation (22) is the Gompertz function $\sigma_{\alpha,\beta}(t)$ (18) [6]). As shown in [28], equation (22) can be interpreted as $y' = ksy$, wherein $s = s(t)$ is the nutrient substrate used for the growth of the population; one see that s is a decay exponential function in the Gompertz model (a similar interpretation can be found in [21]), [40]). For other interpretations see [6]), [8], [20].

VI. Conclusion

In this paper we discuss several computational, modelling and approximation issues related to two familiar classes of sigmoid functions—the logistic (Verhulst) and the Gompertz functions. Both classes find numerous applications in various fields of life sciences, ecology, medicine, artificial neural networks, fuzzy set theory, etc.

bigskip

We study the uniform and Hausdorff approximation of the cut functions by logistic functions. We demonstrate that the best uniform approximation between a cut function and the respective logistic function is an absolute constant not depending on the (largest) slope k. On the other side we show that the Hausdorff distance (H-distance) depends on the slope k and tends to zero with $k \to \infty$. We also discuss the limiting case when the cut function tends to the Heaviside step function in Hausdorff sense, thereby extending a related previous result [4].

The approximation of the cut function by a family of squashing functions induced by the logistic function is also discussed. We propose a new estimate for the H-distance between a cut function and its best approximating squashing function. Our estimate extends a known result stating that the cut function can be approximated arbitrary well by squashing functions [18]. Our estimate is also extended to cover the limiting case of the Heaviside step function.

Finally we study the approximation of the cut and step functions by the family of Gompertz functions. New estimates for the H-distance between a cut function and its best approximating Gompertz function are obtained.

References

[1] A. Akanuma, *Parameter Analysis of Gompertz Function Growth Model in Clinical Tumors*, European J. of Cancer 14 (1978) 681–688.

[2] G. Albano and V. Giorno, *On the First Exit Time Problem for a Gompertz-type Tumor Growth*, Lecture Notes in Computer Science 5717 (2009) 113–120, http://dx.doi.org/10.1007/978-3-642-04772-5_16

[3] R. Alt and S. Markov, *Theoretical and Computational Studies of some Bioreactor Models*, Computers and Mathematics with Applications 64(3) (2012) 350–360, http://dx.doi.org/10.1016/j.camwa.2012.02.046

[4] R. Anguelov and S. Markov, *Hausdorff Continuous Interval Functions and Approximations*, LNCS (SCAN 2014 Proceedings), to appear.

[5] I. A. Basheer and M. Hajmeer, *Artificial Neural Networks: Fundamentals, Computing, Design, and Applications*, Journal of Microbiological Methods 43(1) (2000) 3–31, http://dx.doi.org/10.1016/S0167-7012(00)00201-3

[6] Z. Bajzer and S. Vuk-Pavlovic, *New Dimensions in Gompertz Growth*, J. of Theoretical Medicine 2(4) (2000) 307–315, http://dx.doi.org/10.1080/10273660008833057

[7] D. E. Bentila, B. M. Osei, C. D. Ellingwood and J. P. Hoffmann, *Analysis of a Schnute Postulate-based Unified Growth Mode for Model Selection in Evolutionary Computations*, Biosystems 90(2) (2007) 467–474, http://dx.doi.org/10.1016/j.biosystems.2006.11.006

[8] R. D. Berger, *Comparison of the Gompertz and Logistic Equation to Describe Plant Disease Progress*, Phytopathology 71 (1981) 716–719, http://dx.doi.org/10.1094/Phyto-71-716

[9] M. Carrillo and J. M. Gonzalez, *A New Approach to Modelling Sigmoidal Curves*, Technological Forecasting and Social Change 69(3) (2002) 233–241, http://dx.doi.org/10.1016/S0040-1625(01)00150-0

[10] M. E. Cayre, G. Vignolob and O. Garroa, *Modeling Lactic Acid Bacteria Growth in Vacuum-packaged Cooked Meat Emulsions Stored at Three Temperatures*, Food Microbiology 20(5) (2003) 561–566, http://dx.doi.org/10.1016/S0740-0020(02)00154-5

[11] Y. Chalco-Cano, H. Roman-Flores and F. Gomida, *A New Type of Approximation for Fuzzy Intervals*, Fuzzy Sets and Systems 159(11) (2008) 1376–1383, http://dx.doi.org/10.1016/j.fss.2007.12.025

[12] Z. Chen and F. Cao, *The Approximation Operators with Sigmoidal Functions*, Computers & Mathematics with Applications 58(4) (2009) 758–765, http://dx.doi.org/10.1016/j.camwa.2009.05.001

[13] Z. Chen and F. Cao, *The Construction and Approximation of a Class of Neural Networks Operators with Ramp Functions*, Journal of Computational Analysis and Applications 14(1) (2012) 101–112.

[14] Z. Chen, F. Cao and J. Hu, *Approximation by Network Operators with Logistic Activation Functions*, Applied Mathematics and Computation 256 (2015) 565–571, http://dx.doi.org/10.1016/j.amc.2015.01.049

[15] E. S. Chumerina, *Choice of Optimal Strategy of Tumor Chemotherapy in Gompertz Model*, J. Comp. and Syst. Sci. Int. 48(2) (2009) 325–331, http://dx.doi.org/10.1134/S1064230709020154

[16] D. Costarelli and R. Spigler, *Approximation Results for Neural Network Operators Activated by Sigmoidal Functions*, Neural Networks 44 (2013) 101–106, http://dx.doi.org/10.1016/j.neunet.2013.03.015

[17] D. Costarelli and R. Spigler, *Constructive Approximation by Superposition of Sigmoidal Functions*, Anal. Theory Appl. 29(2) (2013) 169–196, http://dx.doi.org/10.4208/ata.2013.v29.n2.8

[18] J. Dombi and Z. Gera, *The Approximation of Piecewise Linear Membership Functions and Lukasiewicz Operators*, Fuzzy Sets and Systems 154(2) (2005) 275–286, http://dx.doi.org/10.1016/j.fss.2005.02.016

[19] H. Enderling and M. A. J. Chaplain, *Mathematical Modeling of Tumor Growth and Treatment*, Curr. Pharm. Des. 20(30) (2014) 4934–4940, http://dx.doi.org/10.2174/13816128196661131125150434

[20] R. I. Fletcher, *A General Solution for the Complete Richards Function*, Mathematical Biosciences 27(3-4) (1975) 349–360, http://dx.doi.org/10.1016/0025-5564(75)90112-1

[21] J. France, J. Dijkstra and M. S. Dhanoa, *Growth Functions and Their Application in Animal Science*, Annales de Zootechnie 45(Suppl 1) (1996) 165–174.

[22] B. Gompertz, *On the Nature of the Function Expressive of the Law of Human Mortality, and on a New Mode of Determining the Value of the Life Contingencies*, Philos. Trans. R. Soc. London 115 (1825) 513–585.

[23] J. Han and C. Morag, *The Influence of the Sigmoid Function Parameters on the Speed of Backpropagation Learning*, In: Mira, J., Sandoval, F. (Eds) From Natural to Artificial Neural Computation 930 (1995) 195–201, http://dx.doi.org/10.1007/3-540-59497-3_175

[24] F. Hausdorff, *Set Theory* (2 ed.), New York, Chelsea Publ. (1962 [1957]) (Republished by AMS-Chelsea 2005), ISBN: 978–0–821–83835–8.

[25] M. Kodaka, *Requirements for Generating Sigmoidal Time-course Aggregation in Nucleation-dependent Polymerization Model*, Biophys. Chem. 107(3) (2004) 243–253, http://dx.doi.org/10.1016/j.bpc.2003.09.013

[26] M. Kodaka, *Interpretation of Concentration-dependence in Aggregation Kinetics*, Biophys. Chem. 109(2) (2004) 325–332, http://dx.doi.org/10.1016/j.bpc.2003.12.003

[27] M. Koivula, M. Sevon-Aimonen, I. Stranden, K. Matilainen, T. Serenius, K. Stalder and E. Mantysaari, *Genetic (Co)Variances and Breeding Value Estimation of Gompertz Growth Curve Parameters in Finish Yorkshire Boars, Gilts and Barrows*, J. Anim. Breed. Genet. 125(3) (2008) 168–175, http://dx.doi.org/10.1111/j.1439-0388.2008.00726.x

[28] N. Kyurkchiev and S. Markov, *Sigmoidal Functions: Some Computational and Modelling Aspects*, Biomath Communications 1(2), (2014), http://dx.doi.org/10.11145/j.bmc.2015.03.081

[29] N. Kyurkchiev and S. Markov, *On the Hausdorff Distance Between the Heaviside Step Function and Verhulst Logistic Function*, J. Math. Chem., to appear.

[30] N. Kyurkchiev and S. Markov, *Sigmoid Functions: Some Approximation and Modelling Aspects. Some Moduli in Programming Environment Mathematica*, LAP (Lambert Acad. Publ.) (2015), ISBN: 978–3–659–76045–7.

[31] T. P. Labuza and B. Fu, *Growth Kinetics for Shelf-life Prediction: Theory and Practice*, Journal of Industrial Microbiology 12(3–5) (1993) 309–323, http://dx.doi.org/10.1007/BF01584208

[32] A. K. Laird, *Dynamics of Tumor Growth*, Br. J. Cancer 18(3) (1964) 490–502.

[33] D. Lin, Z. Shkedy, D. Yekutieli, D. Amaratunda and L. Bijnens (Eds.), *Modeling Dose Responce Microarray Data in Early Drug Development Experiments Using R*, Springer (2012), ISBN: 978–3–642–24006–5.

[34] S. Markov, *Cell Growth Models Using Reaction Schemes: Batch Cultivation*, Biomath 2(2) (2013), 1312301, http://dx.doi.org/10.11145/j.biomath.2013.12.301

[35] A. G. McKendrick and M. Kesava Pai, *The Rate of Mul-*

tiplication of Micro-organisms: A Mathematical Study, Proc. of the Royal Society of Edinburgh 31 (1912) 649–653, http://dx.doi.org/10.1017/S0370164600025426

[36] N. Radchenkova, M. Kambourova, S. Vassilev, R. Alt and S. Markov, *On the Mathematical Modelling of EPS Production by a Thermophilic Bacterium*, Biomath 3(1) (2014), 1407121, http://dx.doi.org/10.11145/j.biomath.2014.07.121

[37] F. J. Richards, *A Flexible Growth Function for Empirical Use*, J. Exp. Bot. 10 (1959) 290–300, http://dx.doi.org/10.1093/jxb/10.2.290

[38] R. Rickles and A. Scheuerlein, *Biological Imlications of the Weibull and Gompertz Models of Aging*, J. of Gerontology: Biologica Sciences 57(2) (2002) B69–B76, http://dx.doi.org/10.1093/gerona/57.2.B69

[39] A. Sas, H. Snieder and J. Korf, *Gompertz' Survivorship Law as an Intrinsic Principle of Aging*, Medical Hypotheses 78(5) (2012) 659–663, http://dx.doi.org/10.1016/j.mehy.2012.02.004

[40] M. A. Savageau, *Allometric Morphogenesis of Complex Systems: Derivation of the Basic Equations from the First Principles*, Proc. of the National Academy of Sci. USA 76(12) (1979) 6023–6025.

[41] B. Sendov, *Hausdorff Approximations*, Kluwer (1990), ISBN: 978–94–010–6787–4, e-ISBN: 978–94–009–0673–0, http://dx.doi.org/10.1007/978-94-009-0673-0

[42] C. J. Stannard, A. P. Williams and P. A. Gibbs, *Temperature/growth Relationship for Psychrotrophic Food-spoilage Bacteria*, Food Microbiol. 2(2) (1985) 115–122, http://dx.doi.org/10.1016/S0740-0020(85)80004-6

[43] J. H. Van der Walt, *The Linear Space of Hausdorff Continuous Interval Functions*, Biomath 2(2) (2013), 1311261, http://dx.doi.org/10.11145/j.biomath.2013.11.261

[44] P.-F. Verhulst, *Notice Sur la Loi Que la Population Poursuit dans Son Accroissement*, Correspondance Mathematique et Physique 10 (1838) 113–121.

K. van't Riet, *Evaluation of Data Transformations and Validation of a Model for the Effect of Temperature on Bacterial Growth*, Appl. Environ. Microbiol. 60(1) (1994) 195–203.

[45] P.-F. Verhulst, *Recherches Mathematiques sur la Loi D'accroissement de la Population (Mathematical Researches into the Law of Population Growth Increase)*, Nouveaux Memoires de l'Academie Royale des Sciences et Belles-Lettres de Bruxelles 18 (1845) 1–42.

[46] P.-F. Verhulst, *Deuxieme Memoire sur la Loi D'accroissement de la Population*, Memoires de l'Academie Royale des Sciences, des Lettres et des Beaux-Arts de Belgique 20 (1847) 1–32.

[47] D. L. Wilson, *The Analysis of Survival (Mortality) Data: Fitting Gompertz, Weibull, and Logistic Functions*, Mech. Ageing Devel. 74(1–2) (1994) 15–33, http://dx.doi.org/10.1016/0047-6374(94)90095-7

[48] C. P. Winsor, *The Gompertz Curve as a Growth Curve*, Proceedings of the National Academy of Sciences 18(1) (1932) 1–8.

[49] M. H. Zwietering, I. Jongenburger, F. M. Rombouts and K. van't Riet, *Modeling of the Bacterial Growth Curve*, Appl. Envir. Microbiol. 56(6) (1990) 1875–1881.

[50] M. H. Zwietering, H. G. Cuppers, J. C. de Wit and

APPENDIX 1.

The Module "Computation of the distance d and visualization of the cut function c_Δ and squashing function $s_\Delta^{(\beta)}$" in *CAS MATHEMATICA*.

```
Print["Calculation of the value of the distance d (see Eq. (16)) and graphical visualization
of the generalized cut function [x]_{0,r} and the interval squashing function S_{0,r}^k"];

k = Input[" k"];(*5 *)
Print["The parameter -  k = ", k];

r = Input[" r"];(*5 *)
Print["The parameter -  r = ", r];

Print["The following nonlinear equation is used to determination of the distance d: "];

m = Log[(1 + Exp[k * (r + d)]) / (1 + Exp[d * k])] - k * r * (1 - d);

Print[m, " = 0"];
Print["The unique positive root of the equation is the searched value of d:"];
FindRoot[m == 0, {d, 0}];
Print[TableForm[%]];

Print["Graphical visualization of the generalized cut function [x]_{0,r}:"];
pw = Piecewise[{{0, x ≤ 0}, {x/r, 0 < x < r}, {1, r ≤ x}}]
g1 = Plot[pw, {x, -1, 7}]

Print["Graphical visualization of the the interval squashing function S_{0,r}^k:"];
g2 = Plot[1/r * Log[((1 + Exp[k * t]) / (1 + Exp[k * (t - r)]))^(1/k)], {t, -1, 7}]

Print["Comparing of both graphical visualizations:"];
Show[g1, g2]
```

Fig. 7.   Module in programming environment *MATHEMATICA*.

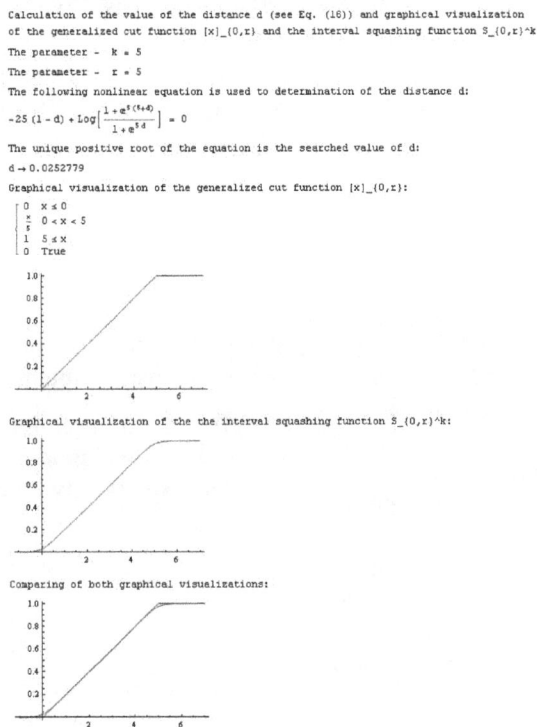

Fig. 8.   The test provided on our control example.

4

Biological control of sugarcane caterpillar (Diatraea saccharalis) using interval mathematical models

José Renato Campos *, Edvaldo Assunção †, Geraldo Nunes Silva ‡ and Weldon Alexander Lodwick §
* Area of Sciences
Federal Institute of Education, Science and Technology of São Paulo, Votuporanga, SP, Brazil
jrcifsp@ifsp.edu.br, jrcifsp@gmail.com
† Department of Electrical Engineering
UNESP - Univ Estadual Paulista, Ilha Solteira, SP, Brazil
edvaldo@dee.feis.unesp.br
‡ Department of Applied Mathematics
UNESP - Univ Estadual Paulista, São José do Rio Preto, SP, Brazil
gsilva@ibilce.unesp.br
§ Department of Mathematical and Statistical Sciences
University of Colorado, Denver, Colorado, USA
weldon.lodwick@ucdenver.edu

Abstract—**Biological control is a sustainable agricultural practice that was introduced to improve crop yields and has been highlighted among the various pest control techniques. However, real mathematical models that describe biological control models can have error measurements or even incorporate lack of information. In these cases, intervals may be feasible for indicating the lack of information or even measurement errors. Therefore, we consider interval mathematical models to represent the biological control problem. Specifically, in the present paper, we illustrate the solution of a discrete-time interval optimal control problem for a practical application in biological control. To solve the problem, we use single-level constrained interval arithmetic [9] and the dynamic programming technique [3]** along with the idea proposed in [23] for the solution of the interval problem.

Keywords-Interval optimal control problem; interval mathematical models; single-level constrained interval arithmetic; dynamic programming; biological control.

I. INTRODUCTION

Sugarcane culture plays an important role in the Brazilian economy. It is estimated that the country has more than 8 million hectares of cultivated area [1] and that sugarcane is responsible for over 4.5 million jobs [38]. In addition to the production of sugar, ethanol and various other byproducts, it is also used to produce electricity with the use of biomass (bagasse and straw). Thus, sustainable management of this culture is fundamental. Among the various types of management that can be implemented (control of pests and weeds, soil handling, etc.) and the various methods of manufacture (biological control, use of insecticide and herbicide, manual and mechanical control, etc.), pest control through biological control stands out.

Biological control is sustainable because it does not affect the environment. For the culture of sugarcane, the control involves the caterpillar and wasp. The caterpillar (Diatraea Saccharalis) is an insect that causes damage to the crop, and its natural predator, Cotesia Flavipes, is a wasp that deposits its eggs on the caterpillar and inhibits the development of the caterpillar. Hence, the caterpillar dies without completing its life cycle and without causing economic loss to the crop.

The spread of the caterpillar can cause damage to the crop such as weight loss and reduction in germination, leading to the death of germinating plants, which directly reflects on the costs of production. Thus, the biological control of pests is a good alternative to the feasibility of such crops for the country. In addition, the biological control process is part of the integrated crop protection [11] that is a benchmark for sustainable farming practices.

Control theory study began in the USA in the 1930s with studies of problems in electrical engineering and mechanical engineering [8]. In the 1950s, with optimization methods developed by Bellmann in 1957 (see [2]) and Pontryagin in 1958 (see [29], [30]), modern control theory or optimal control theory was born. Such theory brought advances in several areas such as Agriculture, Biology, Economics, Engineering and Medicine.

In Agriculture or Biology, deterministic optimal control problems are widely studied, and some biomathematical models illustrating deterministic models can be found in [7], [15], [16], [19], [37]. In these studies, conventional models were assumed with fixed coefficients.

For problems with uncertain parameters, the optimal control problem usually utilizes stochasticity

[4], [16] or, more recently, fuzzy set theory [12], [10], [28]. In the two cases, the coefficients are viewed as random variables or as fuzzy sets, and it is assumed that their probability distributions or membership functions, respectively, are known.

In biological problems, uncertainty arises frequently because it is inherent to the determination of biological data; for example, uncertainty arises due to measurement errors, inaccuracies in the equipment, climatic factors, and lack of specification, among many others. Thus, we propose interval uncertainty to describe the uncertainty in obtaining data in biological problems. We can represent a parameter of the model, such as the mortality rate of predators, as an interval. This is relevant because we can model an environment with several variations in the mortality rates of predators and not have to consider a unique rate for all the predators, especially if this information has been obtained imprecisely.

Optimal control problems involving uncertain systems are described in [6], [13], [14], [39]. However, in these approaches, the functional is a real number and thus differs from the approach proposed in this paper. Additionally, the problem discussed here does not include state feedback. References on control problems that present interval uncertainty but still differ from that proposed in this paper can be found in [20], [17], [32].

Thus, in this work, we consider a new kind of problem called the *interval optimal control problem*. The interval arithmetic used in this approach is described in [9], [21], [22] and is different from the standard interval arithmetic proposed in [24]. To solve the interval optimal control problem, we choose single-level constrained interval arithmetic [9] because it eliminates certain problems related to other types of interval arithmetic, such as the existence of the additive inverse or the distributive law property. Single-level constrained interval arithmetic also has properties closer to the space of real numbers. Therefore, we study the discrete time interval optimal control problem with the interval initial condition or interval parameters in the dynamic equation.

The paper is arranged as follows. Section II presents the application in Biology, and some biological aspects will be demonstrated. We also present the deterministic and interval optimal control problem. In Section III, we present the solutions of the discrete time interval optimal control problems previously proposed. The discussion of the results is provided in Section IV.

II. THE BIOMATHEMATICAL MODEL

The biological situation studied is a problem encountered in sugarcane culture. According to Silva and Bergamasco [35], the environmental management of sugarcane culture requires performance prediction in production and environmental risk at various levels of control in sugarcane production because manipulation of the soil, planting depth and density, pest and diseases, among other factors, and biological control have proven to be effective in operational management of the culture.

Thus, the problem studied corresponds to a model of competition between the wasp (Cotesia Flavipes) and the caterpillar (Diatraea Saccharalis) in terms of sugarcane, represented using the Lotka-Volterra two-species model.

Tusset and Rafikov in [37] ran a simulation of the dynamics of the system without application control and showed that the system begins to stabilize at 350 days and that during this period, economic losses are experienced. Thus, we need to apply control in previous periods, and the application of control corresponds to the introduction of wasps in sugarcane culture.

Tusset and Rafikov [37] solve the continuous deterministic optimal control problem using the Riccati equation. Campos [7] also solved the deterministic and discrete problem using dynamic programming, and the results are similar for the two approaches.

The goal here is to present the interval optimal control problem and solve the biological control problem encountered in sugarcane culture. We analyze the biological situation and describe the biomathematical model. According to Tusset and Rafikov [37], the Lotka-Volterra two-species model used in the problem of sugarcane culture is given by

$$\begin{cases} \dot{x} = x\,(a - \gamma\,x - c\,y) \\ \dot{y} = y\,(-d + r\,x) + u^* + u \end{cases}, \qquad (1)$$

where $x(t)$ is the number of preys and $y(t)$ is the number of predators for $t \geq 0$. Here, u^* is the control that carries the system to the desired equilibrium point, and u is the control that stabilizes the system at this point.

The dynamic model (1) is a Lotka-Volterra model for the case of the caterpillar that is the sugarcane parasitoid, where the coefficient a represents the interspecific growth of the preys, the coefficient d represents the mortality of the predators, c represents the capture rate, r is the maximum rate of growth of the predator population, and γ is the self-inhibition coefficient of growth of the preys due to restriction of food.

According to [37], the parameter a is calculated assuming the absence of predators in (1). Then, we obtain

$$\dot{x} = x\,(a - \gamma\,x), \qquad (2)$$

where we suppose that $\gamma = a/k$. Solving the differential equation (2) and isolating the value of the parameter a, we obtain

$$a = -\frac{1}{t}\left[ln\left(\frac{\frac{k-x}{x}}{\frac{k-x^0}{x^0}} \right) \right].$$

Assuming $k = 25000$ and considering that the caterpillar lives on average 70 days and after mating lays on average 300 eggs (see [27]), we find that $t = 70$ days with $x(70) = 300$ caterpillars per hectare. Assuming an initial number of preys equal to $x^0 = 2$ caterpillars per hectare, it follows that the interspecific growth of the caterpillar is $a = 0.0716$ caterpillars per hectare per day.

The calculation of the other parameters of the dynamic equation of problem (1) can be found in [37], following a similar analysis.

Thus, in this work, we obtain the numerical coefficients a, γ, c, d and r in [37] as well as the expression for the functional of the optimal control problem. The problem proposed in [37] with a quadratic objective function subject to nonlinear restrictions is given by

$$minC = \frac{1}{2} \int_0^{t_f} 8(x-x^*)^2 + 0.2841(y-y^*)^2 + u^2 dt$$

subject to

$$\begin{cases} \dot{x} = x\,(0.0716 - 0.0000029\,x - 0.0000464\,y) \\ \dot{y} = y\,(-1 + 0.000520235\,x) + u^* + u \end{cases},$$

(3)

where t_f is the final time, the initial conditions are $x_0 = 5000$ and $y_0 = 1500$, and the final conditions are the desired equilibrium point (x^*, y^*). From a biological point of view, Segato et al. [33] show that when the number of preys (Diatraea Saccharalis) reaches 5000 caterpillars per hectare, application of control u corresponds to the release of predators (Cotesia Flavipes).

The calculations for the numerical coefficients of the states x and y in the functional of problem (3) are extensive and can be found in [37]; such calculations are based on [34], [31]. Furthermore, Tusset and Rafikov consider in [37] a positive semidefinite and symmetric quadratic functional in order to take the system to the desired equilibrium point the fastest way possible when considering only small oscillations in the path of the system. This is important for the biological control problem studied.

To solve problem (3), Tusset and Rafikov in [37] considered a problem with a linear dynamic equation. The linearization of the model is feasible because we suppose that the linear and nonlinear dynamic system behaviors are qualitatively equivalent in the vicinity of the equilibrium point (see [25], Grobman-Hartman Theorem). Thus, the dynamic equation of problem (3) is linearized (see [25]) assuming that the initial conditions are near the equilibrium point (2000, 1418.10). In a real system, this is possible when we apply a value several times that of the control.

According to Botelho and Macedo [5] for the sugarcane crop, greater than or equal to 2500 caterpillars per hectare causes damage to the

culture. We fix $x^* = 2000$ (a value that does not cause damage) and hence obtain the value y^* using the equation $f(x^*, y^*) = 0$, where $f(x, y) = 0.0716 - 0.0000029\,x - 0.0000464\,y$. Therefore, the desired equilibrium point for the prey and the predator is represented by $(x^*, y^*) = (2000, 1418.10)$ and used in the final condition of the problem.

Finally, the optimal control problem with a quadratic objective function subject to linear restrictions proposed in [37] is given by

$$min\,C = \frac{1}{2} \int_0^{t_f} 8 z_1^2 + 0.2841 z_2^2 + u^2 \, dt$$

subject to

$$\dot{z} = \begin{bmatrix} -0.0058 & -0.0928 \\ 0.7386 & 0.0405 \end{bmatrix} z + \begin{bmatrix} 0 \\ 1 \end{bmatrix} u, \quad (4)$$

with initial conditions $z_{10} = 3000$ and $z_{20} = 80.17$ due to translation to the equilibrium point. Note that $z = (z_1, z_2)^T = (x-x^*, y-y^*)^T$, where z is the translation of the point of equilibrium (x^*, y^*) to the origin and T denotes the transposed vector. In particular, the change in coordinates to problem (4) is performed assuming that we are close to the fixed point; furthermore, the change in coordinates facilitates the computational implementation.

To find the solution of the problem of biological control of the sugarcane caterpillar (4) with a discrete dynamic programming method, Campos [7] discretized problem (4).

The discrete model (and match) proposed in [7] is

$$min\,C = \frac{h}{2} \sum_{k=0}^{N} 8 z_{1\,k}^2 + 0.284 z_{2\,k}^2 + u_k^2$$

subject to

$$z_{k+1} = \begin{bmatrix} 0.960 & -0.093 \\ 0.743 & 1.006 \end{bmatrix} z_k + \begin{bmatrix} -0.047 \\ 1.009 \end{bmatrix} u_k,$$

(5)

where $z_k = (z_{1k}, z_{2k})^T$ and the initial conditions are $z_{10} = 3000$ and $z_{20} = 80.17$. Here, k denotes the discrete iterations in days for the problem. Furthermore, for problem (5), the simulation period equals $N = 18$ days, and hence, $t_f = hN = 18$ days.

Campos in [7] used the zero-order hold method (function $c2d$ in MATLAB 7.4) to discretize the dynamic equation of problem (4). Thus, the zero-order hold method provides an exact match between the continuous dynamic system of problem (4) and the discrete dynamic system of problem (5). For the biological analysis of the optimal control problem, we are assuming that the control decision u_k, introduction of predators, occurs only once a day.

The discretization of the functional of problem (4) introduces an error because it is approximated using a numerical quadrature. However, the error in the discretization of the functional does not change the behavior of the dynamic equations of problems (4) and (5). Furthermore, the weight assigned to the coefficient of control u_k in the functional of problem (5) can be modified and adapted according to the costs involved in the operations.

Next, we illustrate the formulation of interval control problems for two distinct situations. The first involves the problem with the interval initial condition. The second formulation considers an interval coefficient in the dynamic equation.

A. Uncertainty in the Initial Condition

Suppose that the model (5) uses the interval initial condition because we consider there to be inaccurate information in the data. We use an interval initial condition of $Z_{10} = [2970, 3030]$, which represents an error of 2%. The second initial condition used is $Z_{20} = 80.17$ and represents a degenerate interval.

Therefore, the problem with the interval initial condition is described below. It is given by

$$min\,\mathbf{C} = \frac{h}{2} \otimes \sum_{k=0}^{N} 8 \otimes Z_{1k}^2 \oplus 0.284 \otimes Z_{2k}^2 \oplus U_k^2$$

subject to

$$\begin{cases} Z_{1k+1} = 0.960 \otimes Z_{1k} \ominus 0.093 \otimes Z_{2k} \ominus 0.047 \otimes U_k \\ Z_{2k+1} = 0.743 \otimes Z_{1k} \oplus 1.006 \otimes Z_{2k} \oplus 1.009 \otimes U_k \end{cases}$$
(6)

where Z_{1k}, Z_{2k}, U_k and $\mathbf{C}$ are intervals and the initial conditions are $Z_{10} = [2970, 3030]$ and $Z_{20} = 80.17$. For the interval problem, the symbols $\oplus, \ominus, \otimes$ and $\oslash$ represent the sum, subtraction, multiplication and division of intervals, respectively, according to single-level constrained interval arithmetic. This model is presented in [7]; however, here it is presented as an interval problem. In particular, the initial condition is also an interval. Problem (6) is called the interval optimal control problem. Furthermore, we emphasize that the functional is an interval and that its optimality is given by the order relation of single-level constrained interval arithmetic (see [18]). According to Leal [18], given two intervals $A = [\underline{a}, \bar{a}]$ and $B = [\underline{b}, \bar{b}]$, the order relation between them is given by

$$A \leq_{SL} B \text{ iff } A(\lambda) \leq B(\lambda) \text{ for all } \lambda \in [0, 1],$$

where $\leq_{SL}$ denotes the inequality between intervals according to single-level constrained interval arithmetic and $A(\lambda)$ and $B(\lambda)$ are the convex constraint functions associated with A and B, respectively. Note that $A(\lambda) = (1 - \lambda)\underline{a} + \lambda \bar{a}$, $0 \leq \lambda \leq 1$.

Initially, the interval optimal control problem (6) can be transformed into a real classic problem using single-level constrained interval arithmetic. Thus, the interval optimal control problem (6), rewritten as the single-level constrained interval arithmetic [9], is given by

$$min\,C = \frac{h}{2} \sum_{k=0}^{N} 8 Z_{1k}^2(\lambda) + 0.284 Z_{2k}^2(\lambda) + U_k^2(\lambda)$$

subject to

$$Z_{k+1}(\lambda) = \begin{bmatrix} 0.960 & -0.093 \\ 0.743 & 1.006 \end{bmatrix} Z_k(\lambda) + \begin{bmatrix} -0.047 \\ 1.009 \end{bmatrix} U_k(\lambda),$$
(7)

where $Z_k(\lambda) = (Z_{1k}(\lambda), Z_{2k}(\lambda))^T$ and the initial conditions are $Z_{10}(\lambda) = 2970 + 60\lambda$ and $Z_{20}(\lambda) = 80.17$, $0 \leq \lambda \leq 1$. Here, $Z_k(\lambda)$ and $U_k(\lambda)$ are the convex constraint functions associated with intervals Z_k and U_k, respectively. Furthermore, we also suppose $Z_k(\lambda)$ and $U_k(\lambda)$ to have the appropriate dimensions.

Now, problem (7) is a classic optimal control problem for all fixed $\lambda \in [0, 1]$. Therefore, we use dynamic programming as our solution technique for the discrete time optimal control problem. The advantage of dynamic programming is that it determines the optimal solution of a multistage problem by breaking it into stages, where each stage is a subproblem. Solving a subproblem is a simpler task in terms of calculation than dealing with all the stages simultaneously. Moreover, a dynamic programming model is a recursive equation that links the different stages of the problem, ensuring that the optimal solution at each stage is also optimal for the entire problem (see [36]). Details on dynamic programming can be found in [3].

Finally, we solve problem (7) for all fixed $\lambda \in [0, 1]$, and we present the solution in the interval space in accordance with the ideas proposed in [9] and [23], i.e., we return the solution to the interval space using the minimum and maximum of the values obtained for each stage of the problem, provided that the minimum and maximum exist.

B. Uncertainty in the Dynamic Equation

For the interval problem with uncertainty in the dynamic equation, we consider again the biomathematical model (5) described previously. Suppose that, due to some biological factors, the first parameter of the first dynamic equation is an interval. Specifically, consider that due to some inaccuracy in obtaining the data for the model, the interval optimal control problem represents the first parameter of the dynamic equation as an interval, that is, the value 0.960 is substituted by the interval $[0.760, 1.160]$. This interval represents 41.67% of the error in relation to the deterministic value.

Therefore, the interval optimal control problem is

$$min \, \mathbf{C} = \frac{h}{2} \otimes \sum_{k=0}^{N} 8 \otimes Z_{1k}^2 \oplus 0.284 \otimes Z_{2k}^2 \oplus U_k^2$$

subject to

$$\begin{cases} Z_{1k+1} = [0.760, 1.160] \otimes Z_{1k} \ominus 0.093 \otimes Z_{2k} \ominus 0.047 \otimes U_k \\ Z_{2k+1} = 0.743 \otimes Z_{1k} \oplus 1.006 \otimes Z_{2k} \oplus 1.009 \otimes U_k \end{cases}$$
$$(8)$$

where Z_{1k}, Z_{2k}, U_k and $\mathbf{C}$ are intervals and the initial conditions are $Z_{10} = 3000$ and $Z_{20} = 80.17$ (degenerate intervals) due to translation of the equilibrium point.

Similar to Subsection II-A, we rewrite the interval problem according to single-level constrained interval arithmetic [9]. We then solve the corresponding problem using dynamic programming [3]. According to the methodology proposed in [23] and [9], we find the solution interval.

The numerical solution to the problems (6) and (8) will be presented in the next section.

III. NUMERICAL ANALYSIS AND SIMULATIONS

The implementation and adaptation of the dynamic programming algorithm to solve problems (6) and (8) were performed using MATLAB 7.4. Furthermore, problems (6) and (8) were solved using a microcomputer with a Dual-Core AMD E 300 processor and 3 GB of memory. For the interval problems, we chose $N = 18$ days. The computational time to solve problem (6) was approximately 4.5 minutes, and the computational time required to solve problem (8) was approximately 26 minutes.

The interval cost found in the solution is called the optimal interval cost. The interval state obtained is called the optimal interval state, and the interval control obtained for each iteration in the interval optimal control problem is called the optimal interval control.

The following figures represent the numerical results of problems (6) and (8). The deterministic and discrete solutions are also introduced in the figures. In the solutions presented, the translation of the solution has been reversed. In addition, the points representing the deterministic and interval

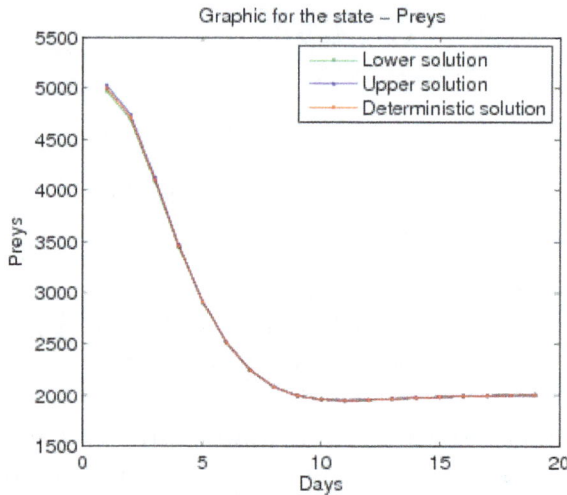

Fig. 1.   Preys for problem (6).

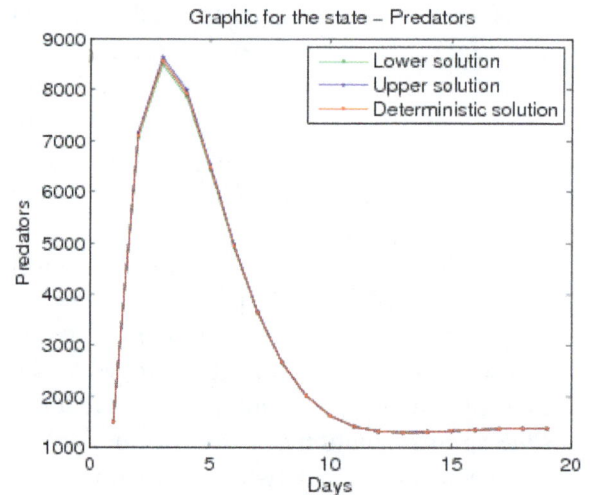

Fig. 2.   Predators for problem (6).

solutions to the problem are connected by line segments for facilitating the visualization of the temporal evolution. The solutions given by the minimum and maximum values correspond to the optimal interval solutions.

Graphical solutions are provided for the situations described in problem (6). Figure 1 illustrates the number of preys for the problem with uncertainty in the initial condition. Figure 2 illustrates the number of predators for the same problem. The predators are introduced in Figure 3, and the negative values that appear in the figure correspond to the number of predators that should be removed using some sustainable agricultural practice.

The optimal cost of the deterministic problem is 1.3716×10^8. The optimal interval cost of problem (6) is $[1.3443 \times 10^8, 1.3992 \times 10^8]$. Thus, the interval uncertainty inserted in the initial condition of the problem results in a variation in the cost of approximately 4.00% compared with the deterministic solution.

The graphical solution to problem (8) is presented below. Figure 4 illustrates the number of preys for this problem. Figure 5 illustrates the number of predators for (8). The values of the control variable are presented in Figure 6.

The optimal interval cost of problem (8) is

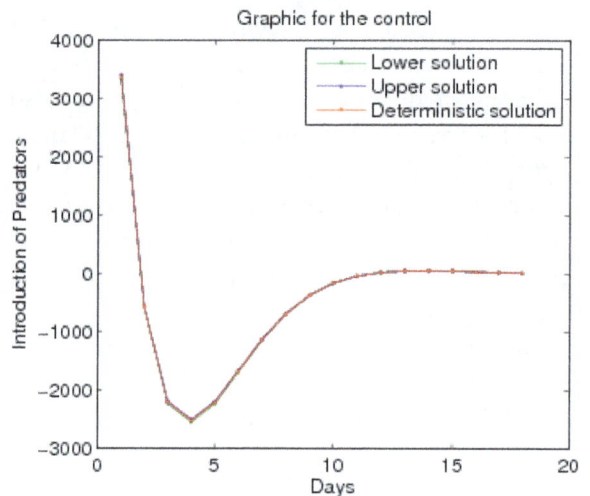

Fig. 3.   Introduction of predators for problem (6).

$[7.8523 \times 10^7, 2.7181 \times 10^8]$. The uncertainty introduced into the dynamic equation generated a variation of approximately 140.92% in the functional in relation to the deterministic solution.

Remark 3.1: The solutions of the interval problems (6) and (8) converge to the desired equilibrium point. The interval solutions converge to the desired equilibrium point if the distance between them tends to zero according to the definition of the distance between intervals given by [9]. Thus, the approximate interval X to x^* means that the

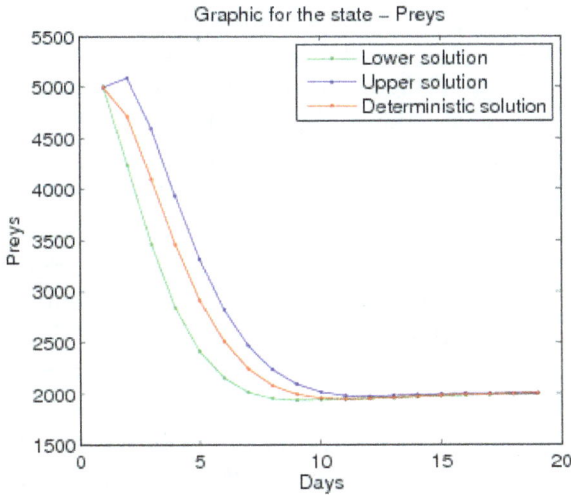

Fig. 4.   Preys for problem (8).

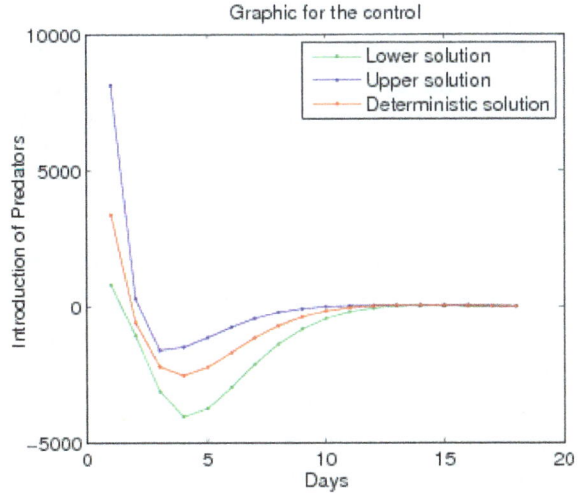

Fig. 5.   Predators for problem (8).

distance between them, given by $\max\limits_{0 \leq \lambda \leq 1} |X(\lambda) - x^*|$ where $X(\lambda)$ is a convex constraint function associated with X, tends to zero. Further, analyzing the interval problems (6) and (8) according to the associated convex constraint functions (see, for example, problem (7)), we have that the corresponding optimal control problems are classical optimal control problems for all fixed $\lambda \in [0, 1]$ and satisfy the stability criterion (see [26], [3]) for optimal control problems with quadratic functional and linear constraints.

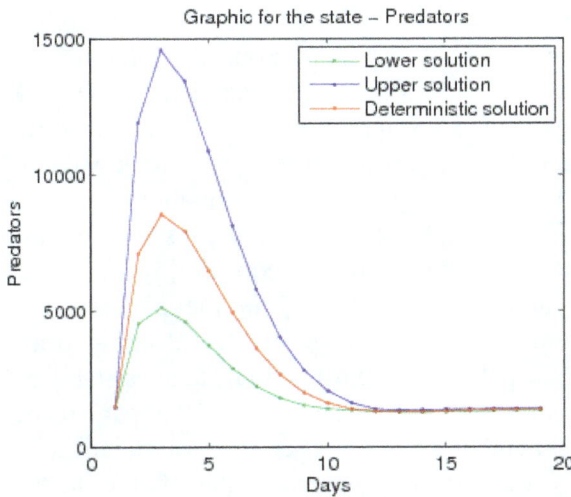

Fig. 6.   Introduction of predators for problem (8).

Remark 3.2: Other interval optimal control problems can be investigated, such as the problem with interval initial conditions and interval parameters in the interval dynamic equation. Thus, considering the interval optimal control problem given by

$$min\, \mathbf{C} = \frac{h}{2} \otimes \sum_{k=0}^{N} 8 \otimes Z_{1\,k}^2 \oplus 0.284 \otimes Z_{2\,k}^2 \oplus U_k^2$$

subject to

$$\begin{cases} Z_{1k+1} = [0.760, 1.160] \otimes Z_{1k} \ominus 0.093 \otimes Z_{2k} \ominus 0.047 \otimes U_k \\ Z_{2k+1} = 0.743 \otimes Z_{1\,k} \oplus 1.006 \otimes Z_{2\,k} \oplus 1.009 \otimes U_k \end{cases} \tag{9}$$

where $Z_{1\,k}$, $Z_{2\,k}$, U_k and $\mathbf{C}$ are intervals and the interval initial conditions are $Z_{1\,0} = [2970, 3030]$ and $Z_{2\,0} = 80.17$, we have that the optimal interval cost is given by $[7.6959 \times 10^7,\, 2.7729 \times 10^8]$. Furthermore, the solution of the interval problem (9) shows basically the same qualitative behavior as that of the solution of the interval problem (8).

IV. DISCUSSION OF THE RESULTS

In the problems studied, the initial condition or the dynamic equation has intervals because the data are generally inaccurate and may be represented by interval uncertainty. Consequently, this implies a variation in the functional, state and control at each iteration (cost, state and control represented by intervals). The decision maker should consider whether it is feasible to run the model for the values obtained in these intervals.

Therefore, to analyze if the number of preys or predators achieves the minimum or maximum values is an important question in the decision-making process of a manager because it can lead to financial loss and environmental damage. Furthermore, the analysis of interval costs is also very important for the company.

We now emphasize some points from the solutions obtained previously.

A. Analysis of the interval problem (6)

In the solution presented for the interval problem (6), we found consistency with the deterministic results as can be seen from Figures 1, 2 and 3. The behaviors of the interval state variable and interval control variable are also quite regular and in accordance with the variation of the deterministic solution. The extremes of the intervals of the state interval solutions X and Y approached the desired value, as was observed with the interval control U. Therefore, the decision maker obtains values close to those found for the deterministic solution; associated with this, we observe only a small variation in the functional. Thus, an error caused by lack of information in obtaining the initial condition generated small variations in cost and did not result in drastic changes for the decision maker.

B. Analysis of the interval problem (8)

For the interval problem (8), the behaviors of the interval state variable X and interval control variable U followed the same trajectory as that of the deterministic solution after the thirteenth day. Thus, for the state variable X (preys) and with the introduction of predators U, there was no large variation in comparison with the deterministic solution after the thirteenth day. However, in the initial periods, the introduction of predators U presented a large variation, with direct implications for agricultural practice of pest control.

We emphasize the large variation of the interval state variable Y, which represents the variation of the predators (Figure 5). For the third period, we obtained a variation of 5.1270×10^3 up to 1.4565×10^4 corresponding to the Y optimal interval state given by the interval $[5.1270 \times 10^3, 1.4565 \times 10^4]$. For this variable, we obtained an approximation of the extremes of the interval, which represents the interval solution, to the deterministic solution after the fifteenth day. Furthermore, the problem presents a large variation in the optimal interval cost.

Finally, we can conclude that the facts described above will certainly influence the company's decision making.

C. Conclusion

In Section III, we perceive that the optimal interval state X was approximately 2000 in problems (6) and (8). The optimal interval state Y (predators) also approximated the desired value. The optimal interval control tends to the value of 16 wasps per day for the two situations.

These values approximated the results presented in [5]. Botelho and Macedo in [5] show that the application of control in the population of caterpillars in the State of São Paulo - Brazil utilizing the parasitoid Cotesia Flavipes stabilized the number of caterpillars to $x = 1900$ per hectare. The number of wasps per hectare stabilized to $y = 1423$ with the average rate of introduction of 16.4 wasps per day.

Thus, considering the deterministic or interval problem, the values that represent the solution to the problem are near the desired values and in accordance with the actual situation practiced in the State of São Paulo.

For the implementation of biological control in practice, the simulation results show us that we should introduce a daily number of predators (Cotesia Flavipes) in the tillage, and this number should be contained in the interval solution. We remark that inserting large numbers of predators does not necessarily guarantee a higher cost compared to the costs that are contained in the optimal interval cost and does not necessarily guarantee a control of the infestation in a shorter time, although this is a likely outcome for both interval problems studied. We only know that independent of the number of predators inserted in tillage, and because this number of predators is contained in the interval solution, we can control the infestation with a cost contained in the optimal interval cost state and control contained in the optimal interval state and optimal interval control, respectively. Furthermore, the daily number of predators inserted in tillage corresponds to the difference, in absolute value, between the number of predators inserted the previous day and the number that will be inserted the day after.

ACKNOWLEDGMENT

The authors wish to express their sincere thanks to the referees for valuable suggestions that improved the manuscript.

The authors also thank the CAPES – Coordination for the Improvement of Higher Education Personnel. The author Edvaldo Assunção was partially supported by the CNPq – Brazilian National Council for Scientific and Technological Development – under Grant number 300703/2013-9. The author Geraldo Nunes Silva was partially supported by the São Paulo State Research Foundation (FAPESP – CEPID) under Grant number 2013/07375-0.

REFERENCES

[1] AGRIANUAL 2012: anuário da agricultura brasileira, São Paulo: FNP, 2012.

[2] R. E. Bellman, *Dynamic programming*, Princeton: University Press, 1957.

[3] D. P. Bertsekas, *Dynamic programming and optimal control*, vol. I, Belmont Massachusetts: Athena Scientific, 1995.

[4] D. P. Bertsekas, *Dynamic programming and stochastic control*, New York: Academic Press, 1976.

[5] P. S. M. Botelho and N. Macedo, *Cotesia Flavipes para o controle de Diatraea Sacharalis*, In: J. R. P. Parra, P. S. M. Botelho, B. S. C. Ferreira and J. M. S. Bento, Controle biológico no Brasil: parasitóides e predadores, São Paulo: Manole, 409–447, 2002.

[6] S. Boyd, L. E. Ghaoui, E. Feron and V. Balakrishnan, *Linear matrix inequalities in system and control theory*, Philadelphia: SIAM, 1994.

[7] J. R. Campos, *Controle ótimo da lagarta da cana-de-açúcar utilizando modelos linearizados e funcional quadrático: uma resolução usando programação dinâmica*, Universitas, 3 (1): 241–252, 2007.

[8] E. Cerdá, *Optimización dinámica*, Madri: Pearson Educación, 2001.

[9] Y. Chalco-Cano, W. A. Lodwick and B. Bede, *Single level constraint interval arithmetic*, Fuzzy Sets and Systems, 2014. http://dx.doi.org/10.1016/j.fss.2014.06.017 http://dx.doi.org/10.1016/j.fss.2014.06.017

[10] M. M. Diniz and R. C. Bassanezi, *Problema de controle ótimo com equações de estado p-fuzzy: programação dinâmica*, Biomatemática, 23: 33–42, 2013.

[11] C. J. Doyle, *A review of the use of models of weed control in integrated crop protection*, Agriculture, Ecosystems and Environment, 64: 165–172, 1997. http://dx.doi.org/10.1016/S0167-8809(97)00035-2

[12] D. Filev and A. Plamen, *Fuzzy optimal control*, Fuzzy Sets and Systems, 47: 151–156, 1992. http://dx.doi.org/10.1016/0165-0114(92)90172-Z

[13] J. C. Geromel, P. L. D. Peres and S. R. Souza, H_2 *guaranteed cost control for uncertain continuous-time linear systems*, Systems & Control Letters, 19: 23–27, 1992. http://dx.doi.org/10.1016/0167-6911(92)90035-Q

[14] J. C. Geromel, P. L. D. Peres and S. R. Souza, H_∞ *control of discrete-time uncertain systems*, IEEE Transactions on Automatic control, 39: 1072–1075, 1994. http://dx.doi.org/10.1109/9.284896

[15] R. Jones and O. J. Cacho, *A dynamic optimisation model of weed control*, Agricultural and Resource Economics, 1: 1–17, 2000.

[16] J. O. S. Kennedy, *Dynamic programming: applications to Agriculture and natural resources*, New York: Elsevier, 1986.

[17] K. Kobayashi and K. Hiraishi, *Analysis and control of hybrid systems with parameter uncertainty based on interval methods*, American Control Conference, 3632–3637, 2009. http://dx.doi.org/10.1109/ACC.2009.5159932

[18] U. A. S. Leal, *Incerteza intervalar em otimização e controle*, 2015, 163 f. Tese (Doutorado em Matemática), Universidade Estadual Paulista, São José do Rio Preto, 2015.

[19] U. A. S. Leal, G. N. Silva and D. Karam, *Otimização dinâmica multiobjetivo da aplicação de herbicida considerando a resistência de plantas daninhas*, Biomatemática, 22: 1–16, 2012.

[20] B. Li, R. Chiong and M. Lin, *A two-layer optimization framework for UAV path planning with interval uncertainties*, Computational Intelligence in Production and Logistics Systems, 120–127, 2014.

[21] W. A. Lodwick, *Constrained interval arithmetic*, CCM Report 138, 1999. http://dx.doi.org/10.1109/CIPLS.2014.7007170

[22] W. A. Lodwick, *Interval and fuzzy analysis: an unified approach*, In: Advances in Imagining and Electronic Physics, Academic Press, 148: 75–192, 2007.

[23] W. A. Lodwick and O. A. Jenkins, *Constrained interval and interval spaces*, Soft Computing, 17: 1393–1402, 2013. http://dx.doi.org/10.1007/s00500-013-1006-x

[24] R. E. Moore, R. B. Kearfott and M. J. Cloud, *Introduction to interval analysis*, Philadelphia: SIAM - Society for Industrial and Applied Mathematics, 2009.

[25] L. H. Monteiro, *Sistemas dinâmicos*, São Paulo: Editora Livraria da Física, 2002.

[26] K. Ogata, *Engenharia de controle moderno*, Rio de Janeiro: LTC, 1998.

[27] J. R. P. Parra, P. S. M. Botelho, B. S. C. Ferreira and J. M. S. Bento, *Controle biológico no Brasil: parasitóides e predadores*, So Paulo: Manole, 2002.

[28] C. M. Pereira, M. S. Cecconello and R. C. Bassanezi, *Controle ótimo em sistemas baseados em regras fuzzy*, Biomatemática, 23: 147–167, 2013.

[29] L. S. Pontryagin, *Optimal regulation processes*, In: Proceedings of International Congresses of Mathematicians, 1958.

[30] L. S. Pontryagin, V. G. Boltyansky, P. V. Gamkrelidze and E. F. Mischenko, *The mathematical theory of optimal processes*, New York: Interscience Publishers, 1962.

[31] M. Rafikov and P. A. P. Borges *Alguns aspectos do controle em um modelo dinâmico de objetos de produção de interações conforme uma funcional quadrática*, Análise Econômica, 19: 103–121, 1993.

[32] A. Rauh, J. Minisini and E. P. Hofer, *Interval techniques for design of optimal and robust control strategies*, Proceedings of the Twelfth GAMM-IMACS International Symposium on Scientific Computing, Computer Arithmetic and Validated Numerics, 1–9, 2006. http://dx.doi.org/10.1109/SCAN.2006.27

[33] S. V. Segato, A. S. Pinto, E. Jendiroba and J. C. M. Nóbrega *Atualização em produção de cana-de-açúcar*, Piracicaba: Livroceres, 2006.

[34] Y. P. Shih and C. J. Chen *On the weighting factores of the quadratic criterion in optimal control*, International Journal of Control, 19: 947–955, 1974. http://dx.doi.org/10.1080/00207177408932688

[35] F. C. Silva and A. F. Bergamasco, *Levantamento de modelos matemáticos descritos para a cultura da cana-de-açúcar*, Embrapa Informática Agropecuária, 2001.

[36] H. A. Taha, *Pesquisa operacional*, São Paulo: Pearson Prentice Hall, 2008.

[37] A. M. Tusset and M. Rafikov, *Controle ótimo da lagarta da cana-de-açúcar, utilizando modelos linearizados e funcional quadrático*, In: Anais do 3° Congresso Temático de Dinâmica e Controle da SBMAC, Ilha Solteira, 1572–1577, 2004.

[38] UNICA: União da Indústria de Cana-de-açúcar. Available at: http://unica.com.br .

[39] L. Yu, J. M. Xu and Q. L. Han, *Optimal guaranteed cost control of linear uncertain systems with input constraints*, Proceedings of the Fifth World Congress on Intelligent Control and Automation, 553–557, 2004. http://dx.doi.org/10.1109/WCICA.2004.1340636

5

Multiple regulation mechanisms of bacterial quorum sensing

Peter Kumberger *, Christina Kuttler †, Peter Czuppon ‡ and Burkhard A. Hense §
* Center for Modeling and Simulation in the Biosciences
BioQuant-Center, Heidelberg University, Germany
peter.kumberger@bioquant.uni-heidelberg.de
† Zentrum Mathematik
Technische Universität München, Germany
kuttler@ma.tum.de
‡Mathematisches Institut: Abt. f. Math. Stochastik
Universität Freiburg, Germany
czuppon@stochastik.uni-freiburg.de
§ Institute of Computational Biology
Helmholtz-Zentrum München, Germany
burkhard.hense@helmholtz-muenchen.de

Abstract—Many bacteria have developed a possibility to recognise aspects of their environment or to communicate with each other by chemical signals. The so-called Quorum sensing (QS) is a special case of this kind of communication. Such an extracellular signalling via small diffusible compounds (called autoinducers) is known for many bacterial species, including pathogenic and beneficial bacteria. Using this mechanism allows them to regulate their behaviour, e.g. virulence. We will focus on the typical QS system of Gram negative bacteria of the so-called *lux* type, based on a gene regulatory system with a positive feedback loop.

There is increasing evidence that autoinducer systems themselves are controlled by various factors, often reflecting the cells' nutrient or stress state. We model and analyse three possible interaction patterns. Typical aspects are e.g. the range of bistability, the activation threshold and the long term behaviour. Additionally, we aim towards understanding the differences with respect to the biological outcomes and estimating potential ecological or evolutionary consequences, respectively.

Keywords-Quorum Sensing, ODE system, bifurcations, nutrients, qualitative behaviour

I. INTRODUCTION

Extracellular signalling via small diffusible compounds (autoinducers) is known for an increasing number of bacterial species, including pathogenic and human health promoting bacteria. Briefly, bacteria release autoinducers and simultaneously regulate target gene expression dependent on the environmental autoinducer concentration. Regulated behaviour often includes critical life style switches, e.g. from non-virulent to virulent. Thus mechanistic understanding of autoinducer regulation and its ecological significance is of high relevance for the development of treatment strategies. Autoinducer regulation was originally assumed to be a strategy enabling coordinated responses of whole bacteria populations dependent on the cell density (Quorum sensing) [15]. The later detected influence of other aspects such as mass transfer properties of the environment and cell distribution led to the alternative concept of diffusion sensing (assuming that the mass transfer properties of the environment around a cell - including diffusion conditions - are estimated by autoinducers) and of the unifying efficiency sensing [31, 19]. The autoinducer mechanism was first described in the gram-negative *Vibrio fischeri*, which possesses an autoinducer system of *lux*-type with an AHL (acylhomoserine lactone) acting as signal. The signal is produced by the synthase LuxI. It binds to a receptor molecule (LuxR). Dimers of the AHL-LuxR complex bind to the *lux* box in the *lux* operon, where the autoinducer synthase (LuxI) and luminescence genes are up-regulated (Fig. 1), but also to other target genes of the regulon [15]. This AHL system, including the positive feedback loop, represents an archetypal example for the architecture of autoinducer mediated gene regulation of many gram negative bacteria. Autoinducer systems in other bacteria often follow similar design principles, although details may vary.

There is increasing evidence that autoinducer systems themselves are controlled by various factors, often reflecting the cells' nutrient or stress state [12, 27]. Recently it has been suggested that such controls allow for integrating the demand of the cells for the regulated behaviour into the signal strength, generating a kind of hybrid push/pull control [20]. Here, "demand" reflects the strength of the potential benefit a group of cells could have from this behaviour under the current environmental conditions. For example, the demand for the release of an exoprotease might be low as long as available essential amino acids abound in the environment, but increase when the amino acids deplete. Integration of the demand into signal strength can be realised by tuning Quorum sensing dependent on the environmental conditions. The factors have been shown to interfere with the

autoinducer regulation pathway in various ways. The reasons for this variety remain largely unclear. We hypothesise that different ecological and/or evolutionary impacts emerge. A number of fine-tuning strategies with respect to autoinducer systems are realised or at least possible, including e.g. degradation of autoinducers, control of the availability or activity of autoinducers, control of the activity of autoinducer synthases or receptors, or a combination of these. The existence of multiple regulation systems within the same species, controlled by different environmental or cellular factors, respectively, has been reported (e.g. [27]). In this study, we focus on three basic interaction principles affecting the signal synthase and receptor by control of production and degradation:

1) Regulation of the LuxR-type signal receptor (termed LuxR)
2) Regulation of the LuxI-type signal synthase (termed LuxI)
3) Regulation of LuxI and LuxR

Different scenarios are analysed by mathematical modelling. Our aim is to understand the differences with respect to resulting regulation dynamics and the reached equilibria, and to estimate potential ecological and evolutionary consequences, respectively. Relevant aspects are the range of bistability, the activation threshold and long term behaviour. From a mathematical point of view, bifurcation analysis can help to answer these questions. We mainly study single effects on single cells using deterministic models; nevertheless combinations of effects are also possible.

However, small numbers of cellular molecules in the regulatory system or spatial inhomogeneity of environmental factors controlling the regulatory system may cause stochastic differences between cells. We therefore consider shortly the potential relevance of stochasticity in the regulation systems on a small population. Remark that we neglect any spatial structure itself, as our goal here is to understand the basic principles of the regulation system and its qualitative behaviour. For the same reason, but also due to the differences between species or even strains and the general lack of

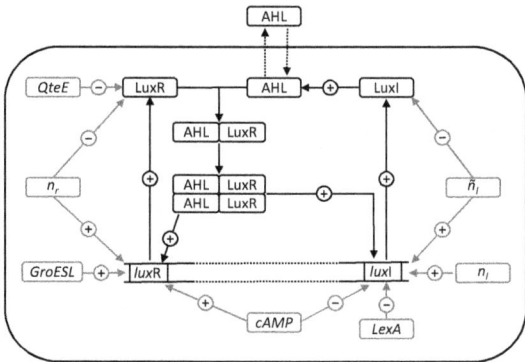

Fig. 1: Scheme of the *lux*-type Quorum sensing system with potential influences of regulators

available experimentally derived quantitative data, we do not emphasize on real parameter values, which are realised in a specific species.

The paper is organised as follows: We start in section II by introducing the basic model for Quorum sensing of LuxI-LuxR type and explain the influences by nutrient-governed regulators. To focus on the signal dynamics, we assume that all other processes not involving AHL are fast and thus in equilibrium, including concentration of the regulators of Quorum sensing ([28]). The qualitative behaviour of these modified systems is examined in section III, e.g. by considering bifurcation diagrams. Some stochastic influences caused possibly by small numbers of molecules are simulated in section IV. As an example we consider coupled influences of different regulators in the stochastic case

II. THE BASIC MATHEMATICAL MODEL AND ITS MODIFICATIONS

In order to focus on the basic qualitative behaviour of our system we neglect any spatial structure and assume a homogeneous intracellular distribution of all involved regulators and substances. Also in the extracellular space, spatial structure is neglected, which is a reasonable assumption, e.g. for well stirred batch cultures or continuous cultures. For the typical Quorum sensing system of LuxI/LuxR type, basic ODE models were introduced, e.g. in [9, 28]. We start with the following

Name	Variable
x_e	extracellular AHL concentration
x_c	intracellular AHL concentration
l	concentration of LuxI
r	concentration of LuxR
y_1	concentration of the LuxR-AHL complex
y_2	concentration of the dimer of LuxR-AHL complexes

TABLE I: Model variables of the basic Quorum sensing model

ODE system for a single cell which distinguishes between intracellular and extracellular AHL (x_c resp. x_e), including equations for LuxR, LuxR-AHL complex, the corresponding dimer and LuxI:

$$\dot{x}_e = d_c x_c - d_e x_e - \gamma_e x_e \quad (1)$$

$$\dot{x}_c = \beta_l l - \gamma_c x_c - d_c x_c + d_e x_e \quad (2)$$
$$\quad - \pi_1^+ r x_c + \pi_1^- y_1$$

$$\dot{r} = \alpha_r + \pi_1^- y_1 - \pi_1^+ r x_c - \gamma_r r \quad (3)$$

$$\dot{y}_1 = \pi_1^+ r x_c - \pi_1^- y_1 + 2\pi_2^- y_2 - 2\pi_2^+ y_1^2 \quad (4)$$

$$\dot{y}_2 = \pi_2^+ y_1^2 - \pi_2^- y_2 \quad (5)$$

$$\dot{l} = \alpha_l - \gamma_l l + \beta_y \frac{y_2}{1 + (\beta_y/\kappa_y)y_2}. \quad (6)$$

For the meaning of all variables and parameters see Tables I and II. The model assumes the typical positive feedback which leads to a Hill function in the equation for LuxI (the AHL producing enzyme, denoted by l) with Hill coefficient $n = 2$, assuming that LuxR-AHL dimers (denoted by y_2) are relevant for the increased LuxI production. Exchange of AHL between intracellular and extracellular space is described by rates d_e and d_c. For LuxR (r), a constitutive basic production is assumed. The notation of the model terms is chosen in a similar way as in previous publications (e.g. [28, 23]), to keep it comparable to the simpler models.

Even though *V. fischeri* possesses at least two Quorum sensing systems, we restrict ourselves to the well-known *lux* system, i.e., there is only one positive feedback via LuxI. Degradation of LuxR

Name	Parameter
α_l	Basal/Background production rate of LuxI
α_r	Basal/Background production rate of LuxR
β_c	Maximum increase of/Slope of increase of LuxR-production by cAMP
β_l	Production rate of AHL by LuxI
β_y	Maximum increase of/Slope of increase of LuxI-production by AHL-LuxR dimer
γ_c	Degradation rate of AHL in the cytoplasm
γ_e	Degradation rate of AHL outside of the cell
γ_l	Degradation rate of LuxI
γ_r	Degradation rate of LuxR
κ_r	Asymptotics of increase of LuxR-production (high cAMP concentration)
κ_y	Asymptotics of increase of LuxI-production (high AHL-LuxR dimer concentration)
μ_l	Production rate of LuxI induced by regulator n_l, $\tilde{n}_l$
μ_r	Production rate of LuxR induced by GroESL
π_1^+	Rate of AHL binding to LuxR (complex association)
π_1^-	Rate of AHL-LuxR complex dissociation
π_2^+	Rate of AHL-LuxR dimer association (binding of two AHL-LuxR complexes)
π_2^-	Rate of AHL-LuxR dimer dissociation
d_c	Diffusion rate of AHL from the cell to the extracellular space
d_e	Diffusion rate of AHL from the extracellular space into the cell
n_l, $\tilde{n}_l$	Regulator n_l, $\tilde{n}_l$ which influences the LuxI-production
n_r	GroESL, a regulator, which influences the LuxR-production
a	LexA, a regulator, which inhibits binding of the AHL-LuxR dimer to the LuxI-operon
b	Affinity of a regulator (LexA or cAMP) to the *lux* operon compared to the AHL-LuxR dimer
c	cAMP, which influences LuxI as well as LuxR
$n_{l,thr}$	Michaelis constant for destabilisation of LuxI by regulator $\tilde{n}_l$
$n_{r,thr}$	Michaelis constant for destabilisation of LuxR by GroESL
p_{n_l}	Strength of destabilisation of LuxI by regulator $\tilde{n}_l$
p_{n_r}	Strength of destabilisation of LuxR by GroESL
p_q	Strength of destabilisation of LuxR by QteE
q	QteE, which destabilises LuxR
q_{thr}	Michaelis constant for destabilisation of LuxR by QteE

TABLE II: Model parameters of the basic and the modified Quorum sensing models

is for simplicity only assumed to take place in the state of a single LuxR, not within the LuxR-AHL complex and not within the dimer.

In order to derive the model, we essentially assume that all dynamics of the more detailed model (Eq.(1) - (6)) are fast but that of x_c and x_e. E.g. complex association or dissociation is faster than the production of a larger molecule. This results in

$$\dot{x}_e = d_c x_c - d_e x_e - \gamma_e x_e$$
$$\dot{x}_c = \beta_l l - \gamma_c x_c - d_c x_c + d_e x_e - \pi_1^+ r x_c + \pi_1^- y_1$$
$$\varepsilon \dot{r} = \alpha_r + \pi_1^- y_1 - \pi_1^+ r x_c - \gamma_r r$$
$$\varepsilon \dot{y}_1 = \pi_1^+ r x_c - \pi_1^- y_1 + 2\pi_2^- y_2 - 2\pi_2^+ y_1^2$$
$$\varepsilon \dot{y}_2 = \pi_2^+ y_1^2 - \pi_2^- y_2$$
$$\varepsilon \dot{l} = \alpha_l - \gamma_l l + \beta_y \frac{y_2}{1 + (\beta_y/\kappa_y)y_2}.$$

This mathematical assumption is valid as considering the whole system shows qualitatively the same behaviour as the reduced system.

For $\varepsilon \to 0$ we obtain a function for l, only depending on x_c,

$$l = \frac{\alpha_l}{\gamma_l} + \frac{\beta_y}{\gamma_l} \frac{x_c^2}{\frac{\pi_2^-}{\pi_2^+}\left(\frac{\pi_1^-\gamma_r}{\pi_1^+\alpha_r}\right)^2 + x_c^2\beta_y/\kappa_y}.$$

Hence we obtain the simplified model

$$\dot{x}_c = \beta_l\left(\frac{\alpha_l}{\gamma_l} + \frac{\beta_y}{\gamma_l} \frac{x_c^2}{\frac{\pi_2^-}{\pi_2^+}\left(\frac{\pi_1^-\gamma_r}{\pi_1^+\alpha_r}\right)^2 + x_c^2\beta_y/\kappa_y}\right)$$
$$\quad -(\gamma_c + d_c)x_c + d_e x_e$$
$$\dot{x}_e = d_c x_c - d_e x_e - \gamma_e x_e$$

or, lumping parameters together,

$$\dot{x}_c = f(x_c) - d_c x_c + d_e x_e$$
$$\dot{x}_e = d_c x_c - d_e x_e - \gamma_e x_e \qquad (7)$$
$$f(x_c) := \alpha + \frac{\beta x_c^2}{x_{thresh}^2 + x_c^2} - \gamma_c x_c.$$

In a further step we introduce some typical additional influences to the mathematical models.

A. Influences on the dynamics of LuxR

Increase of the LuxR production: It was reported, e.g. [1] that the protein GroESL in *V. fischeri* appears in high numbers, when there are insufficient nutrients available. Although the mechanisms behind this are not fully understood, GroESL seems to cause, besides a stabilisation of LuxR, an up-regulation of the gene expression. Production of LuxR-type autoinducers by environmental factors has been reported also for other species such as *Pseudomonas aeruginosa* [32]. Focusing on the regulation of LuxR production, we change the equation, which describes the dynamics of LuxR, to

$$\dot{r} = \mu_r n_r + \alpha_r + \pi_1^- y_1 - \pi_1^+ r x_c - \gamma_r r, \qquad (8)$$

where n_r describes the available concentration of e.g. GroESL. As the copy number of the protein in the cell is low, we neglect saturation effects.

Using this equation instead of the basic equation 3 for LuxR and applying again the idea of different time scales yields

$$\dot{x}_c = B\alpha_l + B\beta_y \frac{A_r x_c^2}{1 + (\beta_y/\kappa_y)A_r x_c^2} \qquad (9)$$
$$\quad - \gamma_c x_c - d_c x_c + d_e x_e,$$

where $B := \frac{\beta_l}{\gamma_l}$ and $A_r := \frac{\pi_2^+}{\pi_2^-}\left(\frac{\pi_1^+(\alpha_r+\mu_r n_r)}{\pi_1^-\gamma_r}\right)^2$.

Destabilisation of LuxR: The protein QteE destabilises the LuxR-homologue LasR in *Pseudomonas aeruginosa* resulting in a faster degradation of LasR [35]. Although the regulation of *qteE* expression yet needs to be investigated in detail, environmental factors seem to be involved [40]. This extension can be described by a slight modification of the LuxR-governing equation

$$\dot{r} = \alpha_r + \pi_1^- y_1 - \pi_1^+ r x_c - \left(1 + \frac{p_q q}{q + q_{thr}}\right)\gamma_r r. \quad (10)$$

Proceeding in the same way as done for GroESL results in

$$\dot{x}_c = B\alpha_l + B\beta_y \frac{A_q x_c^2}{1 + (\beta_y/\kappa_y)A_q x_c^2} \qquad (11)$$
$$\quad - \gamma_c x_c - d_c x_c + d_e x_e,$$

where

$$A_q := \frac{\pi_2^+}{\pi_2^-}\left(\frac{\pi_1^+\alpha_r(q + q_{thr})}{\pi_1^-(pq + q + q_{thr})\gamma_r}\right)^2.$$

Increase of the LuxR production and destabilisation of LuxR: Typically a number of mechanisms regulating Quorum sensing systems occur in the same species (see e.g. [3]). As a hypothetical example, we assume that both mechanisms analysed before, i.e., up-regulation of LuxR production and destabilisation of the LuxR protein, are induced at the same time by environmental triggers, in our case by a single regulator. Such a combination can be assumed to help the bacteria to react faster to environmental changes. The equation of LuxR has the following form:

$$\dot{r} = \mu_r n_r + \alpha_r + \pi_1^- y_1 - \pi_1^+ r x_c$$
$$\quad - \left(1 + \frac{p_{n_r} n_r}{n_r + n_{r,thr}}\right)\gamma_r r. \qquad (12)$$

Using the same mathematical tools as in the paragraphs above yields

$$\dot{x}_c = B\alpha_l + B\beta_y \frac{A_r^{new} x_c^2}{1 + (\beta_y/\kappa_y)A_r^{new}x_c^2} \quad (13)$$
$$- \gamma_c x_c - d_c x_c + d_e x_e,$$

where
$$A_r^{new} := \frac{\pi_2^+}{\pi_2^-}\left(\frac{\pi_1^+}{\pi_1^-}\right)^2\left(\frac{n_r+n_{r,thr}}{(1+p_{n_r})n_r+n_{r,thr}}\right)^2\left(\frac{\mu_r n_r+\alpha_r}{\gamma_r}\right)^2.$$

LuxR feedback: LuxR type receptors may be able to induce the expression of their own gene after binding to its autoinducer [34]. Considering the possibility of a self-induced positive feedback of LuxR leads to qualitatively similar results as the addition of GroESL into our model. We thus omit the analysis in this study for the reason of brevity.

B. Influences on the dynamics of LuxI

Increase of the LuxI production: Stress factors as starvation have been reported to up-regulate the transcription of the lux operon in *V. fischeri*, including the *luxI* gene, via $\sigma 32$ [38]. AHL synthase genes in other species also are known to be controlled in an environment dependent way (e.g. [8]). Regulation of AHL synthase can be incorporated in two different ways: Either only the basal synthase expression (and correlated with this the basal autoinducer production) is increased by the addition of a regulator n_l, or both, the basal and the induced production, are increased. Unfortunately, experimental studies usually do not allow to discriminate between both variants. However as the qualitative behaviour is the same in both approaches, we will only consider the second in this study. This modification leads to the following governing equation for LuxI:

$$\dot{l} = \left(\alpha_l + \beta_y\frac{y_2}{1+(\beta_y/\kappa_y)y_2}\right)(1+\mu_l n_l) - \gamma_l l. \quad (14)$$

Assuming again different time scales and reducing the system to a two component model changes the governing equation for the intracellular concentration of AHL accordingly (equation not shown here for the reason of brevity).

Inhibition of the LuxI production: LexA is a repressor enzyme, which usually acts on SOS response genes. In *V. fischeri*, it has been reported to act antagonistically with LuxR-AHL dimers by competing for the same binding site on the *lux* operon. LexA binding does not induce transcription of the *lux* operon, the transcription is not increased [37]. Repressors of AHL synthase genes have also been shown in other species (see e.g. [43]). Neglecting the details about the binding mechanism, we follow a non-classic approach (as used in [23]): The percentage of present molecules determines if transcription is possible and the grade of transcription is determined as usual by the Monod term. The corresponding modified equation for LuxI reads

$$\dot{l} = \alpha_l - \gamma_l l + \beta_y\frac{y_2}{1+(\beta_y/\kappa_y)y_2} \cdot \frac{by_2}{by_2+a}. \quad (15)$$

The modified equation for x_c is left out again. The influence of oxygen concentration on the expression of the *lux* operon, which is mediated via ArcA, may act similarly [5].

Increase of the LuxI production and destabilisation of LuxI: Although much more evidence exists for regulation of stability of LuxR type AHL receptors, similar behaviour was also reported for LuxI type AHL synthase. In *P. aeruginosa*, the half-life of LasI is controlled by the LON protease, which itself has been reported to be induced by environmental stress due to certain antibiotics [36, 25]. Analogue to the analysis of effects on LuxR, we thus analyse a combination of a destabilising effect on LuxI and an increased LuxI-production by a single regulator. This changes the equation for LuxI in a similar way as in the corresponding regulation of LuxR:

$$\dot{l} = \left(\alpha_l + \beta_y\frac{y_2}{1+(\beta_y/\kappa_y)y_2}\right)(1+\mu_l\tilde{n}_l)$$
$$- \left(1+\frac{p_{n_l}\tilde{n}_l}{\tilde{n}_l+n_{l,thr}}\right)\gamma_l l. \quad (16)$$

The resulting governing equation for the intracellular AHL concentration reads

$$\dot{x}_c = \frac{\tilde{n}_l + n_{l,thr}}{(1 + p_{n_l})\tilde{n}_l + n_{l,thr}}(1 + \mu_l\tilde{n}_l) \cdot$$
$$\left(B\alpha_l + B\beta_y\frac{A_l x_c^2}{1 + (\beta_y/\kappa_y)A_l x_c^2}\right)(17)$$
$$-\gamma_c x_c - d_c x_c + d_e x_e.$$

C. Influence on the dynamics of LuxI and LuxR

Regulation factors can have pleiotropic effects on different target molecules. Starvation induces an increased occurrence of 3':5'-cyclic AMP (cAMP) in bacteria such as V. fischeri [12]. This molecule is able to bind to the cAMP receptor protein (CRP). The so-formed complex influences the lux system in V. fischeri on two different sites. On the one hand it amplifies the production of LuxR. On the other hand cAMP inhibits the LuxI-production using a similar mechanism as LexA. We analysed the effect of cAMP as an example for more complex regulation mechanisms. From now on for the reason of simplicity the (cAMP-CRP)-complex will be referred to as cAMP. Adding cAMP to the model yields a change in the dynamics of LuxI and LuxR resulting in

$$\dot{r} = \alpha_r + \beta_c\frac{c}{1 + (\beta_c/\kappa_r)c} + \pi_1^- y_1 - \pi_1^+ r x_c - \gamma_r r$$
$$\dot{l} = \alpha_l - \gamma_l l + \beta_y\frac{y_2}{1 + (\beta_y/\kappa_y)y_2}\frac{by_2}{by_2 + c}.$$

The reduction of the so modified system obviously affects the governing equation of x_c. Those changes lead to the following equation:

$$\dot{x}_c = B\alpha_l + B\beta_y\frac{PA_c^2 x_c^2}{1 + (\beta_y/\kappa_y)PA_c^2 x_c^2}\frac{bPA_c^2 x_c^2}{bPA_c^2 x_c^2 + c}$$
$$- \gamma_c x_c - d_c x_c + d_e x_e,$$
$$(18)$$

where $P := \frac{\pi_2^+}{\pi_2^-}\left(\frac{\pi_1^+}{\pi_1^-}\right)^2$ and $A_c := \frac{\beta_c}{\gamma_r}\frac{c}{1 + (\beta_c/\kappa_r)c}$.

III. MODEL ANALYSIS AND RESULTS

In this section we analyse the effects of different strengths of the regulation impact on Quorum sensing signals. Therefore we take a look at simulations made with the above derived models for the different influences of regulators. The variables (listed in Table III) and parameters are used in a non-dimensional form. The values of the parameters in the simulations are shown in Table IV. We aim in this study at comparing the potential qualitative consequences of different regulators on the function of the AHL-type Quorum sensing system in a generic approach. The parameter values were chosen in a way to disclose the full complexity of such a system, including e.g. the maximum number of stationary states. We assume that evolution of a system enabling complex behaviour suggests that the bacterium at least under certain conditions exploits this complexity. Using an experimentally derived parameter set of a specific bacterium, which was gained under certain environmental conditions, was thus not meaningful, and would have been difficult due the lack of such data and variability of parameters in response to changes of the environmental conditions ([18]). Note that other parameter values might cause more simple behaviour, including absence of multistationarity. However, the qualitative messages in the results with respect to time and strength of Quorum sensing induction will hold. As the basal production rate of the autoinducer synthase, which is critical for induction dynamics, may vary between different species, we use two different parameter values. Changes due to the variation of the basal LuxR-production rate are not subject to this study and hence the same value was used throughout. In the following solid lines in the bifurcation diagrams represent stable stationary states whereas dashed lines represent unstable stationary states.

For the time courses in this section an initial condition of zero intra/extracellular AHL was assumed.

The numerical analyses were done with XPPAUT Version 5.41 [13].

A. Influences on LuxR

Increase of the LuxR production: For low basal production rates of LuxI (α_l) we observe a bistable behaviour of the lux system (Fig. 2(a)), when assuming the strength of the regulator (e.g. GroESL) to be the bifurcation parameter. This means that for

	Regulator	Influence	Eq.	Fig.
LuxR	n_r	Increased production	9	2(a)+(b)
LuxR	QteE	Destabilisation	11	2(c)+(d)
LuxR	$\tilde{n}_r$	Increased production & Destabilisation	13	2(e)+(f)
LuxI	n_l	Increased production	14*	3(a)+(b)
LuxI	LexA	Inhibited production	15*	3(c)+(d)
LuxI	$\tilde{n}_l$	Increased production & Destabilisation	17	3(e)+(f)
LuxI/LuxR	cAMP	Increased LuxR production & Inhibited LuxI production	18	4

TABLE III: Different scenarios. *Equation referenced has to be inserted into Eq.(2) assuming a quasi-steady state.

Parameter	Value	Parameter	Value
α_l	refer to figures	α_r	0.1
β_c	1	β_l	0.1
β_y	1	γ_c	0.03
γ_e	0.03	γ_l	0.1
γ_r	0.1	κ_r	1
κ_y	1	μ_l	0.1
μ_r	0.1	π_1^+	1
π_1^-	1	π_2^+	1
π_2^-	10	d_c	0.5
d_e	0.5	b	1
$n_{l,thr}$	1	$n_{r,thr}$	1
p_{n_l}	10	p_{n_r}	10
p_q	5	q_{thr}	1

TABLE IV: Values of the dimensionless parameters for the simulations

concentrations of GroESL (or similar acting regulators) larger than a certain threshold (the bifurcation point, here at about $n_r = 1.9$) the system will always be induced in the used parameter value setting. For concentrations of regulator beneath this threshold the final AHL-concentration within the cells depends on the extra- and intracellular concentration of AHL at the beginning of the simulation. The unstable stationary state (dashed region of the black line in Fig. 2(a)) marks the threshold: a starting AHL-concentration lower than the threshold causes the system to stay non-induced, while higher AHL-concentrations lead to considerably higher stationary AHL-concentrations, i.e.

an induction of the whole system.

Assuming higher basal rates α_l shifts the bifurcation diagram to the left and hence bistability is lost. In this case independent of the starting AHL-concentration the system always gets activated (Fig.2(a) red line).

Fig. 2(b) shows the time course of the extracellular AHL-concentration for different GroESL-concentrations in the low basal production case, corresponding to the black line in Fig. 2(a). In case the GroESL concentration is above the bifurcation point, increasing the GroESL concentration results in an earlier induction of the cell. The final AHL-concentration of the induced cells does not depend

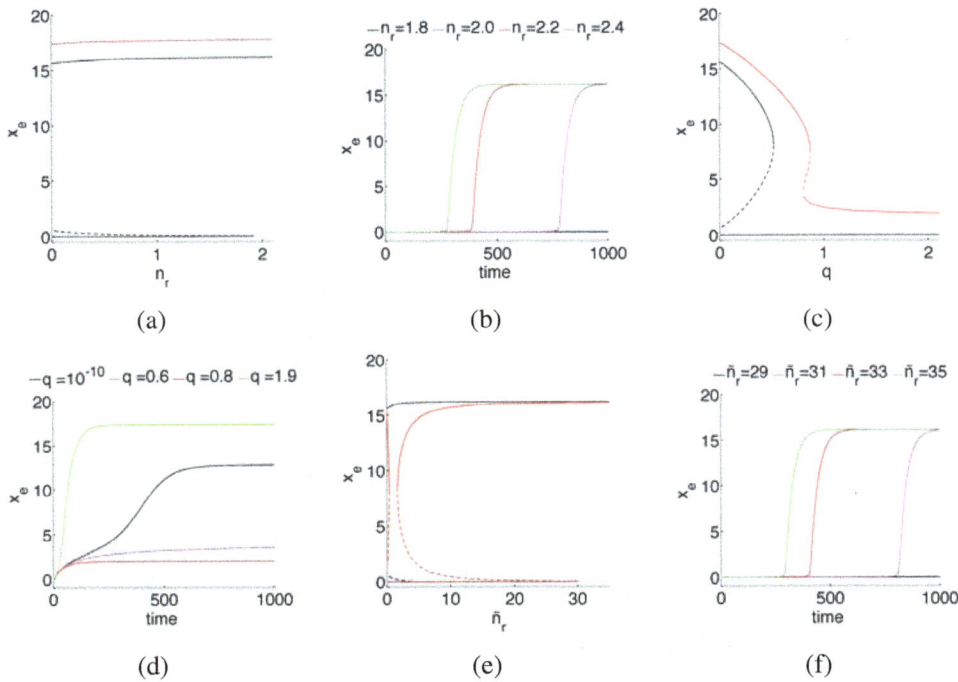

Fig. 2: Influences on LuxR. (a) Bifurcation diagram for the extracellular AHL-concentration, where only the increase in the LuxR-production by GroESL was considered. The basal LuxI-production rate α_l is 0.001 (black line) resp 0.1 (red). (b) Time courses for the extracellular AHL-concentration, which were generated by cells with different GroESL-concentrations. Those time courses correspond to the bifurcation diagram shown in (a) by the black line, i.e., $\alpha_l = 0.001$. (c) Bifurcation diagram for the extracellular AHL-concentration, where the influence of QteE on the system is examined. QteE destabilises LuxR and hence leads to a faster degradation of LuxR. Basal LuxI-production rate is assumed to be 0.001 (black) resp $\alpha_l = 0.1$ (red). (d) Time courses for the extracellular AHL-concentration, which were generated by cells with different QteE-concentrations. Those time courses correspond to the bifurcation diagram shown in the red line in (c), i.e., $\alpha_l = 0.1$. (e) Bifurcation diagram, where both, an up-regulation of LuxR-production, and a destabilising effect on the LuxR protein is assumed. Basal LuxI-production rate is 0.001. The destabilising effect is $p_{n_r} = 1$ (black) or $p_{n_r} = 10$ (red) (f) Time courses for the extracellular AHL-concentration, which were generated by cells with different concentrations of the GroESL-like regulator. Those time courses correspond to the bifurcation diagram shown in (e) with $p_{n_r} = 10$.

on the amount of GroESL.

Destabilisation of LuxR: Here (Fig. 2(c)-(d)) we choose the concentration of a QteE-like regulator as the bifurcation parameter. The bifurcation diagrams show that a high level of QteE completely prevents an activation of the Quorum sensing system. Even induced systems will switch to the non-induced state after some time, when there is a high concentration of QteE present. In addition the basic production rate of LuxI (α_l) also plays a significant role: if α_l is large, q must be large as well to prevent an induction of the *lux* system.

However, if α_l is very low, a non-induced system will never, i.e., independent of the concentration of QteE, be able to activate itself. The presence of QteE may shift the potential stationary states, but typically keeps the bistable behaviour with the possibility to switch on for growing bacteria, see Fig. 2(c). Fig. 2(d) shows the time course for cells which are provided with different amounts of QteE. It is evident that the time of induction and the height of the final AHL-concentration depend on the amount of present QteE. The more QteE available the lower is the final AHL-concentration.

The moment of induction - in case the system is induced - is late if the concentration of QteE is close to the QteE-concentration at which the bifurcation occurs.

Increase of the LuxR production and destabilisation of LuxR: The results of a combined impact on LuxR, i.e., an increase of the LuxR-production and a faster degradation of LuxR, which could be interpreted as a combined effect of GroESL- and QteE-like regulators, is shown in Figures 2(e) - 2(f). As we have already discussed different basal production rates α_l, we now focus on changing the ratio between the strength of degradation of LuxR and the increase of the LuxR-production by varying the latter. As a bifurcation parameter we use GroESL concentration.

For weak effects of GroESL on LuxR stability the bifurcation diagram is similar to the one, where no influence on the degradation of LuxR was assumed (compare Figures 2(a) and 2(e) black lines). When assuming a stronger destabilisation of LuxR, an intermediate range of GroESL-concentrations exists for which the system is never able to get activated in our parameter setting (Fig. 2(e) red line). By increasing the destabilisation strength, the bistable range increases.

As already seen for Fig. 2(b) the moment of induction depends on the GroESL-concentration. The closer it is to the bifurcation point the later the system gets activated (Fig. 2(f)).

B. Influences on LuxI

Increase of the LuxI production: In contrast to a regulator which acts by increasing the production of LuxR, introducing regulator n_l into the system changes the concentration of AHL in the stationary phase (Fig. 3(a)). While the system acts bistable when a small basal LuxI-production rate α_l is assumed, this bistability is lost for high basal rates. In Fig. 3(b) the time courses of AHL concentration for different amounts of regulator n_l are shown. Increasing n_l does not only result in higher maximum concentrations of AHL, but - similar to a factor up-regulating the production of LuxR - promotes an earlier induction

Inhibition of the LuxI production: Inhibition of LuxI production by LexA results in similar effects as described for the LuxR destabilising regulator above (Figures 3(c) - 3(d)), including a decrease of maximum AHL concentration in stationary phase, and a delay of activation for higher LexA concentrations. A similar effect takes place if one considers LuxI destabilisation only, due to the "simple" production of AHL by LuxI, formulated as a linear term, no further non-trivial effects appear in that context.

Increase of the LuxI production and destabilisation of LuxI: The results are shown in Figures 3(e) - 3(f) (Please note the logarithmic axes in Fig. 3(e)). Similar to the corresponding regulation of LuxR, in our parameter setting using a stronger destabilisation effect, there is an intermediate range of concentrations of regulator $\tilde{n}_l$, in which the system cannot be activated. As a main difference between regulation of LuxR and LuxI, the maximum concentration of AHL in an activated state increases significantly with increasing concentrations of regulator $\tilde{n}_l$. Again similar to LuxR regulation, the intermediate range vanishes for small values of p_{n_l}.

C. Influence on the dynamics of LuxI and LuxR

Increasing the affinity of cAMP to the *lux* operon (parameter b) stretches the bifurcation diagram, but keeps its shape (compare the black with the red lines in Figures 4(a),(c) and (e), respectively).

In Figures 4(a) - 4(b) a low basal LuxI-production rate was assumed. With this assumption and our parameter setting a system which starts in a non-activated state is not able to get induced (Fig. 4(b)). The four different curves are all close to zero (thus indistinguishable from each other).

Using our set of parameters and an intermediate basal production rate α_l the bacteria will always get activated as long as they are neither starving nor drowned with nutrients, i.e., an intermediate amount of cAMP is present (Fig. 4(c)). Contrarily, the system is never activated with very low or very high amounts of cAMP. Regions of bistability exist, i.e., dependent on the initial concentration,

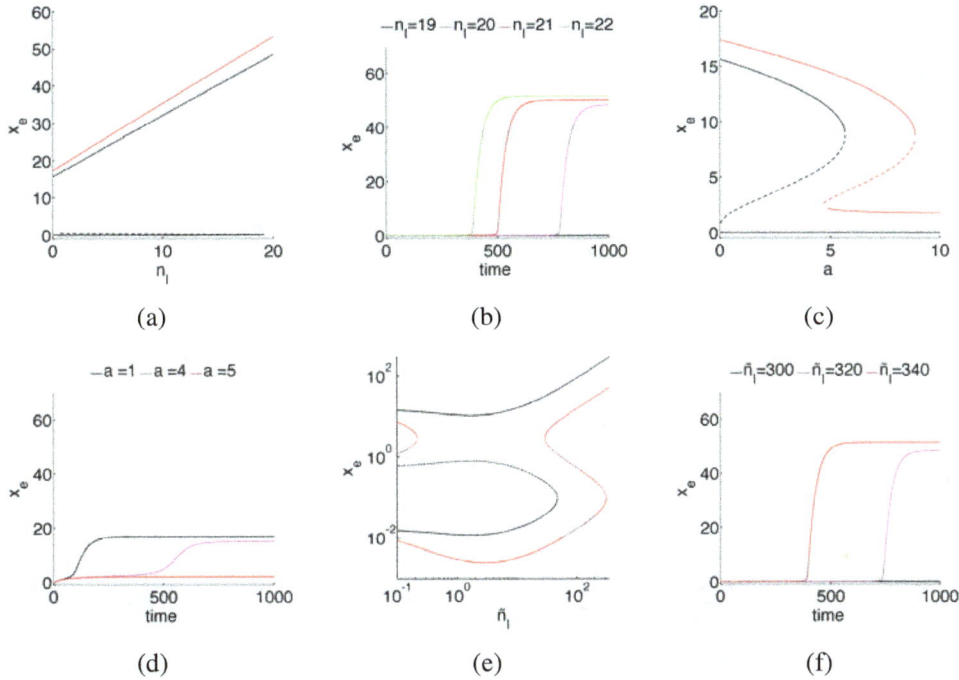

(a) (b) (c)

(d) (e) (f)

Fig. 3: Influences on LuxI. (a) Bifurcation diagram for the extracellular AHL-concentration, where only the increase in the LuxI-production by regulator n_l was considered. The basal LuxI-production rate α_l is 0.001 (black line) or 0.1 (red line), respectively. (b) Time courses for the extracellular AHL-concentration, which were generated by cells with different concentrations of regulator n_l. Those time courses correspond to the bifurcation diagram shown in (a), with $\alpha_l = 0.001$. (c) Bifurcation diagram for the extracellular AHL-concentration, where the influence of LexA on the system is examined. LexA destabilises LuxI and hence leads to a faster degradation of LuxI. Basal LuxI-production rate is assumed to be $\alpha_l = 0.001$ (black line) or $\alpha_l = 0.1$ (red line), respectively. (d) Time courses for the extracellular AHL-concentration, which were generated by cells with different concentrations of regulator n_l. Those time courses correspond to the bifurcation diagram shown in (c), with $\alpha_l = 0.1$. (e) Bifurcation diagram, where in addition to the increased LuxI-production by regulator $\tilde{n}_l$, a destabilising effect of regulator $\tilde{n}_l$ on LuxI is assumed. Basal LuxI-production rate is 0.1. The destabilising effect is $p_{n_l} = 1$ (black line) or $p_{n_l} = 10$ (red line). (f) Time courses for the extracellular AHL-concentration, which were generated by cells with different concentrations of regulator $\tilde{n}_l$. Those time courses correspond to the bifurcation diagram shown in (e), with $p_{n_l} = 10$.

the system will either be activated or not. The time course, which is shown in Fig. 4(d), displays similar effects as already seen for LexA-, QteE- and n_l-type regulators. Depending on the proximity of the cAMP-concentration to the bifurcation point the *lux* system is induced at different time points. The final AHL-concentration in an activated system changes with different concentrations of cAMP.

The bistability behaviour of the previous figures is lost, when assuming a high basal production rate α_l. The system is induced independently of

the added cAMP-concentration (Fig. 4(e)). All systems are induced at about the same time (Fig. 4(f)). They only differ in the final AHL-concentrations.

IV. STOCHASTIC INFLUENCES

So far, any stochasticity was neglected in our modelling approach. Nevertheless, as e.g. some parts of the intracellular regulation system may consist only of few molecules, and the regulation system acts non-linearly, the behaviour of individual cells might significantly differ from the

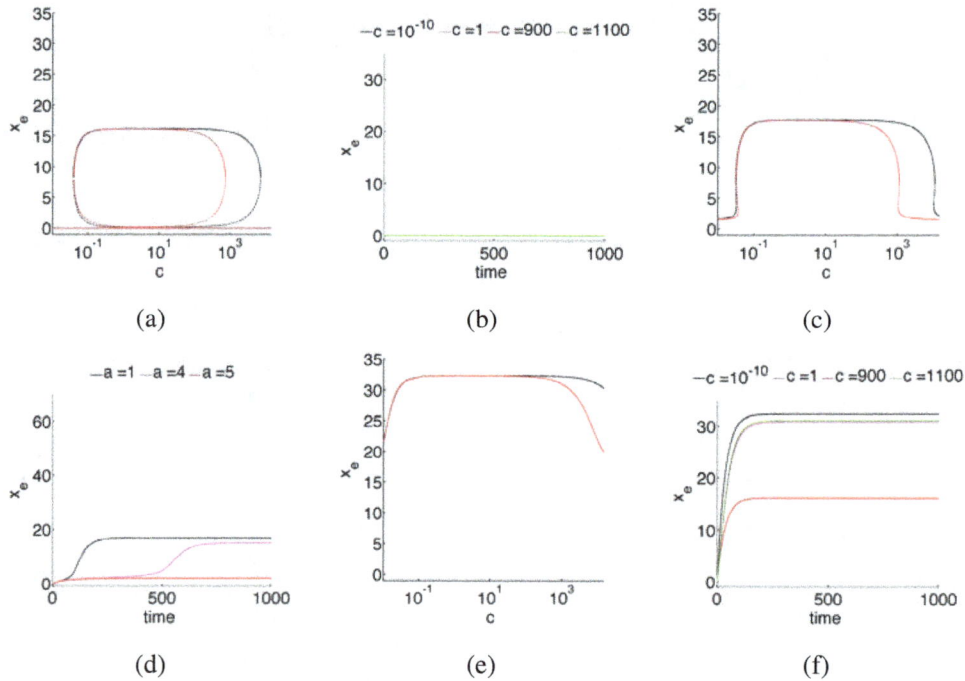

Fig. 4: Influences on LuxI and LuxR simultaneously. (a) Bifurcation diagram for the extracellular AHL-concentration, where the effect of cAMP on the competitive inhibition of the LuxI-production and the increase of the LuxR-production is considered. The basal LuxI-production rate α_l is 0.001. Affinity of cAMP to the *lux* operon compared to the AHL-LuxR dimer is assumed to be equal. This is achieved by setting $b = 1$ (black line) or by assuming the affinity of cAMP to the *lux* operon to be stronger compared to the AHL-LuxR dimer by setting $b = 10$ (red line). (b) Time courses for the extracellular AHL-concentration, which were generated by cells with different cAMP-concentrations. Those time courses correspond to the bifurcation diagram shown in (a), with $b = 1$ and $\alpha_l = 0.001$. (c) Same figure as seen in (a), only the basal LuxI-production rate is increased to 0.1. (d) Time courses for the extracellular AHL-concentration, which were generated by cells with different cAMP-concentrations. Those time courses correspond to the bifurcation diagram shown in (c), with $b = 1$ and $\alpha_l = 0.1$. (e) Same figure as seen in (a), only $\alpha_l = 1$. (f) Time courses for the extracellular AHL-concentration, which were generated by cells with different concentrations of regulator n_l. Those time courses correspond to the bifurcation diagram shown in (e), with $b = 1$ and $\alpha_l = 1$

bulk behaviour. This is also the case for nutrient-dependent regulators, as nutrients often are heterogeneously distributed under natural conditions. As an example we will consider nutrient-dependent influences in this section. Of course, the dynamic behaviour itself is the same as in the deterministic setting. But this stochastic approach allows us to track a number of cells with typical variations in molecule numbers and hence, leads to a better understanding of how realistic cell populations could behave.

The numerical analyses were done with MATLAB Version R2010a [26], using the solver ode45 with its standard precision.

A. Influence of a single regulator on the system

We start by considering the influence of stochasticity of a single regulator on the whole Quorum sensing system. For the number of regulator molecules per cell we assume a normal distribution with a fixed expected value and variance. This can be interpreted as a normal distributed nutrient availability under natural conditions, which then transfers to the nutrient-dependent regulator.

For the simulations, we set the number of cells to ten. For higher numbers of cells the results are qualitatively the same (not shown). A fixed cell number can be realised experimentally e.g. in a chemostat-like setting. Above, a deterministic single cell model was introduced. Now we slightly alter this model in order to obtain a model with n cells and a random distribution of regulators, i.e., we focus on the influence of stochasticity by the regulators but neglect other stochastic effects on the Quorum sensing system. This means that we still assume AHL production in each cell to be deterministic but dependent on the random number of regulator molecules in each cell. Assuming once again different time scales, we reduce the model to a two component hybrid model. While the basic equation for the intracellular AHL-concentration in the reduced model (Eq. (7)) stays the same, the governing equation for x_e changes to be

$$\dot{x}_e = \sum_{k=1}^{n} d_c x_c^{(k)} - n d_e x_e - \gamma_e x_e, \qquad (19)$$

where the superscript describes the k-th cell, as now, each cell may have an individual intracellular AHL-concentration, dependent on its available regulators. When regarding the above mentioned assumptions, the governing equations for x_c are modified only slightly. As an example we show how the equation for the intracellular AHL-concentration under the influence of a regulator controlling LuxR production in a way as reported for GroESL (Eq. (9)) changes:

$$\dot{x}_c^{(k)} = B\alpha_l + B\beta_y \frac{A_r \left(x_c^{(k)}\right)^2}{1 + (\beta_y/\kappa_y) A_r \left(x_c^{(k)}\right)^2} - \gamma_c x_c^{(k)} - d_c x_c^{(k)} + d_e x_e, \qquad (20)$$

for $k = 1, ..., n$ and $A_r := \frac{\pi_2^+}{\pi_2^-} \cdot \left(\frac{\pi_1^+}{\pi_1^-}\right)^2 \cdot \left(\frac{\alpha_r + \mu_r N_r}{\gamma_r}\right)^2$.

Note that the only difference to the non-stochastic equation concerns the superscript k, which describes the k-th cell, and the random variable N_r instead of the fixed n_r. This random variable

N_r is, as stated above, normally distributed with an expected value $E[N_r]$ and a variance $Var[N_r]$. In the following we choose the variances relatively high such that the effects due to the randomness in the regulator distribution become visible. The realisations of N_r will be different concentrations of GroESL-like regulators in different cells. All the other equations are altered in a similar way but omitted here for the reason of brevity. For the hybrid model, which includes a higher number of cells, the diffusion constants d_c and d_e are changed. This helps to identify the studied effects better. In the simulations the diffusion constants are set to $d_c = d_e = 0.05$ in contrast to 0.5 in the simulations without a stochastic distribution of regulators in order to keep the extracellular concentration of AHL comparable to the single cell scenario, i.e., we implicitly assume that the extracellular volume of n cells is n times the extracellular volume of one cell.

Taking these changes into account, the bifurcation points in the simulations with multiple cells are considerably lower, i.e., lower regulator-concentrations - in the case of regulator n_l and GroESL - are bifurcation points than the ones identified in the single cell simulations (results not shown). Introducing LexA- or QteE-like regulators into the equations and assuming that the basal production rate of LuxI (α_l) to be 0.001 obviously never leads to an activation of the bacteria in the ten cell setting, under the given conditions, when starting with zero AHL and an arbitrary concentration of LexA or QteE (Figures 2(d) and 3(d)).

From here on it is important to keep the differences between the following figures - especially Figures 6, 8, 9 and 10 - and the time courses in Figures 2, 3 and 4 in mind. While the single cells were not able to influence each other in the previous sections, there is now an influence between the different cells within one colony.

Running 1000 simulations with the amount of a regulator near the bifurcation point in each run, results in large differences of the final intra- and extracellular AHL concentrations due to non-linearity (Fig.5(a) for extracellular AHL concentration). We use the same amount of regulators for each run as we only want to examine the effect of the distribution of the regulator on the final AHL-concentration. Each data point in the box plot can be interpreted as one colony, where each colony has the same size and the same amount of regulator available. The only difference between the runs is the distribution of the regulator over the cells. This result gives rise to the idea that the distribution of a regulator is to some extent responsible for the activation of the system, neglecting the time course for a moment which also might be influenced by the stochastic regulatory effects. The same result was attained for the other effects of regulators on the system, but they are omitted here.

Effects due to GroESL-variation: When running a simulation with one colony, one can compare the cell with the highest intracellular AHL-concentration at the end of a simulation ($t_{end} = 1000$) within the colony with the one having the lowest final intracellular AHL-concentration. Subtracting those concentrations from each other gives information about variation between cells within a colony. Doing this for one thousand colonies - again assuming the same size of the colonies - leads to the box plot shown in Figure 5(b). Most cells within a colony - when assuming an inhomogeneous distribution of GroESL-like regulators - have a similar final intracellular AHL-concentration as the difference between the cells is low compared to the relative deviation of regulator n_l of approximately 40% (Fig. 5(c)). However, some outliers occur in Figure 5(b) (red crosses). A possible interpretation for those is that the distribution of regulators within one colony might influence the time of activation as some cells are already activated while others are not yet. This idea will be supported in section IV-B (see below).

Effects due to n_l-variation: Proceeding with regulator n_l in the same way as with GroESL leads to the box plot shown in Fig. 5(c). There, one can see that on the one hand the system as a whole is always either induced or non-induced as the difference of the AHL-concentrations of the

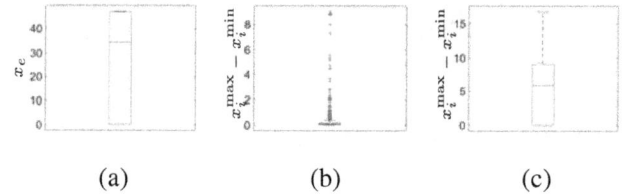

(a)   (b)   (c)

Fig. 5: Box plots of AHL-concentrations under different conditions. The colours mean the following: red line is the median, blue box is the $25 - 75\%$-quantile, black limiters (whiskers) extend to the most extreme values which are no more than $1.57/\sqrt{1000} \cdot (75\%-\text{quantile} -25\%-\text{quantile})$ away from the box and red crosses show outliers not belonging to the region limited by the whiskers. (a) Extracellular AHL-concentration within one colony. Each colony had the same amount of regulator n_l available and the same number of cells. However, the concrete distribution of regulator n_l amongst the individuals is different in each colony. The AHL-concentration is measured at the end of the simulation at time $t_{end} = 1000$. The values of regulator n_l are simulated with $E[N_l] = 6.5$ and $Var[N_l] \approx 5.2$. (b) Box plot, where each data point is obtained by subtracting the cell with the lowest intracellular AHL-concentration at the end of the simulation ($t_{end} = 1000$) from the cell with the highest intracellular AHL-concentration within one colony. The difference of the intracellular AHL-concentration between the cells is - in this subfigure - due to the influence of GroESL on the LuxR-production. One thousand colonies were simulated to create this box plot. $E[N_r] = 6$ and $Var[N_r] \approx 0.055$. (c) Box plot was created in the same way as in (b), only the influence of regulator n_l on the LuxI-production is varying this time. $E[N_l] = 6.5$ and $Var[N_l] \approx 5.2$. End of simulation at time $t_{end} = 1000$.

two cells is considerably lower than between an activated and a non-activated state (compare to Fig. 5(a)). This means that regulator n_l has no relevant effect on the time of activation within one colony in our parameter setting. On the other hand the AHL-concentration level at the end of the simulation ($t_{end} = 1000$) depends on the amount of n_l (Fig. 5(c)), which is different compared to the influence of GroESL where the distance between the cells containing most and fewest AHL-molecules is considerably lower than here (Fig. 5(b)).

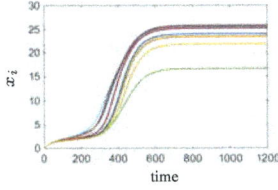

Fig. 6: Time course of the intracellular AHL-concentration of ten cells within one colony with a inhomogeneous distribution of cAMP (i.e. $E[c] = 0.03, Var[c] = 0.00005; \alpha_l = 0.05$)

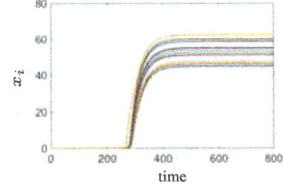

Fig. 7: Time course of the intracellular AHL-concentration of ten cells which all belong to the same colony. The distribution of regulator n_l fixed and given by the vector $(1, 2, 3, 6, 7, 8, 9, 12, 13, 15)$.

Effects due to cAMP-variation: The simulations on cAMP are done with $\alpha_l = 0.05$ since otherwise the system never gets activated in our parameter setting. In this case the lower bifurcation point of cAMP is lower than in the (deterministic) one cell setting (see Fig. 4(d)), whereas the upper bifurcation point is even higher (results not shown). The effect which cAMP has on the system is a combination of the effects of GroESL and regulator n_l, similar as in the deterministic model system. On the one hand cells with low cAMP-concentration will get activated later than cells with intermediate cAMP concentration. A low concentration of cAMP on the other hand leads to a lower final intracellular concentration of AHL than an intermediate concentration (Fig. 6). In Fig. 6 the expected value for cAMP is $c = 0.03$ and the variance is 0.00005, i.e., quite small.

Here, as well as below, one can see that the system is quite stable with respect to the variation of the different regulators, i.e., the resulting relative deviation in the AHL-concentration was below 10% even though the coefficient of variation ($\sqrt{Var}/E$) of c was 24% approximately, in our parameter setting. Nevertheless the figures are included to see the possible effects of the different regulators on the Quorum sensing system. In contrast to this, Fig. 7 was included, in which a range of concentrations of regulator n_l was distributed over the different cells within one colony. This means that the different regulators may yield different resulting variability in the system, due to their non-linear influences.

B. Combining several regulators

So far only the influence of a single regulator on the system has been studied. In the following we investigate the effect of several regulators influencing the Quorum sensing system at the same time. We show the governing equation for x_c in the reduced model for an influence of the exemplarily chosen regulators GroESL, LexA and n_l on the Quorum sensing system:

$$
\dot{x}_c^{(k)} = (1 + \mu_l N_l) \cdot
$$
$$
\left(B\alpha_l + B\beta_y \frac{A_r \left(x_c^{(k)}\right)^2}{1 + (\beta_y/\kappa_y) A_r \left(x_c^{(k)}\right)^2} \frac{bA_r x_c^{(k)}}{bA_r x_c^{(k)} + A} \right)
$$
$$
- \gamma_c x_c^{(k)} - d_c x_c^{(k)} + d_e x_e, \qquad (21)
$$

where $A_r := \frac{\pi_2^+}{\pi_2^-} \left(\frac{\pi_1^+}{\pi_1^-}\right)^2 \left(\frac{\alpha_r + \mu_r N_r}{\gamma_r}\right)^2$.

Again N_l, N_r and A are random variables representing the different amounts of regulator n_l, GroESL and LexA in the cells. To get a deeper understanding of the functionality of the respective regulators, we only examine the influence of two regulators on the system at a time, one with a fixed value, the other one with the usual variation. The expected values and variances of the simulations in this section are given in Table V. The values of expectation and variance are chosen such that all regulators have the same coefficient of variation (0.5). Note however that we had to increase the values of N_l significantly in Figure 9. Elsewise the system would not activate which is due to the inhibition of LexA even though it seems negligible.

In the deterministic model approach, we guessed that GroESL affects the time of activation of our

	N_l		N_r		LexA	
	Expected value	Variance	Expected value	Variance	Expected value	Variance
Fig. 8 (a)	3	2.25	1.5	0	0	0
Fig. 8 (b)	3	0	1.5	0.5625	0	0
Fig. 9 (a)	20	100	0	0	0.01	0
Fig. 9 (b)	20	0	0	0	0.01	0.000025
Fig. 10 (a)	0	0	1.5	0.5625	0.01	0
Fig. 10 (b)	0	0	1.5	0	0.01	0.000025

TABLE V: Expected values and variances for the different regulators in the Figures 8 (a) - 10 (b)

system. This behaviour can also be found in the stochastic approach under the additional presence of n_l, see Fig. 8(b). Additionally, note that the final concentrations are basically indistinguishable. The earlier expressed assertion that n_l changes the final concentration, but has no impact on the time-point of activation, is visible in Fig. 8(a).

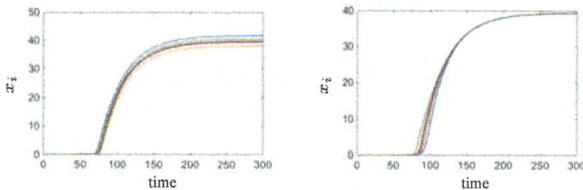

(a) constant GroESL, varying N_l

(b) constant N_l, varying GroESL

Fig. 8: Time course of the intracellular AHL-concentration of ten cells within one colony influenced by GroESL and Regulator n_l

Regarding the inhibitor LexA, a connection between the time of activation and the LexA-concentration becomes visible now. The lower the concentration of LexA, the earlier the cell is activated (Fig. 9(b)). Fig. 9(a) shows qualitatively the same behaviour as Fig. 8(a) suggesting that the qualitative impact of LexA and N_r is similar, at least once the colony gets activated.

This fact is confirmed by Figures 10(a) and (b) which show the effect of the coupled influence of LexA and GroESL. The pictures are similar, the only slight difference being that the time-point of activation with varying GroESL leads to a rather homogeneous distribution of activation

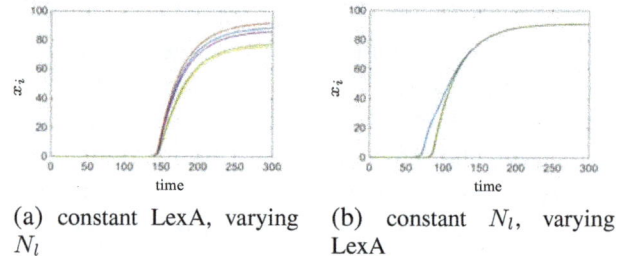

(a) constant LexA, varying N_l

(b) constant N_l, varying LexA

Fig. 9: Time course of the intracellular AHL-concentration of ten cells within one colony influenced by LexA and Regulator n_l

time-points, whereas varying LexA only favours a single cell to activate and afterwards the bulk is induced. This means: some regulators, especially inhibitors may affect first mainly single cells and later the whole colony.

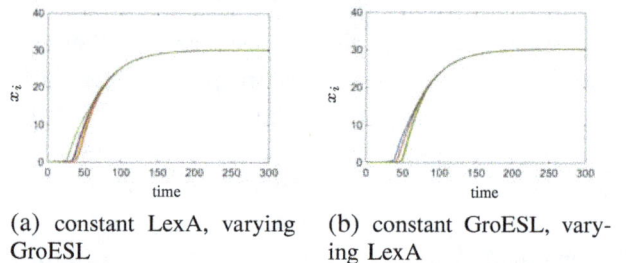

(a) constant LexA, varying GroESL

(b) constant GroESL, varying LexA

Fig. 10: Time course of the intracellular AHL-concentration of ten cells within one colony influenced by GroESL and LexA

V. DISCUSSION AND CONCLUSION

Although a number of studies use mathematical models to investigate traits of different Quorum sensing systems, little is known about the impact of external regulation factors (see e.g. [22, 41, 39]). To our knowledge, we present the first comparison of different mechanisms affecting the typical basic motif of AHL based communication systems.

Modelling the full gene regulatory system for Quorum sensing of *lux* type, including all mentioned influencing regulators and mechanisms, leads to a large system of ODEs in the classical deterministic approach, containing a vast amount of (quantitatively unknown) parameters. The application of singular perturbation on the resulting mathematical model can shrink down the system essentially, and allows for a clearer analysis of the system, e.g. concerning bifurcations. Especially the possibility of bistable regions is of great interest in this context, as it allows (via a kind of hysteretic behaviour) the stabilisation of the system against perturbations [28].

Our results indicate that depending on the mode of action some regulators mainly affect the time of induction (e.g. Fig. 2(b),(f), Fig. 3(b)), which is connected with a critical cell density (plankton) or cell number (colonies). Others change the maximum signal concentration (e.g. Fig. 2(d), Fig. 3(d)) or both (e.g. 4(d)). The potential ecological and/or evolutionary benefit of these different regulator effects depends on the context, in which the population lives. For example, under spatially structured conditions such as populations living in microcolonies which support development of heterogeneity between cells, synchronicity of responses on a population level could be supported by higher induced AHL production. Hense and Schuster [18] argue that the fitness benefit of Quorum sensing regulated activity typically is not only a function of its potential strength, influenced by the cell density and some other factors, but also of the cells which demand it. Furthermore, it is highly desirable to control the timing of induction as a function of environmental conditions. Bistability can be interpreted as a simple kind of memory. The underlying positive autoregulation of components of Quorum sensing often seems to be heritable

and can thus be understood as an epigenetic control [30]. It supports stability of the population e.g. against environmental fluctuations. When a Quorum sensing controlled switch between two cellular states is costly, such stability helps to minimise costs, however, at the expense of adaptation rate. Shifts of range of stability enable the cells to optimise trade-off between these opposing aspects. By combinations of various regulators, or multiple effects of one regulator on Quorum sensing via different mechanisms the cell can realise complex reaction patterns such as maximum or minimum Quorum sensing at an intermediate strength of the environmental control factor.

For example our model predicts that under certain conditions environmental factors acting via cAMP show such an intermediate maximum. cAMP is connected with starvation strength. [37] showed experimentally an intermediate maximum in a dilution series of culture medium for *V. fischeri*. Although the biochemical mechanisms behind were not fully clarified and their experimental design did not exactly reflect our model, their experiments show that such complex regulation patterns are relevant in vivo. This intermediate peak is interesting, as usually a more monotone relation between environmental factors and Quorum sensing systems has been reported (see e.g. Hense and Schuster, and citations therein.). Unfortunately, quantitative information about dose-response relations over a larger range of the strength of these factors are largely lacking, which impedes statements about the prevalence of such intermediate peaks. An exception is the well-studied *Bacillus subtilis*, in which mild starvation induce sacrifice of a fraction of the population ([24]). The purpose of this highly cooperative activity seems to be to supply nutrients for the remaining cells which might help to delay a costly sporulation. However, if starvation increases even more, the population induces sporulation. Induction of sacrifice thus peaks at intermediate starvation levels. Similarly, *B. subtilis* induces competence in a certain window of environmental intermediate stress conditions ([33]). Quorum sensing is involved in the control of these processes. However the architecture of Quorum sensing in *B. subtilis* differs strongly from that of the AHL-type. The Com Quorum sensing system of *B. subtilis* and the influence of stress act rather in parallel in regulation of their target competence in *B. subtilis*. In contrast, in the scenario analysed in our study, cAMP impacts the Quorum sensing system directly. In more abstract terms, Quorum sensing usually induces cooperative behaviour, often as a stress response ([18]). Stress as a promoter of cooperation is a well-established concept also in other areas of ecology ([21]). [21] state that extreme stress does no longer support cooperation, but other aspects like competition tend to become dominant. As a consequence, under very severe stress conditions cells may induce other phenotypes like persistence or motility to escape from stress. Intermediate stress levels as optimal activator conditions for Quorum sensing fits to this concept. Based on these hints we speculate that such a regulation strategy may occur more frequently. More experimental dose-response studies investigating the relation between environmental conditions and regulation of Quorum sensing systems are thus desirable. Although a number of external regulators have been experimentally identified for an increasing number of Quorum sensing systems, the effect of most of these regulators on dynamics of Quorum sensing is usually unclear. RsaL in *P. aeruginosa* acts by suppressing the expression of the LuxI homologue and thereby delays the induction in experiments, which fits to our results for LexA-like regulators [11, 7]. LitR promotes the expression of LuxR in *V. fischeri*. In accordance to what we predict for GroESL-like regulators, *litR* mutants show delayed expression of Quorum sensing regulated phenotypes (Lupp and Ruby, 2005). However, for both, RsaL and LitR, effects on maximum AHL production and the potential ecological relevance of it have not been determined experimentally yet. In a second step, we combined the deterministic behaviour of a single cell with a stochastic distribution of regulators in a number of cells, allowing for simulations of more realistic populations with some individual variations. Our study indicates that, depending on how a regulator of Quorum sensing systems acts on the molecular level, such a stochastic distribution may have effects on timing of induction and/or strength of induction, due to the non-linearity and the interaction of the single cells via the signalling molecule AHL.

As Quorum sensing regulation has been regarded as a source of synchronous responses of cell populations, the existence of stochastic heterogeneity on Quorum sensing systems of isogenic populations has only recently been recognised ([16]). Underlying mechanisms, as well as ecological effects and potential benefits are far from being understood. Generally, Quorum sensing systems are thought to be prone to fluctuations due to often low numbers of receptors and signals. However, mechanisms to suppress dominance of stochasticity and hence making the system more reliable have been described (e.g. [29, 42]). There are hints that heterogeneity of expression in QS genes and/or QS regulated target genes may be a common phenomenon even in isogenic populations [2, 6]. Stochastic differences between cells play a stronger role if only a few cells are involved in the autoinducer based decision making process, e.g. in extreme if a single cell is induced by highly limited mass transfer in a pore (diffusion sensing) [17]. Our study investigates, how regulators of Quorum sensing can cause heterogeneity in Quorum sensing dynamics.

Such a heterogeneity can be an unavoidable side effect. However, if it causes significant phenotypic differences, it might have an ecological purpose, as it is often interpreted in terms of division of work [6]. The benefit of division of work strongly depends on the environmental conditions. It thus seems probable that stochastic heterogeneity of environment-dependent regulators are involved in the emergence of molecular heterogeneities between cells. Therefore, cells may not just suppress noise in their Quorum sensing systems, but rather control its level or its impact on the Quorum sensing regulation [10].

Our results indicate how stochastic variations in the concentration of factors regulating Quorum sensing influence inter-cell heterogeneity of Quorum sensing response. Dependent on the mode of action of the regulator respectively the combination of different regulators, both timing and/or strength of the response can vary. Stochastic differences in timing of Quorum sensing induced mobility resulting in a removal of single cells from colonies has been reported for *Pseudomonas putida* ([6]). In other bacteria rather the expression levels of Quorum sensing regulated genes seem to vary ([16]), although the design of the experimental studies often impedes a clear discrimination. In almost all cases both the causes of the heterogeneity and the ecological or evolutionary benefit of heterogeneity are unknown yet. By investigating the potential impact of regulators on heterogeneity, our study aims to shed some light on these questions. The differences of the Quorum sensing response between the cells caused by the regulators, i.e., the strength of heterogeneity, was limited in our simulations. However, they might be larger in the real world, as they depend on the variability of the regulator concentration, and on the degree of coupling between cells. The latter, which is mediated by the Quorum sensing signal, has been predicted to be controlled by the cells dependent on the environmental conditions ([14]). Interestingly, [14] predicted in a mathematical model that fluctuations on the molecular level, which are regulated by environmental factors, cause a switch between all-or-none and graded responses of Quorum sensing systems on a population level. Stochastic heterogeneities between cells can also impact the functionality of Quorum sensing systems, e.g. on the induction threshold on a population level [42]. It is thus highly desirable to get a deeper understanding of sources and outcome of Quorum sensing associated stochastic heterogeneity.

Our analysis focuses on typical AHL based Quorum sensing systems, but also Quorum sensing systems with other architectures exist. The ex-act net effect of different regulation mechanisms depends on the design of the complete cellular regulation network (see e.g. [3]). As most pathogens and many other bacteria relevant from a human perspective use Quorum sensing to regulate virulence or factors beneficial for human health, the qualitative and quantitative understanding of the underlying mechanisms are critical for the development of adequate treatment strategies. Furthermore, knowledge of the behaviour of such motifs is required in the growing field of synthetic biology (see e.g. [4]). Thus, the qualitative and quantitative impact of regulators in QS systems should be investigated in more depth, both experimentally and theoretically.

VI. Supplementary Information

Figures 11 and 12 allow for the comparison of the qualitative behaviour of the full basic model system with the basic system with quasi-steady state assumption. Please note that the large initial differences are due to the fact that we continued to take our "standard initial values", which are not close to the quasi-steady state and needs some adaptation first.

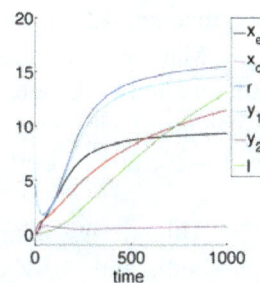

Fig. 11: Simulation of the basic model (Eq.(1)-(6)) with parameters from Table IV, $\alpha_l = 0.001$ and initial conditions $x_e = 10$ and $x_c = r = y_1 = y_2 = l = 0$.

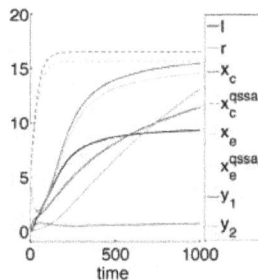

Fig. 12: Simulation of the basic model (Eq.(1)-(6)) with parameters from Table IV, $\alpha_l = 0.001$ and initial conditions $x_e = 10$ and $x_c = r = y_1 = y_2 = l = 0$. The variables x_c^{qssa} and x_e^{qssa} correspond to the basic system with quasi-steady state assumption.

References

[1] Y.Y. Adar, M. Simaan, and S. Ulitzur. Formation of the LuxR protein in the *Vibrio fischeri lux* system in controlled by HtpR through the GroESL. *J. Bacteriol.*, 174:7138–7143, 1992.

[2] C. Anetzberger, U. Schell, and K. Jung. Single cell analysis of *Vibrio harveyi* uncovers functional heterogeneity in response to quorum sensing signals. *BMC Microbiol.*, 12: 209, 2012. doi: 10.1186/1471-2180-12-209.

[3] D. Balasubramanian, L. Schneper, H. Kumari, and K. Mathee. A dynamic and intricate regulatory network determines *Pseudomonas aeruginosa* virulence. *Nucl. Acids Res.*, pages 1—20, 2012. doi: 10.1093/nar/gks1039.

[4] Y. Borg, E. Ullner, A. Alagha, A. Alsaedi, D. Nesbeth, and A. Zaikin. Complex and unexpected dynamics in simple genetic regulatory networks. *Int. J. Mod. Phys. B*, 28:1430006, 2014. doi: 10.1142/S0217979214300060.

[5] J.L. Bose, U. Kim, W. Bartkowski, R.P. Gunsalus, A.M. Overley, N.L. Lyell, K.L. Visick, and E.V. Stabb. Bioluminescence in *Vibrio fischeri* is controlled by the redox-responsive regulator ArcA. *Mol. Microbiol.*, 65:538–553, 2007. doi: 10.1111/j.1365-2958.2007.05809.x.

[6] G. Carcamo-Oyarce, P. Lumjiaktase, R. Kümmerli, and L. Eberl. Quorum sensing triggers a stochastic escape of individual cells from *Pseudomonas putida* biofilms. *Nat. Communication*, 6:5945, 2015. doi: 10.1038/ncomms6945.

[7] T. De Kievit, P.C. Seed, J. Nezezon, L. Passador, and B. Iglewski. RsaL, a novel repressor of virulence gene expression in *Pseudomonas aeruginosa. J. Bacteriol.*, 181: 2175–2184, 1999.

[8] G. Dieppois, V. Ducret, O. Caille, and K. Perron. The transcriptional regulator CzrR modulates antibiotic resistance and quorum sensing in *Pseudomonas aeruginosa. PLOS One*, page e38148, 2012. doi: 10.1371/journal.pone.0038148.

[9] J.D. Dockery and J.P. Keener. A Mathematical Model for Quorum sensing in *Pseudomonas aeruginosa. Bull. Math. Biol.*, 63: 95–116, 2001. doi: 10.1006/bulm.2000.0205.

[10] B. Drees, M. Reiger, K. Jung, and I.B. Bischofs. A modular view of the diversity of cell-density-encoding schemes in bacterial quorum-sensing systems. *Biophys. J.*, 107: 266–277, 2014. doi: 10.1016/j.bpj.2014.05.031.

[11] J.-F. Dubern, B.J.J. Lugtenberg, and G.V. Bloemberg. The *ppuI-rsaL-ppuR* quorum-sensing sysem regulates biofilm formation of *Pseudomonas putida* PCL1445 by controlling biosynthesis of the cyclic lipopeptides Putisolvins I and II. *J. Bacteriol.*, 188:2898–2906, 2006.

[12] P.V. Dunlap. Quorum Regulation of Luminescence in *Vibrio fischeri. J. Molec. Microbiol. Biotechnol.*, 1:5–12, 1999.

[13] B. Ermentrout. XPPAUT Version 5.41. http://www.math.pitt.edu/~bard/bardware/, February 2003.

[14] K. Fujimoto and S. Sawai. A design principle of group-level decision making in cell populations. *PLOS Comp. Biology*, 2013. doi: 10.1371/journal.pcbi.1003110.

[15] W. Fuqua, S. Winans, and E. Greenberg.

Quorum sensing in bacteria: The LuxR-LuxI family of cell density-responsive transcriptional regulators. *J. Bacteriol.*, 176:269–275, 1994.

[16] J. Grote, D. Krysciak, , and W.R. Streit. Phenotypic heterogeneity, a phenomenon that may explain why quorum sensing does not always result in truly homogenous cell behavior. *Appl. Environm. Microbiol.*, 81:5280–5289, 2015.

[17] S.J. Hagen, M. Son, J.T. Weiss, and J.H. Young. Bacterium in a box: sensing of quorum and environment by the LuxI/LuxR gene regulatory circuit. *J. Biol. Physics*, 36:317–327, 2010. doi: 10.1007/s10867-010-9186-4.

[18] B.A. Hense and M. Schuster. Core principles of bacterial autoinducer systems. *Microbiol. Mol. Biol. Rev.*, 79:153–169, 2015. doi: 10.1128/MMBR.00024-14.

[19] B.A. Hense, C. Kuttler, J. Müller, M. Rothballer, A. Hartmann, and J.U. Kreft. Does efficiency sensing unify diffusion and quorum sensing? *Nat. Rev. Microbiol.*, 5:230–239, 2007. doi: 10.1038/nrmicro1600.

[20] B.A. Hense, J. Müller, C. Kuttler, and A. Hartmann. Spatial heterogeneity of autoinducer regulation systems. *Sensors*, 12:4156–4171, 2012. doi: 10.3390/s120404156.

[21] M. Holmgren and M. Scheffer. Strong facilitation in mild environments: the stress gradient hypothesis revisited. *J. Eco.*, 98:1269–1275, 2010.

[22] S. Jabbari, J.T. Heap, and J.R. King. Mathematical modelling of the sporulation-initiation network in *Bacillus subtilis* - revealing the dual role of the putative quorum-sensing signal molecule PhrA. *Bull. Math. Biol.*, 73:181–211, 2011. doi: 10.1007/s11538-010-9530-7.

[23] C. Kuttler and B. Hense. Interplay of two quorum sensing regulation systems of *Vibrio fischeri*. *J. Theor. Biol.*, 251:167–180, 2008.

[24] D. Lopez and R. Kolter. Extracellular signals that define distinct and coexisting cell fates in bacillus subtilis. *FEMS Microbiol. Rev.*, 34:134–149, 2010.

[25] A.K. Marr, J. Overhage, M. Bains, and R.E. Hancock. The Lon protease of *Pseudomonas aeruginosa* is induced by aminoglycosides and is involved in biofilm formation and motility. *Microbiology*, 153:474–482, 2007.

[26] The MathWorks. MATLAB Version R2010a. http://www.mathworks.de/, March 2010.

[27] B.L. Mellbye and M. Schuster. A physiological framework for the regulation of quorum-sensing dependent public goods in *Pseudomonas aeruginosa*. *J. Bacteriol.*, 196:1155–1164, 2014.

[28] J. Müller, C. Kuttler, B.A. Hense, M. Rothballer, and A. Hartmann. Cell-cell communication by quorum sensing and dimension-reduction. *J. Math. Biol.*, 53:672–702, 2006. doi: 10.1007/s00285-006-0024-z.

[29] J. Müller, C. Kuttler, and B.A. Hense. Sensitivity of the quorum sensing system is achieved by low pass filtering. *BioSystems*, 92:76–81, 2008. doi: 10.1016/j.biosystems.2007.12.004.

[30] E.M. Nelson, V. Kurz, N. Perry, D. Kyrouac, and G. Timp. Biological noise abatement: Coordinating the responses of autonomous bacteria in a synthetic biofilm to a fluctuation environment using a stochastic bistable switch. *ACS Synth. Biol.*, 3:286–297, 2014. doi: 10.1021/sb400052f.

[31] R. Redfield. Is quorum sensing a side effect of diffusion sensing? *Trends Microbiol.*, 10:365–370, 2002.

[32] C. Reimmann, M. Beyeler, A. Latifi, H. Winteler, M. Foglino, A. Lazdunski, and D. Haas. The global activator GacA of *Pseudomonas aeruginosa* PAO positively controls the production of the autoinducer N-butyryl-homoserine lactone and the formation of the virulence factors pyocyanin, cyanide, and lipase. *Mol. Micriobiol.*, 24:309–319, 1997.

[33] D. Schultz, P.G. Wolynes, E. Ben Jacob, and J.N. Onuchica. Deciding fate in adverse times: Sporulation and competence in bacillus subtilis. *Proc. Nat. Acad. Sci. USA*, 106:

21027–21034, 2009.

[34] G.S. Shadel and T.O. Baldwin. The *Vibrio fischeri* LuxR protein is capable of bidirectional stimulation of transcription and both positive and negative regulation of the *luxR* gene. *J. Bacteriol.*, 173:568–574, 1991.

[35] R. Siehnel, B. Traxler, D.D. An, M.R. Parsek, A.L. Schaefer, and P.K. Singh. A unique regulator controls the activation threshold of quorum-regulated genes in *Pseudomonas aeruginosa*. *Proc Natl Acad Sci USA*, 107:7916–7921, 2010. doi: 10.1073/pnas. 0908511107.

[36] A. Takaya, F. Tabuchi, H. Tsuchiya, E. Isogai, and T. Yamamoto. Negative regulation of quorum-sensing systems in *Pseudomonas aeruginosa* by ATP-dependent Lon protease. *J. Bacteriol.*, 190:4181–4188, 2008.

[37] S. Ulitzur. The Regulatory Control of the Bacterial Luminescence System-A New View. *J. Biolumin. Chemilumin.*, 4:317–325, 1989. doi: 10.1002/bio.1170040144.

[38] S. Ulitzur and J. Kuhn. The Transcription of Bacterial Luminescence is Regulated by Sigma 32. *J. Biolumin. Chemilumin.*, 2:81–93, 1988. doi: 10.1002/bio.1170020205.

[39] A.U. Viretta and M. Fussenegger. Modeling the quorum sensing regulatory network of human-pathogenic *Pseudomonas aeruginosa*. *Biotechnol. Prog.*, 20:670–678, 2004.

[40] D. Wang, C. Seeve, L.S. Pierson, and E.A. Pierson. Transcriptome profiling reveals links between ParS/ParR, MexEF-OprN, and quorum sensing in the regulation of adaptation and virulence in *Pseudomonas aeruginosa*. *BMC Genomics*, 14:618, 2013. doi: 10.1186/ 1471-2164-14-618.

[41] J.P. Ward, J.R. King, A.J. Koerber, J.M. Croft, R.E. Sockett, and P. Williams. Cell-signalling repression in bacterial quorum sensing. *Math. Med. Biol.*, 21:169–204, 2004.

[42] M. Weber and J. Buceta. Dynamics of the quorum sensing switch: stochastic and non-stationary effects. *BMC Systems Biology*, 7, 2013. doi: 10.1186/1752-0509-7-6.

[43] P. Williams and M. Camara. Quorum sensing and environmental adaptation in *Pseudomonas aeruginosa*: a tale of regulatory networks and multifunctional signal molecules. *Curr. Opin. Microbiol.*, 12:182–191, 2009. doi: 10.1016/j.mib.2009.01.005.

6

A mathematical model of absorbing Markov chains to understand the routes of metastasis

David H. Margarit [*†] and Lilia Romanelli [*†]

* Instituto de Ciencias, Universidad Nacional de General Sarmiento, Buenos Aires, Argentina
† Consejo Nacional de Investigaciones Científicas y Técnicas (CONICET), Buenos Aires, Argentina
dmargari@ungs.edu.ar, lili@ungs.edu.ar

Abstract—**Metastasis is a complex and multi-step stochastic process. The study of the probabilities of generating a tumor from a primary site in another organs is the aim of this work. Based on statistics of National Institute of Cancer of Argentina (INC), a characterization of the routes of metastasis for the principal organs is presented by using Absorbing Markov chains. The metastasis propagation from different primary sites towards secondary and tertiary sites is also shown, emphasizing the relation and analysis about absorbing states.**

Keywords-**Metastasis; Complex Systems; Mathematical Modelling; Absorbing Markov chains**

I. INTRODUCTION

Cells have a specific and stipulated time of death (apoptosis) and reproduction rate to maintain cell balance. A tumor is the result of an uncontrolled growth of abnormal cells or when cells lose the ability to die. These cells form accumulations that affect the normal functioning of the organs, and can spread to other organs so as to cause metastasis. Metastasis is the spread of circulating tumor cells (CTC) from a primary site to near or distant locations by different ways. This depends on the organ and its initial localization through either the bloodstream or the lymphatic system (a collection of vessels that carry fluid and immune cells)[1], [2].

Markov chains are used to model different natural systems based on statistics and applications ([3], [4]), making the dynamic and continuous processes where the states depend only on the actual state and not as a result of previous events[5]. Recently, some works ([6], [7]) give different approaches by using Markov chains in metastasis processes from the lung, where mechanisms of progression and time scales of systemic disease are quantized. In the present work, we use absorbing Markov chains to analyse the metastasis transmission of solid tumors of different organs: from the primary site to a secondary site (called metastasis from primary site), and from there to a tertiary site (metastasis of secondary site) for the principal cancers in Argentina. The probabilities of having a tumor in a tertiary or secondary site from a primary site and their differences depending on each organ is analysed, as well as the probabilities that from a primary site ends in those organs that have very low probability of spreading CTCs are calculated. Finally, the steps (meaning the stages between metastasis from a primary to secondary or tertiary site) based in absorbing states (in our case, organs with low probability of generating metastasis) are found. The aim of the present work is to understand how the organs are related to each other giving a characterization of the routes of metastasis.

The metastasis are different for males and females. However the analysis performed here is similar in both cases, although different organs are affected (ovaries, vagina or uterus in females and prostate or testicles in males). In this paper we refer only the metastasis in males as a case study.

This work is organized as follow: In Section II a brief description of the methodology is given; Section III is devoted to describe transition matrices and absorbing states; The transition matrix for tertiary sites is shown in Section IV; An analysis of expected number of steps and probability of absorption by absorbing states is found in Section V; Finally, comments and conclusions are drawn in Section VI.

II. METHODOLOGY

The Markov chain transition matrix P was assembled and it is determined by the number of organs with higher probabilities of developing metastasis. This matrix shows the probability that an organ can be reached by CTCs[8] from another organ and has been built under the assumption of developing metastasis. Up to now, three leading routes of metastasis are known: *Hematogenous (blood circulation), lymphatic and transcoelomic*. With this information, the statistics of National Institute of Cancer of Argentina[9] of the principal tumors and quantitative data on the main organs affected depending on the primary site (obtained from National Cancer Institute at the National Institutes of Health of United States)[10], we performed the graph (depicted in Fig. 1) showing the most common tumors and the principal sites where they can generate metastasis.

The graph was designed by free software *Visone 2.15*[11]. The size of the nodes represents

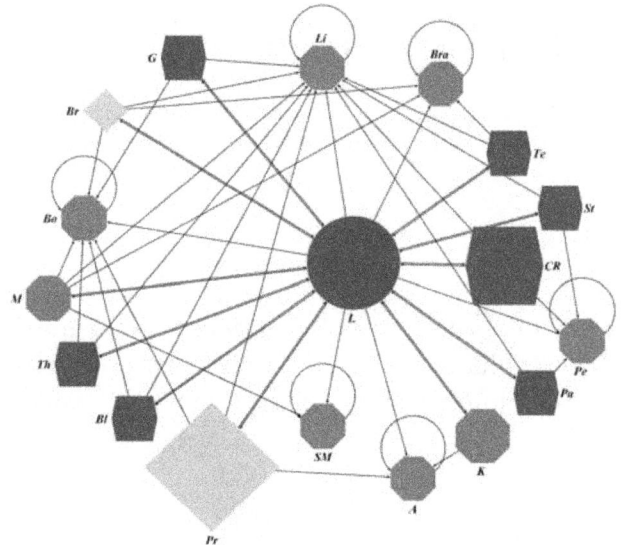

Fig. 1: Graph for the principal links of metastasis for tumors in males based in statistics. The name of the organs were referred by a symbol as depicted in Table I.

the proportion of cases for the main tumors mentioned in reference [9]. The connection of these nodes (the links) is based on the main sites of metastasis, for each specific organ, according to the data of the reference [10]. Besides, the shapes of the nodes depends on the amount of linked organs: Circle, 6 or more organs to propagate metastasis; Rhombuses, 4 or more; Hexagons, 2 or more; and finally Octagons, organs with low probability of propagate metastasis. It is important to note that the organs with low probability of generating metastasis, compared to the rest, are shown in the graph as a link on themselves (not to be confused as to metastasize about themselves).

In Fig. 1, it can be seen that the Lung is the principal link. Although the principal primary sites are Prostate, Colon or Rectum, Lung tumor is the most common but usually comes from some primary site, being the principal secondary site ([12], [13]).

TABLE I: Symbols for organs

ORGAN	NAME
Adrenal Gland	A
Bladder	Bl
Bone	Bo
Brain	Bra
Breast	Bre
Colon/Rectum	CR
Gallbladder	G
Kidney	K
Liver	Li
Lung	L
Melanoma	M
Pancreas	Pa
Peritoneum	Pe
Prostate	Pr
Skin/Muscle	SM
Stomach	St
Testicular	Te
Thyroid	Th

III. TRANSITION MATRIX, PROBABILITIES AND ABSORBING STATES

A. Transition matrix and its probabilities

The characterization of metastatic evolution is developed by using Markov chains, based on a network construction from a primary site.

Let X_0 (primary site) be the organ where the tumor was originated, and X_1 the state of the process where the new tumor is formed coming from the primary site and develop metastasis (Note: The transition time, sub index of X, does not refer to a calendar time, it refers to a general time in which has already been observed a new tumor). The probability that an organ develops metastasis from another one is:

$$p_{ij} = P\left[X_1 = j | X_0 = i\right] \qquad (1)$$

where $i, j = 1, 2, 3..., m$ number of organs.

The values p_{ij} are called *transition probabilities*[14] and have two properties:

- $\sum_{j=1}^{m} p_{ij} = p_{i1} + p_{i2} + \ldots + p_{im} = 1$, since the system must be in one of these states m, the sum of probabilities must be equal to 1.

This means that the elements in any row of the matrix transition must add 1.

- Each element $p_{ij} \geq 0$

Based in the transition probabilities, the *transition matrix* P is given by:

$$P = [p_{ij}], \qquad P \in \mathbb{R}^{m \times m} \qquad (2)$$

The routes for metastasis from one organ to another are known; although, in the literature, no information is available about their relative likelihood on which organs have an advantage over others. Given this slight uncertainty, the qualitative information ([10], [15]), we assume an equal probability that an organ X_0 (primary site), reaches other one X_1 (secondary site), this is under the assumption a metastasis is detected and based in the possibles routes as previously discussed. For other cases, where there are not predominant organs for metastasis from a specific primary site (according references [10], [12] and [13]), we will assume zero probability in order to work only the predominant sites of metastasis.

Looking at matrix P, if a tumor of a primary site, for example Prostate (Fig. 3a), has a non null probability of developing a new tumor in a secondary site, this will be $1/4$ for L, Li, Bo and Pe. The same considering a tumor in Stomach (3b), that probability will be $1/3$ for Li, Pe and L.

The matrix P is given in the Appendix A. Another way to visualize the matrix P is displayed in the Fig. 2, where, in RGB scale, the probabilities expressed in Eq. 2 can been seen.

B. Absorbing states

It is worth to notice the existence of absorbing states in the system, these are states where it is impossible to leave and are found if any row of the matrix satisfies[14]:

$$p_{ii} = 1 \text{ and } p_{ij} \neq 0 \text{ (if } i \neq j) \qquad (3)$$

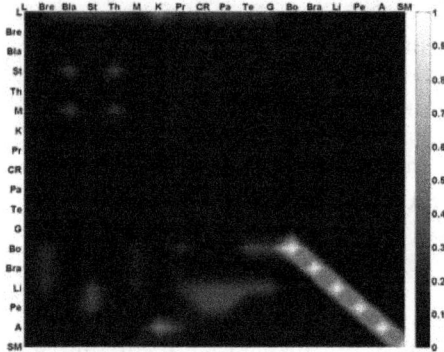

Fig. 2: Visualization of the transition matrix, in gray scale, for secondary sites.

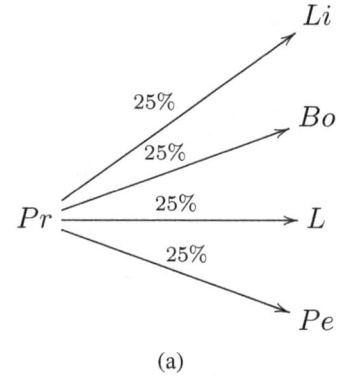

(a)

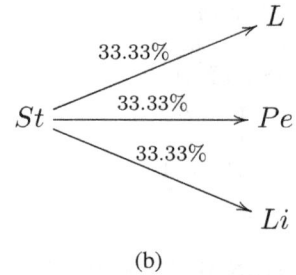

(b)

Fig. 3: Probabilities (in percentage) of metastasis in organs from Prostate (a) and Stomach (b) cancer respectively.

The properties of an Absorbing Markov chain are:

- At least, it has one absorbing state
- The absorption ends in an absorbing state with probability 1

In this work, the transition matrix was performed from the principal organs with higher probability for developing metastasis. However, for some organs, if the tumor is originated there or if it is developed elsewhere (first or secondary sites), do not evolve as metastasis from there. The absorbent states are organs that rarely metastasize, i.e. these are organs do not generate metastasis in the next step of the transition matrix. The Absorbing States of P are: **Bone (Bo), Brain (Bra), Liver (Li), Peritoneum (Pe), Adrenal Gland (A) and Skin/Muscle (SM).**

IV. TRANSITION MATRIX FOR TERTIARY SITES (SECOND METASTASIS)

For the process to move from state i to state j in two steps, it must go through an intermediate state k. If a tumor in a secondary site spreads to a new organ (tertiary site), this will be labelled as X_2. The probability of generate a new tumor in a tertiary site from a primary site is given by:

$$p_{ij}^{(2)} = P\left[X_2 = j | X_0 = i\right] = \sum_{k=1}^{m} p_{ik} \cdot p_{kj} \qquad (4)$$

where $i, j, k = 1, 2, 3..., m$ number of organs. Similarly Eq. 2, the new *transition matrix for Second Step* $P^{(2)}$ is built for males and this is given by:

$$P^{(2)} = [p_{ij}^{(2)}], \qquad P^{(2)} \in \mathbb{R}^{m \times m} \qquad (5)$$

Eq. 5 gives information how to obtain a second metastasis from the original tumor going through the possible connections. Although the metastasis from metastasis is unlikely, some clinical evidence was found in cites [16], [17]. The matrix $P^{(2)}$ is shown Appendix B.

Therefore, $P^{(2)}$ allows to find the probabilities of metastasis in a tertiary site from a primary site. For sake of clarity, examples are shown in Fig. 4a and 4b.

A global analysis of the matrix $P^{(2)}$ shows that if the primary site is the Lung (L), absorbing

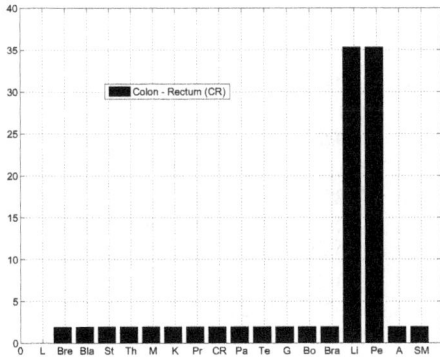

(a)

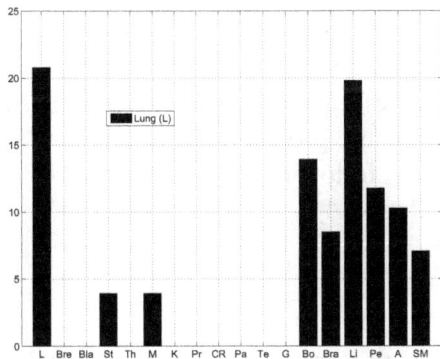

(b)

Fig. 4: Probabilities (in percentage) for tumors in tertiary site from (a) Colon/Rectum (CR) and (b) Lung (L).

states are the most probable tertiary sites as discussed in Sec. 3 (Bo, Bra, Li, Pe, A and SM). This remarks the role of Lung as the principal link between organs[18]. For other organs (disregarding the absorbing states), if the tumor is developed, there is some probability to generate metastasis in a tertiary site.

In sum, according with recent statistics[8], the principal tumors in the population are Lung (L) and Colon/Rectum (CR) tumor. In the P matrix can be observed the probabilities for evolving a tertiary site tumor from those

two. If Colon/Rectum tumor is a primary site, the principal tertiary sites are Liver (Li) and Peritoneum (Pe). And, if there is a Lung tumor, the principal tertiary sites are the absorbing states (Bo, Bra, Li, Pe, A and SM), as it was previously discussed. This can be visualized in the Fig. 5.

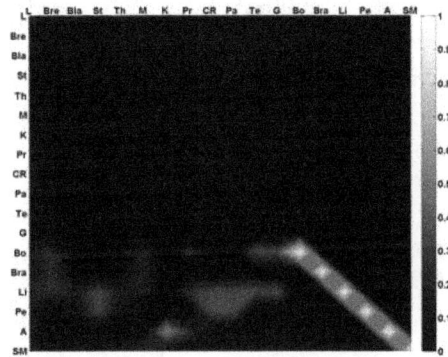

Fig. 5: Visualization of the transition matrix, in gray scale, for tertiary sites.

V. ANALYSIS OF THE EXPECTED NUMBER OF STEPS AND PROBABILITY OF ABSORPTION BY ABSORBING STATES

When the processes are absorbing, the number of steps before the system is absorbed, as well as, the probability of absorption of any absorbing state can be found. In order to find this process, each transition matrix will be represented in its canonical form[14], called J. It is composed by 4 sub-matrices: N (this sub-matrix contains the probabilities are moving from a non-absorbing state to another non-absorbing state), A (sub-matrix that contains the probabilities of going from a non-absorbing state to another absorbing state), O (zero sub-matrix) and I (identity sub-matrix).

$$J = \left(\begin{array}{c|c} N & A \\ \hline O & I \end{array} \right)$$

The matrices with smaller probability contain elements that originate a absorbing states and n non-absorbing states. There are $a + n = m$ states of the system.

Let be the fundamental matrix $F = I + N + N^2 + \cdots = (I-N)^{-1}$[14]. We can calculate from a transient state the expected number of steps before being absorbed by an absorbing state. Let t_i be the expected number of steps before the chain is absorbed when this begins in a transient state i, and let $\bar{t}$ be the column vector whose $i - th$ entry is t_i. Then, the vector $\bar{t}$ can be estimated by the following expression[14]:

$$\bar{t} = F.\bar{c} \qquad (6)$$

where $\bar{c}$ is vector whose entries are all one.

$$\bar{t} = \begin{array}{c} L \\ Bre \\ Bla \\ St \\ Th \\ M \\ K \\ Pr \\ CR \\ Pa \\ Te \\ G \end{array} \begin{bmatrix} \overset{steps}{2.237} \\ 1.559 \\ 2.810 \\ 1.745 \\ 2.810 \\ 1.447 \\ 2.118 \\ 1.559 \\ 1.745 \\ 1.745 \\ 1.745 \\ 1.745 \end{bmatrix}$$

In $\bar{t}$ are shown the tumors with more than one stage of metastasis (given the nearest integer, one stage or a step is the first metastasis in a secondary site, and two steps is the second metastasis in a tertiary site). Bladder and Thyroid have a higher number of steps, Lung and Kidney have the maximal steps around two, this coincides with the main trend of mortality in males due to these tumors[12], [21].

In addition, the probability of absorption of any non-absorbing state by any absorbing states can be calculated, and this is given by[14]:

$$Z = F.A \qquad (7)$$

$Z =$

	Bo	Bra	Li	Pe	A	SM
L	0.190	0.120	0.283	0.169	0.133	0.10
Bre	0.297	0.280	0.320	0.042	0.033	0.025
Bla	0.164	0.128	0.322	0.197	0.068	0.118
St	0.063	0.040	0.427	0.389	0.044	0.033
Th	0.164	0.128	0.322	0.197	0.068	0.118
M	0.238	0.224	0.256	0.033	0.026	0.220
K	0.095	0.060	0.141	0.084	0.566	0.050
Pr	0.297	0.030	0.320	0.042	0.283	0.025
CR	0.063	0.040	0.427	0.389	0.044	0.033
Pa	0.063	0.040	0.427	0.389	0.044	0.033
Te	0.063	0.040	0.427	0.389	0.044	0.033
G	0.063	0.040	0.427	0.389	0.044	0.033

Here, the matrix Z shows the probability that a specific organ (not belonging to the absorbing states) is absorbed by an absorbing specific organ. If we take as a reference the estimated sum of all elements of each column (i.e., each absorbing state), the main probabilities are Liver(Li) and Bones (Bo), following by Peritoneum (Pe) and Adrenal Gland (A).

VI. CONCLUSIONS

In order to search new ways to understand the metastasis process and its interactions among organs sites of possible metastasis, Absorbing Markov chains were used as a mathematical tool to achieve this goal. A characterization of the route of metastasis was developed. The Lung as the main connector between the primary site and the tertiary site, with defined probabilities in emphasized.

The graphs and their connections, in order to develop the transition matrices for the occurrence of tumors, are a good approximation to the reality. These matrices exhibit the connections and the existence of absorbing states in organs with lower probabilities (almost null) to generate metastasis ([19], [20]) in secondary sites, absorbing states represent organs that are not the source of metastasis (sponges as Newton calls in [6]).

By the analysis of the expected steps number and the probability of absorption by absorbing states, it is possible to predict the tertiary sites from the secondary sites[22] (or at least estimated them). Furthermore, a quantized approximation of the transition matrices of second step $P^{(2)}$ can be obtained. It is a result useful for treatments and therapies given its predictive character. Also, we know that as we get more data on statistics of metastasis, our analysis will be more accurate. This can be through statistical methods or by predominant tumor cells (Stems, progenitors or differentiated) prevailing in each organ (or migrated from other organs such as CTCs). The latter is our immediate study object.

On the other hand, by the properties of Absorbing Markov chains, it was found that in no more than 2 steps (second metastasis in tertiary site) any absorbing states are reached: Bo, Bra, Li, Pe, A and SM as can be seen in the references [23], [24], [25], [26] and [27]. This also can be seen as a result from the point of view of our model, the main point is the fact that in 3 steps (P^3), we reach an absorbing state with a high probability (see Appendix C), where the sum of probabilities of each row, in the columns of absorbing organs, is near to 1.

$$
\begin{array}{c}
\sum_{abs=1}^{6} p_{abs} \\
\begin{array}{c}
L \\
Bre \\
Bla \\
St \\
Th \\
M \\
K \\
Pr \\
CR \\
Pa \\
Te \\
G
\end{array}
\left[
\begin{array}{c}
0.845 \\
0.928 \\
0.790 \\
0.905 \\
0.790 \\
0.943 \\
0.857 \\
0.928 \\
0.905 \\
0.905 \\
0.905 \\
0.905
\end{array}
\right]
\end{array}
$$

Where abs = Bo, Bra, Li, Pe, A and SM (The absorbing states).

This analysis is quite similar for females, taking into account the specific organs (Ovaries, Vagina and Uterus).

ACKNOWLEDGEMENTS

This work is partially supported by PIO 14420140100016CO from CONICET Argentina. The authors want to thanks to Dr Marcela Reale for their valuable contributions.

REFERENCES

[1] P. Donald et al., *Modeling Boundary Conditions for Balanced Proliferation in Metastatic Latency*, Clin tumor Res, **19**, 5 (2013), 1063–1070.

[2] P. Donald et al., *The Dormancy Dilemma: Quiescence versus Balanced Proliferation*, Cancer Res., **73**, 13 (2013), 3811–3816.

[3] P. Glaus et al., *Identifying differentially expressed transcripts from RNA-seq data with biological variation*, Bioinformatics, **13** (2010), 1721–1728.

[4] J. Hadfield, S. Nakagawa, *General quantitative genetic methods for comparative biology: phylogenies, taxonomies and multitrait models for continuous and categorical characters*, Journal of evolutionary biology, **23** (2010), 494–508.

[5] Tierney, Luke, Introduction to general state-space Markov chain theory, Ed. Springer US, 1996.

[6] P. Newton, *Spreaders and Sponges Define Metastasis in Lung tumor: A Markov chain Monte Carlo Mathematical Model*, Cancer Research, **73** (2013), 2760–2769.

[7] P. Newton, J. Mason, K. Bethel, L.A. Bazhenova, J. Nieva et al., *A Stochastic Markov chain Model to Describe Lung tumor Growth and Metastasis.* PLoS One, **7**, 4 (2012), 1–18.

[8] J. Aguirre-Ghiso, *On the theory of tumor self-seeding: implications for metastasis progression in humans*, Breast Cancer Research, **12** (2010), 304.

[9] http://www.msal.gov.ar/inc/index.php/acerca-del-cancer/estadisticas

[10] http://www.cancer.gov/cancertopics/what-is-cancer/metastatic-fact-sheet

[11] http://www.visone.info

[12] D. Loria et al., *Tendencia de la mortalidad por cáncer en Argentina, Cuba y Uruguay en un período de 15 años.* Rev. Cubana Salud Pública, **36** (2010), 115–125.

[13] C. Navarro et al., Evaluación externa de registros de cáncer de base poblacional: la Guía REDEPICAN para América Latina, *Rev Panam Salud Pública*, **34** (2013), 336–342.

[14] G. Modica, Poggiolini, A First Course in Probability and Markov chains (United Kindom: John Wiley & Sons, Ltd), 2013, 187–195

[15] J. Talmadge, I. Fidler, *AACR Centennial Series: The Biology of Cancer Metastasis: Historical Perspective. Cancer Research,* **70**, 14 (2010), 5649-5

[16] Stanleya et al., *Clinical Evidence: Metastases can Metastasize*, World J Oncol, **3** (2012), 138–141.

[17] L. Norton, J. Massagu, *Is cancer a disease of self-seeding?*, Nature Medicine, **12** (2006), 875–878.

[18] S. Jeffrey et al., *The Cellular Basis of Site-Specific Tumor Metastasis*, N Engl J Med, **322** (1990), 605–612.

[19] S. Duffy, *Estimation of mean sojourn time in breast tumor screening using a Markov chain model of both entry to and exit from the preclinical detectable phase*, Statistics in Medicine, **14**, 14 (1995), 1531–1543.

[20] A. Mazabraud, *Metastases (secondary tumors of bone)*, Pathology of bone tumours, (1998), 381–390.

[21] L. Weiss, K. Haydock, J. Pickren, W. Lane, *Organ Vascularity and Metastatic Frequency*, Am J Pathol, **101**, 1 (1980), 101-114.

[22] E. Nakamura, *Secondary tumors of the pancreas: Clinicopathological study of 103 autopsy cases of Japanese patients*, Issue Pathology International, **51**, 9 (2001), 686–690.

[23] S. Vanharanta, J. Massague, *Origins of Metastatic Traits*, Tumor Cell, **24**, 4 (2013), 410–421.

[24] M. Cummings et al., *Metastatic progression of breast tumor: insights from 50 years of autopsies*, Tumor Cell, **232**, 1 (2013), 23–31.

[25] C. Voskens et al., *Impact of Bone and Liver Metastases on Patients with Renal Cell Carcinoma Treated with Targeted Therapy*, European Urology, **65**, 3 (2014).

[26] C. Gzell, J. Kench, M. Stockler, G. Hruby, *Biopsy-proven brain metastases from prostate tumor: a series of four cases with review of the literature*, International Urology and Nephrology, **45**, 3 (2013), 735–742.

[27] J. Xue, G. Peng, J. Yang, Q. Ding, J. Cheng, *Predictive factors of brain metastasis in patients with breast tumor*, Medical Oncology, (2013), 30–337.

Appendix

A. Matrix P: Probabilities to generate metastasis in a secondary site from a primary site.

$P =$

	L	Bre	Bla	St	Th	M	K	Pr	CR	Pa	Te	G	Bo	Bra	Li	Pe	A	SM
L	0	$\frac{1}{17}$	$\frac{1}{17}$	$\frac{1}{17}$	$\frac{1}{17}$	$\frac{1}{17}$	$\frac{1}{17}$	$\frac{1}{17}$	$\frac{1}{17}$	$\frac{1}{17}$	$\frac{1}{17}$	$\frac{1}{17}$	$\frac{1}{17}$	$\frac{1}{17}$	$\frac{1}{17}$	$\frac{1}{17}$	$\frac{1}{17}$	$\frac{1}{17}$
Bre	$\frac{1}{4}$	0	0	0	0	0	0	0	0	0	0	0	$\frac{1}{4}$	$\frac{1}{4}$	$\frac{1}{4}$	0	0	0
Bla	$\frac{1}{3}$	0	0	$\frac{1}{3}$	0	$\frac{1}{3}$	0	0	0	0	0	0	0	0	0	0	0	0
St	$\frac{1}{3}$	0	0	0	0	0	0	0	0	0	0	0	0	0	$\frac{1}{3}$	$\frac{1}{3}$	0	0
Th	$\frac{1}{3}$	0	0	$\frac{1}{3}$	0	$\frac{1}{3}$	0	0	0	0	0	0	0	0	0	0	0	0
M	$\frac{1}{5}$	0	0	0	0	0	0	0	0	0	0	0	$\frac{1}{5}$	$\frac{1}{5}$	$\frac{1}{5}$	0	0	$\frac{1}{5}$
K	$\frac{1}{2}$	0	0	0	0	0	0	0	0	0	0	0	0	0	0	0	$\frac{1}{2}$	0
Pr	$\frac{1}{4}$	0	0	0	0	0	0	0	0	0	0	0	$\frac{1}{4}$	0	$\frac{1}{4}$	0	$\frac{1}{4}$	0
CR	$\frac{1}{3}$	0	0	0	0	0	0	0	0	0	0	0	0	0	$\frac{1}{3}$	$\frac{1}{3}$	0	0
Pa	$\frac{1}{3}$	0	0	0	0	0	0	0	0	0	0	0	0	0	$\frac{1}{3}$	$\frac{1}{3}$	0	0
Te	$\frac{1}{3}$	0	0	0	0	0	0	0	0	0	0	0	$\frac{1}{3}$	0	$\frac{1}{3}$	0	0	0
G	$\frac{1}{3}$	0	0	0	0	0	0	0	0	0	0	0	$\frac{1}{3}$	0	$\frac{1}{3}$	0	0	0
Bo	0	0	0	0	0	0	0	0	0	0	0	0	1	0	0	0	0	0
Bra	0	0	0	0	0	0	0	0	0	0	0	0	0	1	0	0	0	0
Li	0	0	0	0	0	0	0	0	0	0	0	0	0	0	1	0	0	0
Pe	0	0	0	0	0	0	0	0	0	0	0	0	0	0	0	1	0	0
A	0	0	0	0	0	0	0	0	0	0	0	0	0	0	0	0	1	0
SM	0	0	0	0	0	0	0	0	0	0	0	0	0	0	0	0	0	1

B. Matrix $P^{(2)}$: Probabilities to generate metastasis in a tertiaryy site from a primary site.

$P^{(2)} =$

	L	Bre	Bla	St	Th	M	K	Pr	CR	Pa	Te	G	Bo	Bra	Li	Pe	A	SM
L	$\frac{53}{255}$	0	0	$\frac{2}{51}$	0	$\frac{2}{51}$	0	0	0	0	0	0	$\frac{71}{510}$	$\frac{29}{340}$	$\frac{101}{510}$	$\frac{2}{17}$	$\frac{7}{68}$	$\frac{6}{85}$
Bre	0	$\frac{1}{68}$	$\frac{1}{68}$	$\frac{1}{68}$	$\frac{1}{68}$	$\frac{1}{68}$	$\frac{1}{68}$	$\frac{1}{68}$	$\frac{1}{68}$	$\frac{1}{68}$	$\frac{1}{68}$	$\frac{1}{68}$	$\frac{9}{34}$	$\frac{9}{34}$	$\frac{9}{34}$	$\frac{1}{68}$	$\frac{1}{68}$	$\frac{1}{68}$
Bla	$\frac{8}{45}$	$\frac{1}{51}$	$\frac{1}{51}$	$\frac{1}{51}$	$\frac{1}{51}$	$\frac{1}{51}$	$\frac{1}{51}$	$\frac{1}{51}$	$\frac{1}{51}$	$\frac{1}{51}$	$\frac{1}{51}$	$\frac{1}{51}$	$\frac{22}{255}$	$\frac{22}{255}$	$\frac{151}{765}$	$\frac{20}{153}$	$\frac{1}{51}$	$\frac{22}{255}$
St	0	$\frac{1}{51}$	$\frac{1}{51}$	$\frac{1}{51}$	$\frac{1}{51}$	$\frac{1}{51}$	$\frac{1}{51}$	$\frac{1}{51}$	$\frac{1}{51}$	$\frac{1}{51}$	$\frac{1}{51}$	$\frac{1}{51}$	$\frac{1}{51}$	$\frac{1}{51}$	$\frac{6}{17}$	$\frac{6}{17}$	$\frac{1}{51}$	$\frac{1}{51}$
Th	$\frac{8}{45}$	$\frac{1}{51}$	$\frac{1}{51}$	$\frac{1}{51}$	$\frac{1}{51}$	$\frac{1}{51}$	$\frac{1}{51}$	$\frac{1}{51}$	$\frac{1}{51}$	$\frac{1}{51}$	$\frac{1}{51}$	$\frac{1}{51}$	$\frac{22}{255}$	$\frac{22}{255}$	$\frac{151}{765}$	$\frac{20}{153}$	$\frac{1}{51}$	$\frac{22}{255}$
M	0	$\frac{1}{85}$	$\frac{1}{85}$	$\frac{1}{85}$	$\frac{1}{85}$	$\frac{1}{85}$	$\frac{1}{85}$	$\frac{1}{85}$	$\frac{1}{85}$	$\frac{1}{85}$	$\frac{1}{85}$	$\frac{1}{85}$	$\frac{18}{85}$	$\frac{18}{85}$	$\frac{18}{85}$	$\frac{1}{85}$	$\frac{1}{85}$	$\frac{18}{85}$
K	0	$\frac{1}{34}$	$\frac{1}{34}$	$\frac{1}{34}$	$\frac{1}{34}$	$\frac{1}{34}$	$\frac{1}{34}$	$\frac{1}{34}$	$\frac{1}{34}$	$\frac{1}{34}$	$\frac{1}{34}$	$\frac{1}{34}$	$\frac{1}{34}$	$\frac{1}{34}$	$\frac{1}{34}$	$\frac{1}{34}$	$\frac{9}{17}$	$\frac{1}{34}$
Pr	0	$\frac{1}{68}$	$\frac{1}{68}$	$\frac{1}{68}$	$\frac{1}{68}$	$\frac{1}{68}$	$\frac{1}{68}$	$\frac{1}{68}$	$\frac{1}{68}$	$\frac{1}{68}$	$\frac{1}{68}$	$\frac{1}{68}$	$\frac{9}{34}$	$\frac{1}{68}$	$\frac{9}{34}$	$\frac{1}{68}$	$\frac{9}{34}$	$\frac{1}{68}$
CR	0	$\frac{1}{51}$	$\frac{1}{51}$	$\frac{1}{51}$	$\frac{1}{51}$	$\frac{1}{51}$	$\frac{1}{51}$	$\frac{1}{51}$	$\frac{1}{51}$	$\frac{1}{51}$	$\frac{1}{51}$	$\frac{1}{51}$	$\frac{1}{51}$	$\frac{1}{51}$	$\frac{6}{17}$	$\frac{6}{17}$	$\frac{1}{51}$	$\frac{1}{51}$
Pa	0	$\frac{1}{51}$	$\frac{1}{51}$	$\frac{1}{51}$	$\frac{1}{51}$	$\frac{1}{51}$	$\frac{1}{51}$	$\frac{1}{51}$	$\frac{1}{51}$	$\frac{1}{51}$	$\frac{1}{51}$	$\frac{1}{51}$	$\frac{1}{51}$	$\frac{1}{51}$	$\frac{6}{17}$	$\frac{6}{17}$	$\frac{1}{51}$	$\frac{1}{51}$
Te	0	$\frac{1}{51}$	$\frac{1}{51}$	$\frac{1}{51}$	$\frac{1}{51}$	$\frac{1}{51}$	$\frac{1}{51}$	$\frac{1}{51}$	$\frac{1}{51}$	$\frac{1}{51}$	$\frac{1}{51}$	$\frac{1}{51}$	$\frac{6}{17}$	$\frac{1}{51}$	$\frac{6}{17}$	$\frac{1}{51}$	$\frac{1}{51}$	$\frac{1}{51}$
G	0	$\frac{1}{51}$	$\frac{1}{51}$	$\frac{1}{51}$	$\frac{1}{51}$	$\frac{1}{51}$	$\frac{1}{51}$	$\frac{1}{51}$	$\frac{1}{51}$	$\frac{1}{51}$	$\frac{1}{51}$	$\frac{1}{51}$	$\frac{6}{17}$	$\frac{1}{51}$	$\frac{6}{17}$	$\frac{1}{51}$	$\frac{1}{51}$	$\frac{1}{51}$
Bo	0	0	0	0	0	0	0	0	0	0	0	0	1	0	0	0	0	0
Bra	0	0	0	0	0	0	0	0	0	0	0	0	0	1	0	0	0	0
Li	0	0	0	0	0	0	0	0	0	0	0	0	0	0	1	0	0	0
Pe	0	0	0	0	0	0	0	0	0	0	0	0	0	0	0	1	0	0
A	0	0	0	0	0	0	0	0	0	0	0	0	0	0	0	0	1	0
SM	0	0	0	0	0	0	0	0	0	0	0	0	0	0	0	0	0	1

C. Matrix $P^{(3)}$

$P^{(3)} =$

	L	Bre	Bla	St	Th	M	K	Pr	CR	Pa	Te	G	Bo	Bra	Li	Pe	A	SM
L	$\frac{16}{765}$	$\frac{5}{409}$	$\frac{5}{409}$	$\frac{5}{409}$	$\frac{5}{409}$	$\frac{5}{409}$	$\frac{5}{409}$	$\frac{5}{409}$	$\frac{5}{409}$	$\frac{5}{409}$	$\frac{5}{409}$	$\frac{5}{409}$	$\frac{18}{113}$	$\frac{55}{522}$	$\frac{43}{186}$	$\frac{1}{7}$	$\frac{41}{356}$	$\frac{33}{364}$
Bre	$\frac{49}{943}$	0	0	$\frac{1}{102}$	0	$\frac{1}{102}$	0	0	0	0	0	0	$\frac{268}{941}$	$\frac{35}{129}$	$\frac{183}{611}$	$\frac{1}{34}$	$\frac{7}{272}$	$\frac{3}{170}$
Bla	$\frac{53}{765}$	$\frac{8}{765}$	$\frac{8}{765}$	$\frac{2}{85}$	$\frac{8}{765}$	$\frac{2}{85}$	$\frac{8}{765}$	$\frac{8}{765}$	$\frac{8}{765}$	$\frac{8}{765}$	$\frac{8}{765}$	$\frac{8}{765}$	$\frac{21}{170}$	$\frac{19}{180}$	$\frac{15}{59}$	$\frac{41}{255}$	$\frac{3}{67}$	$\frac{77}{765}$
St	$\frac{53}{765}$	0	0	$\frac{2}{153}$	0	$\frac{2}{153}$	0	0	0	0	0	0	$\frac{20}{431}$	$\frac{23}{809}$	$\frac{244}{611}$	$\frac{19}{51}$	$\frac{7}{204}$	$\frac{2}{85}$
Th	$\frac{53}{765}$	$\frac{8}{765}$	$\frac{8}{765}$	$\frac{2}{85}$	$\frac{8}{765}$	$\frac{2}{85}$	$\frac{8}{765}$	$\frac{8}{765}$	$\frac{8}{765}$	$\frac{8}{765}$	$\frac{8}{765}$	$\frac{8}{765}$	$\frac{21}{170}$	$\frac{19}{180}$	$\frac{15}{59}$	$\frac{41}{255}$	$\frac{3}{67}$	$\frac{77}{765}$
M	$\frac{18}{433}$	0	0	$\frac{2}{255}$	0	$\frac{2}{255}$	0	0	0	0	0	0	$\frac{18}{79}$	$\frac{28}{129}$	$\frac{98}{409}$	$\frac{2}{85}$	$\frac{7}{340}$	$\frac{91}{425}$
K	$\frac{53}{510}$	0	0	$\frac{1}{51}$	0	$\frac{1}{51}$	0	0	0	0	0	0	$\frac{41}{589}$	$\frac{29}{680}$	$\frac{91}{919}$	$\frac{1}{17}$	$\frac{75}{136}$	$\frac{3}{85}$
Pr	$\frac{49}{943}$	0	0	$\frac{1}{102}$	0	$\frac{1}{102}$	0	0	0	0	0	0	$\frac{268}{941}$	$\frac{10}{469}$	$\frac{183}{611}$	$\frac{1}{34}$	$\frac{75}{272}$	$\frac{3}{170}$
CR	$\frac{53}{765}$	0	0	$\frac{2}{153}$	0	$\frac{2}{153}$	0	0	0	0	0	0	$\frac{20}{431}$	$\frac{23}{809}$	$\frac{244}{611}$	$\frac{19}{51}$	$\frac{7}{204}$	$\frac{2}{85}$
Pa	$\frac{53}{765}$	0	0	$\frac{2}{153}$	0	$\frac{2}{153}$	0	0	0	0	0	0	$\frac{20}{431}$	$\frac{23}{809}$	$\frac{244}{611}$	$\frac{19}{51}$	$\frac{7}{204}$	$\frac{2}{85}$
Te	$\frac{53}{765}$	0	0	$\frac{2}{153}$	0	$\frac{2}{153}$	0	0	0	0	0	0	$\frac{20}{431}$	$\frac{23}{809}$	$\frac{244}{611}$	$\frac{19}{51}$	$\frac{7}{204}$	$\frac{2}{85}$
G	$\frac{53}{765}$	0	0	$\frac{2}{153}$	0	$\frac{2}{153}$	0	0	0	0	0	0	$\frac{20}{431}$	$\frac{23}{809}$	$\frac{244}{611}$	$\frac{19}{51}$	$\frac{7}{204}$	$\frac{2}{85}$
Bo	0	0	0	0	0	0	0	0	0	0	0	0	1	0	0	0	0	0
Bra	0	0	0	0	0	0	0	0	0	0	0	0	0	1	0	0	0	0
Li	0	0	0	0	0	0	0	0	0	0	0	0	0	0	1	0	0	0
Pe	0	0	0	0	0	0	0	0	0	0	0	0	0	0	0	1	0	0
A	0	0	0	0	0	0	0	0	0	0	0	0	0	0	0	0	1	0
SM	0	0	0	0	0	0	0	0	0	0	0	0	0	0	0	0	0	1

7

Numerical solutions of one-dimensional parabolic convection-diffusion problems arising in biology by the Laguerre collocation method

Burcu Gürbüz, Mehmet Sezer
Department of Mathematics,
Manisa Celal Bayar University,
Manisa, Turkey
burcugrbz@gmail.com, mehmet.sezer@cbu.edu.tr

Abstract—**In this work, we present a numerical scheme for the approximate solutions of the one-dimensional parabolic convection-diffusion model problems which arise in biological models. The presented method is based on the Laguerre collocation method used for ordinary differential equations. The approximate solution of the problem in the truncated Laguerre series form is obtained by this method. By substituting truncated Laguerre series solution into the problem and by using the matrix operations and the collocation points, the suggested scheme reduces the problem to a linear algebraic equation system. By solving this equation system, the unknown Laguerre coefficients can be computed. The accuracy and efficiency of the method is studied by comparing with other numerical methods when used to solve some numerical experiments.**

Keywords-**Convection-diffusion equation models, Parabolic problem, Laguerre collocation method.**

I. INTRODUCTION

Diffusion models form a reasonable basis for studying insect and animal dispersal and invasion, which arise from the question of persistence of endangered species, biodiversity, disease dynamics, multi-species competition so on. Convection-diffusion problem is also a form of heat and mass transfer in biological models [1-3].

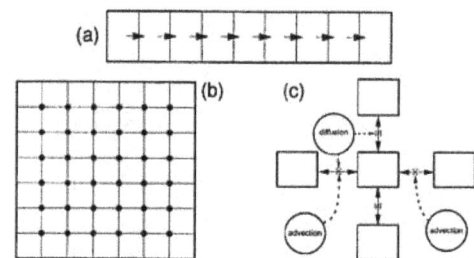

Fig. 1. (a) Flow between imaginary compartments in a continuous one-dimensional system. (b) Discrete grid system used in two-dimensional transport models. (c) A close-up of five grid points showing the similarity to compartment models.

Compartment models are general framework

$$\frac{\partial u(x,t)}{\partial t} = \begin{pmatrix} \text{Convection} \\ \text{In} \end{pmatrix} - \begin{pmatrix} \text{Convection} \\ \text{Out} \end{pmatrix} + \begin{pmatrix} \text{Diffusion} \\ \text{In} \end{pmatrix} - \begin{pmatrix} \text{Diffusion} \\ \text{Out} \end{pmatrix}$$

Fig. 2.   A conceptual rate equation with respect to the convection-diffusion model

that has many applications in biology, ecosystems and enzyme kinetics which can be mostly shown by forrester diagrams. The system is decomposed into flows of material as possibly large number of discrete compartments which are very useful. Conversely, it is also useful for the quantities nominally not flow, for instance, blood or water pressure in animal and plant physiological systems. Furthermore, complex interconnection networks can be addressed by these type of models with respect to link many of them together in many different complicated ways (Fig. 1).

On the other hand, in transport models, we have a physical quantity, such as energy i.e. heat or a quantity of matter, that flows from spatial point to point. There are many forces that could influence the flow of the matter, but the following simplified view uses two that will illustrate the qualitative model formulation. Convection moves the substance with a physical flow of water from point to point (i.e. river flow). Diffusion moves a substance in any direction according to the concentration of the substance around each point (Fig. 2) [4-5].

In this study, we consider the one-dimensional parabolic convection-diffusion problem

$$\frac{\partial u}{\partial t} = \frac{\partial^2 u}{\partial x^2} + A(x)\frac{\partial u}{\partial x} + B(x)u + f(x,t),$$
$$0 \le x \le l, 0 \le t \le T,$$
$$(1)$$

with the initial conditions

$$u(x,0) = g(x), \quad 0 \le x \le l < \infty, \qquad (2)$$

and the boundary conditions

$$u(0,t) = h(t), \quad u(l,t) = K(t), \quad 0 \le t \le T < \infty$$
$$(3)$$

where $f(x,t), A(x), B(x), g(x)$ and $h(t)$ are functions defined in $[0,l] \times [0,T]$; l and T are appropriate constants. In this study, we develop the Laguerre collocation method given in [9,10] and use to obtain the approximate solution of Eq. (1) in the truncated Laguerre series form

$$u(x,t) = \sum_{r=0}^{N}\sum_{s=0}^{N} a_{r,s} L_{r,s}(x,t); \qquad (4)$$
$$L_{r,s}(x,t) = L_r(x)L_s(t)$$

where $a_{r,s}, \; r,s = 0, ..., N$, are the unknown Laguerre coefficients and $L_n(x), \; n = 0,1,2,...,N$ are the Laguerre polynomials defined by [6-8]

$$L_n(x) = \sum_{k=0}^{n} \frac{(-1)^k}{k!}\binom{n}{k}x^k, n \in \mathbb{N}, \; 0 \le x < \infty.$$
$$(5)$$

II. Numerical Method

We first consider the series (4) for $N = 2$, as follows:

$$u(x,t) = \sum_{r=0}^{2}\sum_{s=0}^{2} a_{r,s}L_r(x)L_s(t)$$
$$= a_{00}L_0(x)L_0(t) + a_{10}L_1(x)L_0(t)$$
$$+ a_{20}L_2(x)L_0(t) + a_{01}L_0(x)L_1(t) \quad (6)$$
$$+ a_{11}L_1(x)L_1(t) + a_{21}L_2(x)L_1(t)$$
$$+ a_{02}L_0(x)L_2(t) + a_{12}L_1(x)L_2(t)$$
$$+ a_{22}L_2(x)L_2(t)$$

Then we can generalize the approximate solution (6) for any truncated limit N and can write the obtained series in the matrix form

$$[u(x,t)] = \mathbf{L}(x)\overline{\mathbf{L}}(t)\mathbf{A} \qquad (7)$$

where

$$\mathbf{L}(x) = \begin{bmatrix} L_0(x) & L_1(x) & \cdots & L_N(x) \end{bmatrix},$$

$$\overline{\mathbf{L}}(t) = \begin{bmatrix} \mathbf{L}(t) & 0 & \cdots & 0 \\ 0 & \mathbf{L}(t) & \cdots & 0 \\ \vdots & \vdots & \ddots & \vdots \\ 0 & 0 & \cdots & \mathbf{L}(t) \end{bmatrix}$$

and

$$\mathbf{A} = [a_{0,0}\ a_{0,1} \cdots a_{0,N} \cdots a_{N,0}\ a_{N,1} \cdots a_{N,N}]^T$$

Also, we can put the matrix $\mathbf{L}(x)$ in the matrix form

$$\mathbf{L}(x) = \mathbf{X}(x)\mathbf{H} \qquad (8)$$

where $\mathbf{X}(x)$ and $\mathbf{H}$ are defined as

$$\mathbf{X}(x) = \begin{bmatrix} 1 & x^1 & \cdots & x^N \end{bmatrix}$$

and

$$\mathbf{H} = \begin{bmatrix} \frac{(-1)^0}{0!}\binom{0}{0} & 0 & \cdots & 0 \\ \frac{(-1)^0}{0!}\binom{1}{0} & \frac{(-1)^1}{1!}\binom{1}{1} & \cdots & 0 \\ \vdots & \vdots & \ddots & \vdots \\ \frac{(-1)^0}{0!}\binom{N}{0} & \frac{(-1)^1}{1!}\binom{N}{1} & \cdots & \frac{(-1)^N}{N!}\binom{N}{N} \end{bmatrix}$$

Moreover, it is clearly seen that the relations between the matrix $\mathbf{X}(x)$ and its derivatives $\mathbf{X}'(x)$ and $\mathbf{X}''(x)$ are

$$\mathbf{X}'(x) = \mathbf{X}(x)\mathbf{B} \quad \text{and} \quad \mathbf{X}''(x) = \mathbf{X}(x)\mathbf{B}^2 \qquad (9)$$

where

$$\mathbf{B} = \begin{bmatrix} 0 & 1 & 0 & \cdots & 0 \\ 0 & 0 & 2 & \cdots & 0 \\ \vdots & \vdots & \vdots & \ddots & \vdots \\ 0 & 0 & 0 & \cdots & N \\ 0 & 0 & 0 & \cdots & 0 \end{bmatrix}.$$

Then, by using the expressions (8) and (9) we easily find the matrix relations

$$\mathbf{L}'(x) = \mathbf{X}(x)\mathbf{B}\mathbf{H} \quad \text{and} \quad \mathbf{L}''(x) = \mathbf{X}(x)\mathbf{B}^2\mathbf{H} \ (10)$$

$$\overline{\mathbf{L}}(t) = \overline{\mathbf{X}}(t)\overline{\mathbf{H}} \quad \text{and} \quad \overline{\mathbf{L}}'(t) = \overline{\mathbf{X}}(t)\overline{\mathbf{B}\mathbf{H}} \qquad (11)$$

Now, by means of the relations (7)-(11) we obtain the following matrix forms:

$$[u(x,t)] = \mathbf{L}(x)\overline{\mathbf{L}}(t)\mathbf{A} = \mathbf{X}(x)\mathbf{H}\overline{\mathbf{X}}(t)\overline{\mathbf{H}}\mathbf{A} \quad (12)$$

$$[u_x(x,t)] = \mathbf{L}'(x)\overline{\mathbf{L}}(t)\mathbf{A}$$
$$= \mathbf{X}(x)\mathbf{B}\mathbf{H}\overline{\mathbf{X}}(t)\overline{\mathbf{H}}\mathbf{A} \quad (13)$$

$$[u_{xx}(x,t)] = \mathbf{L}''(x)\overline{\mathbf{L}}(t)\mathbf{A}$$
$$= \mathbf{X}(x)\mathbf{B}^2\mathbf{H}\overline{\mathbf{X}}(t)\overline{\mathbf{H}}\mathbf{A} \quad (14)$$

$$[u_t(x,t)] = \mathbf{L}(x)\overline{\mathbf{L}}(t)\mathbf{A}$$
$$= \mathbf{X}(x)\mathbf{H}\overline{\mathbf{X}}(t)\overline{\mathbf{B}\mathbf{H}}\mathbf{A} \quad (15)$$

By putting the expressions (8), (12), (13), (14) and (15) into Eq. (1), we obtain the matrix equation

$$\{\mathbf{X}(x)\mathbf{H}\overline{\mathbf{X}}(t)\overline{\mathbf{B}} - \mathbf{X}(x)\mathbf{B}^2\mathbf{H}\overline{\mathbf{X}}(t)$$
$$-A(x)\mathbf{X}(x)\mathbf{B}\mathbf{H}\overline{\mathbf{X}}(t) \qquad (16)$$
$$-B(x)\mathbf{X}(x)\mathbf{H}\overline{\mathbf{X}}(\mathbf{t})\}\overline{\mathbf{H}}\mathbf{A} = f(x,t)$$

or briefly,

$$\mathbf{W}(x,t)\mathbf{A} = f(x,t)$$

Besides, by substituting the collocation points defined by

$$x_i = \frac{l}{N}i, \ t_j = \frac{T}{N}j, \ i,j = 0,1,2,...,N,$$

into the Eq.(16), we have the system of the matrix equations $\mathbf{W}(x_i,t_j)\mathbf{A} = f(x_i,t_j)$ or briefly the fundamental matrix equation

$$\mathbf{W}\mathbf{A} = \mathbf{F} \Longrightarrow [\mathbf{W};\mathbf{F}]$$

By using the same procedure for the initial and boundary conditions we obtain the matrix relations for
$i,j = 0,1,...,N$:

$$u(x_i,0) = \mathbf{X}(x_i)\mathbf{H}\overline{\mathbf{X}}(0)\overline{\mathbf{H}}\mathbf{A} = g(x_i) = \lambda_i$$
$$u(0,t_j) = \mathbf{X}(0)\mathbf{H}\overline{\mathbf{X}}(t_j)\overline{\mathbf{H}}\mathbf{A} = h(t_j) = \mu_j$$
$$u(y,t_j) = \mathbf{X}(y)\mathbf{H}\overline{\mathbf{X}}(t_j)\overline{\mathbf{H}}\mathbf{A} = K(t_j) = \gamma_j$$

or briefly,

$$\mathbf{U}\mathbf{A} = [\lambda]; \ [\mathbf{U};\lambda], \mathbf{V}\mathbf{A} = [\mu]; \ [\mathbf{V};\mu], \mathbf{Z}\mathbf{A} = [\gamma]; \ [\mathbf{Z};\gamma].$$

To obtain the approximate solution of Eq. (1) under conditions (2) and (3), we form the augmented matrix

TABLE I

COMPARISON OF THE ABSOLUTE ERRORS WITH TCM AND LCM FOR $N = 25, 50$ IN EXAMPLE 2.

x	TCM E_{25}	LCM E_{25}	TCM E_{50}	LCM E_{50}
0.0	0.69500E-17	7.000000E-13	0.10000E-18	0.0000000000
0.1	0.15886E-03	3.681538E-06	0.15886E-03	1.040134E-07
0.2	0.63428E-03	2.207214E-05	0.63428E-03	4.157153E-08
0.3	0.14208E-02	4.966673E-05	0.14208E-02	9.332459E-08
0.4	0.25078E-02	5.150761E-05	0.25078E-02	1.652982E-05
0.5	0.38799E-02	2.543790E-11	0.38799E-02	2.569605E-15
0.6	0.55168E-02	1.324508E-04	0.55168E-02	3.676180E-06
0.7	0.73934E-02	3.734395E-04	0.73934E-02	4.964259E-06
0.8	0.94802E-02	7.505606E-04	0.94802E-02	6.423988E-06
0.9	0.11744E-01	1.291408E-03	0.11744E-01	8.044241E-07
1.0	0.14146E-01	2.023575E-03	0.14146E-01	9.812752E-07

$$[\tilde{\mathbf{W}}; \tilde{\mathbf{F}}] = \begin{bmatrix} \overline{\mathbf{W}}; \mathbf{F} \\ \mathbf{U}; \lambda \\ \mathbf{V}; \mu \\ \mathbf{Z}; \gamma \end{bmatrix}$$

Hence, the unknown Laguerre coefficients are computed by

$$\mathbf{A} = (\tilde{\tilde{\mathbf{W}}})^{-1} \tilde{\tilde{\mathbf{F}}}$$

where $[\tilde{\tilde{\mathbf{W}}}; \tilde{\tilde{\mathbf{F}}}]$ is obtained by using the Gauss elimination method and then removing zero rows of augmented matrix $[\tilde{\mathbf{W}}; \tilde{\mathbf{F}}]$ [9-11]. By substituting the determined coefficients into Eq. (4), we have the Laguerre series solution

$$u_N(x,t) = \sum_{r=0}^{N} \sum_{s=0}^{N} a_{r,s} L_{r,s}(x,t),$$
$$L_{r,s}(x,t) = L_r(x) L_s(t).$$

III. NUMERICAL RESULTS

Test case[11]

$$\frac{\partial u}{\partial t} = \frac{\partial^2 u}{\partial x^2} + (2x+1)\frac{\partial u}{\partial x} + x^2 u + \frac{e^{x+t}}{\epsilon},$$
$$0 \le x \le 1, 0 \le t \le 1, \qquad (17)$$

with conditions

$$u(x,0) = \frac{e^x}{\epsilon}, \quad 0 \le x \le 1$$
$$u(0,t) = \frac{e^t}{\epsilon}, \quad u(y,t) = \frac{e^{1+t}}{\epsilon} \quad 0 \le t \le 1,$$

with $\epsilon = 2.10^{-4}$ and the exact solution of the problem is $u(x,t) = \frac{e^{x+t}}{\epsilon}$. From Table 1, it is seen that the errors from Laguerre Collocation Method (LCM) are in general less than Taylor Collocation Method (TCM).

Table I. shows the comparison between absolute errors of LCM solutions and TCM solutions for different N values.

IV. CONCLUSION

We have presented and illustrated the Laguerre collocation method is based on computing the coefficients in the Laguerre expansion of solution of a one dimensional parabolic convection-diffusion model problems. A considerable advantage of the method is that the Laguerre polynomial coefficients of the solution are found very easily by using computer programs; Maple and Matlab.

Illustrative example is included to show the validity and applicability of the technique. Shorter computation time and lower operation count results in reduction of cumulative truncation errors and improvement of overall accuracy.

As a result, the method can also be extended to the system of reaction-diffusion-advection model problems with their residual error analysis, but some modifications are required.

ACKNOWLEDGMENT

This work was financially supported by Society for Mathematical Biology for during the Bio-Math 2016, The annual International Conference on Mathematical Methods and Models in Biosciences and it is performed within the "Numerical Solutions of Partial Functional Integro Differential Equations with respect to Laguerre Polynomials and Its Applications" project, Manisa Celal Bayar University Department of Scientific Research Projects, with grant ref. 2014-151.

REFERENCES

[1] J. D. Murray, *Mathematical Biology 1: An Introduction*, Berlin: Springer, 2002.

[2] B. Zduniak, *Numerical analysis of the coupled modified Van Der Pol equations in a model of the heart action*, Biomath Commun. **3**(2014) 1–7. doi:http://dx.doi.org/10.11145/j.biomath.2013.12.281

[3] A. Prieto-Langarica, H. V. Kojouharov, L. Tang, *Constructing one-dimensional continuous models from two-dimensional discrete models of medical implants*, Biomath Commun. **1**(2012) 1–6. doi:http://dx.doi.org/10.11145/j.biomath.2012.09.041

[4] J. W. Haefner, *Modelling Biological Systems*, USA: Springer, 2005.

[5] B. Hannon, M. Ruth, *Modelling Dynamic Biological Systems*, London: Springer, 2014.

[6] G. A. Andrews, R. Askey, R. Roy, *Special Functions*, Cambridge: The University Press, 2000.

[7] J. Dieudonne, *Orthogonal Polynomials and Applications*, Berlin: Springer, 1985.

[8] J. L. Gracia, E. O'Riordan, *Numerical approximation of solution derivatives of singularly perturbed parabolic problems of convection-diffusion type*, Math. Comput. **85**(2016) 581–599. doi:http://dx.doi.org/10.1090/mcom/2998

[9] B. Gürbüz, M. Sezer, *Laguerre polynomial approach for solving Lane-Emden type functional differential equations*, Appl. Math. Comput. **242**(2014) 255–264. doi:http://dx.doi.org/10.1016/j.amc.2014.05.058

[10] B. Gürbüz, M. Sezer, *Laguerre polynomial solutions of a class of initial and boundary value problems arising in science and engineering fields*, Acta Phys. Pol. A **130**(2016) 194–197. doi: 10.12693/APhysPolA.129.194

[11] Ş. Yüzbaşı, N. Şahin, *Numerical solutions of singularly perturbed one-dimensional parabolic convection-diffusion problems by the Bessel collocation method*, Appl. Math. Comput. **174**(2006) 910–920. doi:http://dx.doi.org/10.1016/j.amc.2013.06.027

8

Bifurcations in valveless pumping techniques from a coupled fluid-structure-electrophysiology model in heart development

Nicholas A. Battista[1], Laura A. Miller[2]
[1]Dept. of Mathematics and Statistics, The College of New Jersey
Ewing Township, NJ, USA
battistn@tcnj.edu

[2]Dept. of Biology, Dept. of Mathematics, University of North Carolina at Chapel Hill
Chapel Hill, NC, USA
lam9@unc.edu

Abstract—We explore an embryonic heart model that couples electrophysiology and muscle-force generation to induce flow using a $2D$ fluid-structure interaction framework based on the immersed boundary method. The propagation of action potentials are coupled to muscular contraction and hence the overall pumping dynamics. In comparison to previous models, the electro-dynamical model does not use prescribed motion to initiate the pumping motion, but rather the pumping dynamics are fully coupled to an underlying electrophysiology model, governed by the FitzHugh-Nagumo equations. Perturbing the diffusion parameter in the FitzHugh-Nagumo model leads to a bifurcation in dynamics of action potential propagation. This bifurcation is able to capture a spectrum of different pumping regimes, with dynamic suction pumping and peristaltic-like pumping at the extremes. We find that more bulk flow is produced within the realm of peristaltic-like pumping.

Keywords-valveless pumping; heart development; immersed boundary method; fluid-structure interaction; mathematical biology; biomechanics

I. INTRODUCTION

Various kinds of hearts are found throughout the animal kingdom [1], [2], [3]. In particular many invertebrates have valveless, tubular hearts from their infancy throughout adulthood [4]. These tubular hearts are similar to vertebrate heart morphologies during their first stage of vertebrate heart morphogenesis, e.g., the linear heart tube stage. We begin our discussion of heart tube morphologies by considering the evolution of hearts in the animal kingdom.

Figure 1 shows the evolution of hearts from tunicates to humans. Tunicates have an open circulatory system from infancy through adulthood, in which blood is pushed through out the organism

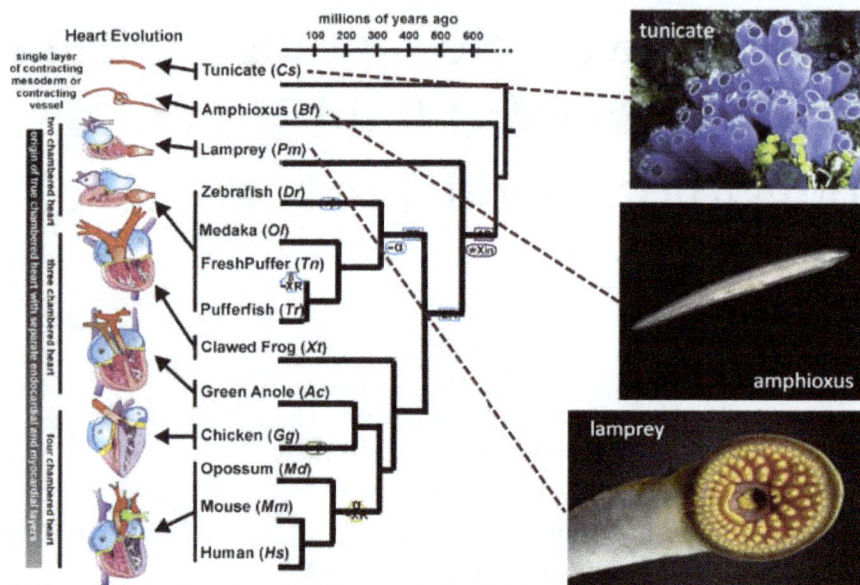

Fig. 1: Figure adapted from Grosskurth *et al.* [5] illustrating the evolution of hearts from the valveless heart tubes in the open circulatory systems of tunicates to the adult multi-chambered-valvular of vertebrates. Tunicate, amphioxus, and lamprey images adapted from [6], [7], [8], respectively.

by a valveless-tubular heart [1], [9], which is composed of only a single layer of myocardial cells. Next on the evolutionary chain is the amphioxus. The amphioxus heart is a rostrocaudally extended tube from its infancy through adulthood [10]. Similar to the tunicate heart, an amphioxus heart consists only of a monolayer of myocardial cells. The amphioxus heart has no chambers, valves, endocardium, epicardium, or other differentiated features of vertebrate hearts. Still, the amphioxus is regarded as the closest living invertebrate relative to vertebrates [11] and appears fish-like.

Furthermore, Figure 1 illustrates an evolutionary morphological change to multi-chambered hearts in a vertebrate - the lamprey. Lampreys are jawless fishes that are a very ancient lineage of vertebrates [12]. The lamprey is considered to have four heart chambers, which are the sinus venosus, atrium, ventricle, and conus arteriosus [13]. This is similar as to the zebrafish heart, which contains four chambers - the sinus venosus, atrium, ventricle, and bulbus arteriosus. Lamprey hearts also are valvular pumping systems, containing valve leaflets between chambers [14]. Moreover,

lampreys are the first organism to develop an endocardial layer in addition to a myocardium, as well as, the first organism to develop cardiac valves [15]. An evolutionary depiction of heart morphology is illustrated in Figure 2, which was adapted from [16]. Note that the additional layer of endocardial cells in lamprey hearts is present during its associated linear heart tube stage and make the heart noticeably more stiff than the tunicate tubular hearts, not including the pericardium.

However, as discussed, the vertebrate embryonic heart begins as a valveless tube, similar to those in various invertebrates, such as urochordates and cephalochordates [17], [18], making invertebrates like sea squirts a possible model for heart development [19]. Historically, the pumping mechanism in these hearts has been described as peristalsis [17], [20], while more recently, dynamic suction pumping (DSP) has been proposed as a novel cardiac pumping mechanism for the vertebrate embryonic heart by Kenner *et. al.* in 2000 [21], and later declared the main pumping mechanism in vertebrate embryonic hearts by Forouhar *et. al.* in 2006 [22]. Debate over which is the actual pump-

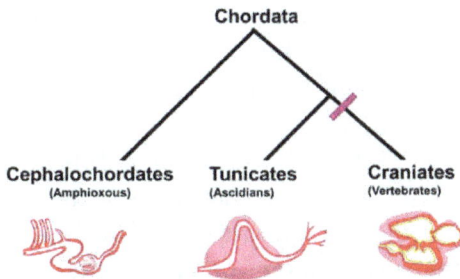

Fig. 2: Figure illustrating the phylogenetic relationship and general heart structure of the Chordate subphyla. Cephalochordates, like amphioxus, have a series of four peristaltic vessels that serve as a pump, while tunicates have a single-chamber pump, which is composed of a single layer of myocardium (red) surrounded by stiff pericardial layer (pink). The earliest vertebrates, e.g., lampreys, have at least a two-chambered heart composed of a layer of cardiac myocardial cells (red), an endocardial cellular layer (yellow), valves that separate distinct chambers, and a surrounding pericardium (pink). Figure adapted from [16].

ing mechanism of the embryonic heart continues today, with the possibility that the mechanism may vary between species or may be some hybrid of both mechanisms [23], [24].

The Liebau pump, a dynamic suction pump, was first described in 1954 [25], and was studied as a novel way to pump water. It has not been until the past 20 years that scientists started looking at the pump as a valveless pumping mechanism in many biological systems and biomedical applications, including microelectromechanical systems (MEMs) and micro-fluidic devices. Direct applications of such pumps include tissue engineering, implantable micro electrodes, and drug delivery [26], [27], [28], [25].

The Womersley Number (Wo) is used to quantify the effects from pulsatile flow in relation to viscous effects. It is a dimensionless number that is traditionally written as

$$Wo = L \left(\frac{2\pi f \rho}{\mu} \right)^{1/2} \qquad (1)$$

where L is a characteristic length scale, f a characteristic frequency, ρ the fluid density, and μ the fluid velocity. For tubular pumping problems the width of the tube is commonly used as the length scale and the pulsation period is used to compute the characteristic frequency.

With extensive industrial applications, dynamic suction pumping has proven to be a suitable means of transport for fluids and other materials in a valveless system, for scales of $Wo > 1$ [29]. DSP can be most simply described by an isolated region of actuation, located asymmetrically along a flexible tube with stiffer ends. Flexibility of the tube is required to allow passive elastic traveling waves, which augment bulk transport throughout the system. The rigid ends of the tube cause the elastic waves to reflect and continue to propagate in the opposite direction, which when coupled with an asymmetric actuation point, can promote unidirectional flow. DSP is illustrated in Figure 3.

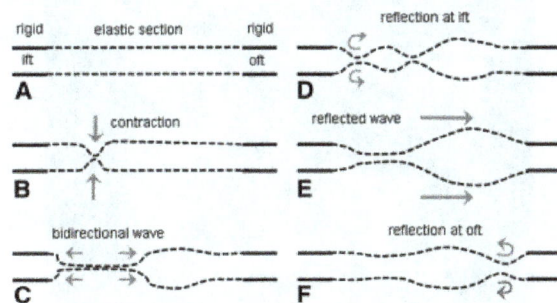

Fig. 3: Schematic diagram illustrating dynamic suction pumping [20]. (A) The flexible tube is at rest, assuming the inflow tract (ifl) is on the left and outflow tract (oft) on the right. (B) Active contraction of the tube in a non-central location along the tube. (C) Contraction induces an elastic passive bidirectional wave to propagate along the tube. (D) Wave reflects off rigid portion of the tube on side nearest to contraction point. (E) The reflected wave travels down the tube. (F) The waves reflect off the rigid section at the far side of the tube. Notice the the reflected wave amplitude is smaller than the reflected wave off the other end.

Due to a coupling between the system's geometry, material properties of the tube wall, and pumping mechanics, there is a complex, nonlinear relationship between volumetric flow rate and pumping frequency [29], [30], [31]. Analytic models of DSP have been developed to address this relationship [32], [33], [34], [35], [30], [36]. Most models use simplifications, such as an inviscid assumption, long wave approximation, small contraction amplitude, and/or one-dimensional flow. Furthermore, no analytical model has described flow reversals, which can occur with changes in the pumping frequency. Relaxing many of these assumptions, physical experiments have been performed to better understand DSP [31], [37], [30], [25], as well as *in silico* investigations [38], [39], [40], [29], [41], [42]. Most of the joint experimental and computational studies focus on the 'high' Wo regime ($Wo \gg 1$), besides studies by Baird *et al.* [41], which also looked at the biologically relevant cases of $Wo \leq 1$.

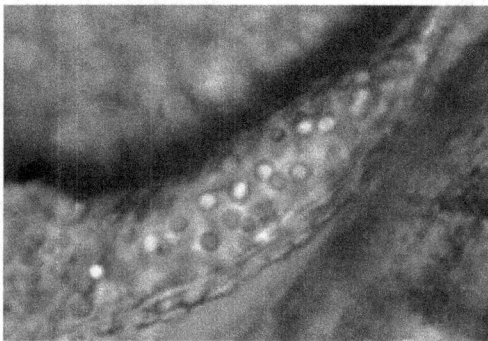

Fig. 4: The embryonic heart tube of a Zebrafish 30 hours post fertilization (hpf), courtesy of [43]. Spherical blood cells are seen within the tubular heart. The heart tube is roughly 5 blood cells thick in diameter.

In this paper we will investigate the pumping phenomena that occurs as a result of a coupled fluid-structure-electrophysiology model [44], and the bulk flow rates thereby induced. The electrophysiology model governs the propagation of action potentials, which then are coupled to muscular contraction, and hence the overall pumping dynamics. We then perturb the diffusion parameter in the electrophysiology model to investigate the bifurcation in pumping dynamics that occurs as

a result of differing action potential propagation. This bifurcation is able to capture a spectrum of different pumping regimes, i.e., dynamic suction pumping to peristaltic-like waves of contraction. Baird et al. 2015 [44] only explored the resulting dynamics for one particular value of the diffusivity within the peristaltic regime; we instead will investigate a spectrum of pumping behaviors. This is the first paper to use an electrophysiology model and demonstrate that a range of pumping behaviors is possible through variations of the action potential diffusivity. The electrophysiology model is governed by the FitzHugh-Nagumo equations [45], [44].

II. METHODS

The immersed boundary method (IB) is a numerical method developed to solve problems involving viscous, incompressible fluid coupled to the movement of an immersed elastic structure [46], [47], [48]. Since its development in the 1970s by Charles Peskin [49], it has been applied to a wide spectrum of biomathematical models, ranging from blood flow through the heart [49], [46], aquatic locomotion [50], [51], insect flight [52], [53], to plant biomechanics [54], [55], and muscle mechanics [56], [48].

The power of this method is that it can be used to describe flow around complicated time-dependent geometries using a regular Cartesian discretization of the fluid domain. The elastic fibers describing the structure are discretized on a moving curvilinear mesh defined in the Lagrangian frame. The fluid and elastic fibers constitute a coupled system, in which the structure moves at the local fluid velocity and the structure applies a singular force of delta-layered thickness to the fluid.

A. *Equations of the IB*

Assume that the immersed boundary is described on a curvilinear, Lagrangian mesh, S, that is free to move. The fluid is described on a fixed Cartesian, Eulerian grid, Ω, that has periodic boundary conditions. Given the size of the domain and the localization of the flow to the tube, the boundary conditions do not significantly affect the fluid motion. The governing equations for the fluid, the Navier-Stokes equations, are given by

$$\rho \left[\frac{\partial \mathbf{u}}{\partial t}(\mathbf{x}, t) + \mathbf{u}(\mathbf{x}, t) \cdot \nabla \mathbf{u}(\mathbf{x}, t) \right] = -\nabla p(\mathbf{x}, t)$$
$$+ \mu \Delta \mathbf{u}(\mathbf{x}, t) + \mathbf{f}(\mathbf{x}, t) \tag{2}$$

$$\nabla \cdot \mathbf{u}(\mathbf{x}, t) = 0. \tag{3}$$

Eqs.(2) and (3) are the Navier-Stokes equations written in Eulerian form, where Eq.(2) is the conservation of momentum for a fluid and Eq.(3) is the conservation of mass, i.e., incompressibility condition. The two constant parameters in these equations are the fluid density, ρ, and the dynamic viscosity of the fluid, μ. The fluid velocity, $\mathbf{u}(\mathbf{x}, t)$, pressure, $p(\mathbf{x}, t)$, and body force, $\mathbf{f}(\mathbf{x}, t)$, are unknown functions of the Eulerian coordinate, $\mathbf{x}$, and time, t. The body force describes the transfer of momentum onto the fluid due to the restoring forces arising from deformations of the elastic structure. It is this term, $\mathbf{f}(\mathbf{x}, t)$, that is unique to the particular model being studied.

The material properties of the structure may be modeled to resist bending, stretching, and displacement from a tethered position. Other forces that can have been modeled include the action of virtual muscles, electrostatic (contact) forces, molecular bonds, porosity, and other external forces [46], [57], [58], [59], [56], [48]. The immersed structure may deform due to bending forces and/or stretching and compression forces. This forces are commonly written in terms of $\mathbf{X}(s, t)$, which gives the position in Cartesian coordinates of the elastic structure at local material point, s, and time t. In this paper, elastic forces are calculated as beams that may undergo large deformations and Hookean springs, i.e.,

$$\mathbf{F}_{beam} = -k_{beam} \frac{\partial^4}{\partial s^4} \left(\mathbf{X}(s, t) - \mathbf{X}_B(s) \right) \tag{4}$$

$$\mathbf{F}_{spring} = -k_{spring} \left(1 - \frac{R_L}{||\mathbf{X}_S - \mathbf{X}_M||} \right) \cdot (\mathbf{X}_M - \mathbf{X}_S). \tag{5}$$

Eq.(4) is the beam equation, which describes forces arising from bending of the elastic structure and Eq.(5) describes the force generated from stretching and compression of the structure. The parameters, k_{beam} and k_{spring}, are the stiffness coefficients of the beam and spring, respectively, and R_L is the resting length of the Hookean spring. The variables $\mathbf{X}_M$ and $\mathbf{X}_S$ give the positions in Cartesian coordinates of the master and slave nodes in the spring formulations, respectively, and $\mathbf{X}_B(s)$ describes the deviation from the preferred curvature of the structure. In all simulations, $\mathbf{X}_B(s) = 0$ along the straight portion of the tube.

A target point formulation can be used to tether the structure or subset thereof in place, holding the Lagrangian mesh in a preferred position that may be time dependent. An immersed boundary point with position $\mathbf{X}(s, t)$ that is tethered to a target point, with position $\mathbf{Y}(s, t)$ undergoes a penalty force that is proportional to the displacement between them. The force that results is given by the equation for a linear spring with zero resting length,

$$\mathbf{F}_{target} = -k_{target} \left(\mathbf{X}(s, t) - \mathbf{Y}(s, t) \right), \tag{6}$$

where k_{target} is the stiffness coefficient of the target point springs. k_{target} can be varied to control the deviation allowed between the actual location of the boundary and its preferred position. The total deformation force that will be applied to the fluid is a sum of the above forces,

$$\mathbf{F}(s, t) = \mathbf{F}_{spring} + \mathbf{F}_{beam} + \mathbf{F}_{target} \tag{7}$$

A more detailed description of existing fiber models can be found in [48]. Once the total force from Eq.(7) has been calculated, it needs to be spread from the Lagrangian frame to the Eulerian grid. This is achieved through an integral transform with a delta function kernel,

$$\mathbf{f}(\mathbf{x}, t) = \int \mathbf{F}(\mathbf{s}, \mathbf{t}) \delta(\mathbf{x} - \mathbf{X}(\mathbf{s}, \mathbf{t})) d\mathbf{s}. \tag{8}$$

Similarly, to interpolate the local fluid velocity onto the Lagrangian mesh, the same delta function transform is used,

$$\mathbf{U}(s, t) = \frac{\partial \mathbf{X}}{\partial t}(s, t) = \int \mathbf{u}(\mathbf{x}, t) \delta(\mathbf{x} - \mathbf{X}(\mathbf{s}, \mathbf{t})) d\mathbf{x}. \tag{9}$$

Eqs.(8) and (9) describe the coupling between the immersed boundary and the fluid, e.g., the communication between the Lagrangian framework and

Eulerian framework. The delta functions in these equations make up the heart of the IB, as they are used to spread and interpolate dynamic quantities between the fluid grid and elastic structure, e.g., forces and velocity. Recall that $\mathbf{X}(s,t)$ gives the position in Cartesian coordinates of the elastic structure at local material point, s, and time t. In approximating these integral transforms, a discretized and regularized delta function, $\delta_h(\mathbf{x})$ [46], is used,

$$\delta_h(\mathbf{x}) = \frac{1}{h^2}\phi\left(\frac{x}{h}\right)\phi\left(\frac{y}{h}\right), \qquad (10)$$

where $\phi(r)$ is defined as

$$\phi(r) = \begin{cases} \frac{1}{4}\left[1 + \cos\left(\frac{\pi r}{2}\right)\right] & |r| \leq 2 \\ \\ 0 & \text{otherwise.} \end{cases} \tag{11}$$

B. Numerical Algorithm

As stated above, we impose periodic boundary conditions on the rectangular domain. To solve Eqs. (2), (3),(8) and (9) we need to update the velocity, pressure, position of the boundary, and force acting on the boundary at time $n+1$ using data from time n. IB does this in the following steps [46]:

Step 1: Find the force density, $\mathbf{F}^n$ on the immersed boundary, from the current boundary configuration, $\mathbf{X}^n$.

Step 2: Use Eq.(8) to spread this boundary force from the curvilinear mesh to nearby fluid lattice points.

Step 3: Solve the Navier-Stokes equations, Eqs.(2) and (3), on the Eulerian domain. In doing so, we are updating $\mathbf{u}^{n+1}$ and p^{n+1} from $\mathbf{u}^n$ and $\mathbf{f}^n$. Note: because of the periodic boundary conditions on our computational domain, we can easily use the Fast Fourier Transform (FFT) [60], [61], to solve for these updates at an accelerated rate.

Step 4: Update the material positions, $\mathbf{X}^{n+1}$, using the local fluid velocities, $\mathbf{U}^{n+1}$ with $\mathbf{u}^{n+1}$ and Eq.(9).

The above steps outline the process used by the IB to update the positions and velocities of both the fluid and elastic structure. A more detailed discussion of the IB can be found in [46].

C. Computational Model

We numerically model a $2D$ closed racetrack where the walls of the tube are modeled as $1D$ fibers. The closed tube is composed of two straight portions, of equal length, connected by two half circles, of equal inner and equal outer radii. The tube, or racetrack, has uniform diameter throughout. This is similar to the racetrack model geometry as in [44]. Furthermore, as in [44], we include the presence of an idealized stiff pericardium surrounding the flexible region of the heart tube.

The tunicate heart consists of a myocardium which is surrounded by a stiff pericardium [62], [63], which provides structural support to the myocardium. Muscle fibers spiral around the heart tube itself, and action potentials propagate to induce myocardial contraction. These action potentials have been previously measured [17]. Myocardial contractions may begin at either end of the heart tube, allowing the propagation of the action potential to occur in either direction [64]. However, we do not concern ourselves with flow reversals in this model. Although the tunicate heart tube has different material properties and physiological properties than the vertebrate embryonic heart, it still is an interesting model for vertebrate heart morphogenesis [19]. However, the conduction properties, e.g., velocities, of action potentials are much more uniform in tunicates than mammalian hearts [65].

The computational model we investigate is shown in Figure 5. Linear springs and beams connect adjacent Lagrangian points in the flexible region of the racetrack geometry. All other Lagrangian points of the boundary are modeled using target points, to hold the stiff portions of the racetrack and pericardium region nearly rigid. The flexible region models the myocardial layer of the tunicate heart, while the pericardium is held nearly rigid. As described below, the myocardial

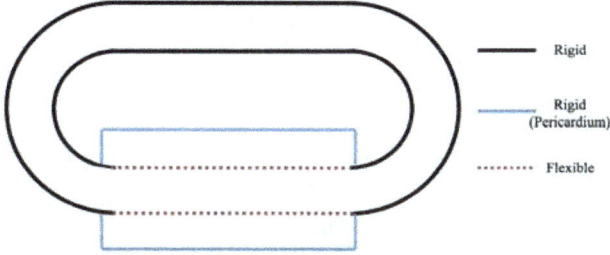

Fig. 5: Computational geometry for the electro-mechanical pumping model. The racetrack is held stiff (black), except for the bottom straight-tube portion, which is flexible (red). There is a stiff pericardium model surrounding the flexible region (blue).

region will actively contract based on an underlying electrophysiology-muscular force generation model. The parameters used in the model are found in Table I below.

Parameter	Value
Length/Width of comp. domain (m)	5.0×10^{-4}
Diameter of tube $[d]$ (m)	3.5×10^{-5}
Outer Radius $[R_o]$ (m)	1×10^{-5}
Inner Radius $[R_i]$ (m)	$d - R_o$
Length of Straight Tube (m)	5.0×10^{-4}
Eulerian Resolution $[dx]$ (m)	8.33×10^{-7}
Lagrangian Resolution $[ds]$ (m)	4.17×10^{-7}
Density of fluid $(\rho) \left[\frac{kg}{m^3}\right]$	1025
Viscosity of fluid $(\mu) \left[\frac{kg}{ms}\right]$	varied
Stretching stiffness of the boundary $(k_{spr}) \left[\frac{kg}{s^2}\right]$	3.24×10^5
Stretching stiffness of target points $(k_{target}) [Nm^2]$	3.24×10^5
Bending coefficient of boundary $(k_{beam}) \left[\frac{kg}{s^2}\right]$	3.24×10^5

TABLE I: Table of the parameters associated with the fluid and the immersed boundary fiber models.

Instead of prescribing contraction, we develop a model for the underlying electrophysiology of the heart, i.e., traveling action potentials arising from a single pacemaker region, to couple to myocardial contraction and hence intracardiac fluid flow. The model of action potential propagation is given by

Parameter	Value
Threshold potential (v_a)	0.1
Strength of blocking (ϵ)	0.1
Diffusive coefficient $(\mathbb{D})$ $\left[m^2/s\right]$	$0.1 - 100$
Resetting rate (γ)	0.5
Current injection $(\mathbb{I})$	0.5
Frequency (f) (Hz)	1.0

TABLE II: Table of the parameters associated with the FitzHugh-Nagumo electrophysiology model.

the FitzHugh-Nagumo equations [45], [44] below,

$$\frac{\partial v}{\partial t} = \mathbb{D}\nabla^2 v + v(v - v_a)(v - 1) - w - \mathbb{I}(t) \quad (12)$$

$$\frac{\partial w}{\partial t} = \epsilon(v - \gamma w), \quad (13)$$

where $v(s,t)$ is the membrane potential, $w(s,t)$ is the blocking mechanism, $\mathbb{D}$ is the diffusion rate of the membrane potential, v_a is the threshold potential, γ is the resetting rate, ϵ is the blocking strength parameter, and $\mathbb{I}(t)$ is an applied current, e.g., an initial stimulus potentially from pacemaker signal activation. Note that v is the action potential and that w can be thought to model a sodium blocking channel. We note that the FitzHugh-Nagumo equations (12)-(13) are a reduced order model of the Hodgkin-Huxley equations, which were the first quantitative model to describe the propagation of an electrical signal across excitable cells [66]. The parameters used in the electrophysiology model are found in Table II.

Next we need to interpolate the information from the electrophysiology model to the fluid-structure interaction solver, i.e., immersed boundary method. Time is scaled in order to match the dynamics of the generated action potentials to the desired active wave of contraction and is given by:

$$dt_f = \frac{dt\mathbb{F}}{\mathbb{T}}, \quad (14)$$

where dt is the time-step associated with the fluid solver, $\mathbb{F}$ is a non-dimensional scaling parameter, and $\mathbb{T}$ is the desired pumping period. When the propagating action potential reaches one of the muscles along the tube, the associated spring stiffness of said muscle model is given by

$$k_e(s,t) = k_m \left(v^4(s,t) \right). \qquad (15)$$

The simplified muscle model is given by a dynamic spring stiffness coefficient, given by $k_e(s,t)$, which is a non-linear function of the traveling action potential, $v(s,t)$. This idea was adapted from Baird *et al.* [41], [44]. The choice of raising $v^4(s,t)$ was adopted to mimic a non-linear stress response using a basic spring-like relation. The force generated by the springs, with non-linear stiffnesses, that connect the bottom and top of the elastic tube can then be computed. These forces represents muscular contraction. The value of k_m is tuned to produce the amount of contraction observed in Ciona hearts, as in [41], [44].

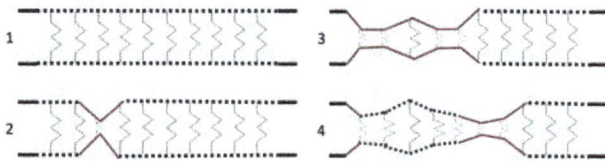

Fig. 6: Schematic of electrodynamical pumping. (1) The tube at rest; the springs connecting the top and bottom of the tube are the muscles. (2) The pacemaker initates an action potential, in which the tube will contract based on the magnitude of the signal (3)-(4) The action potential propagates along the tube, induing contraction.

The idea for electro-dynamic pumping can be seen in Figure 6, which is a schematic for electro-dynamical pumping behavior. First the tube is at rest until a pacemaker initiates a potential signal, which contracts the tube in one singular region. Next the action potential propagates along the tube inducing contraction. Once the action potential passes outside a region on the tube, that location no longer has active contraction, but can return to its resting position.

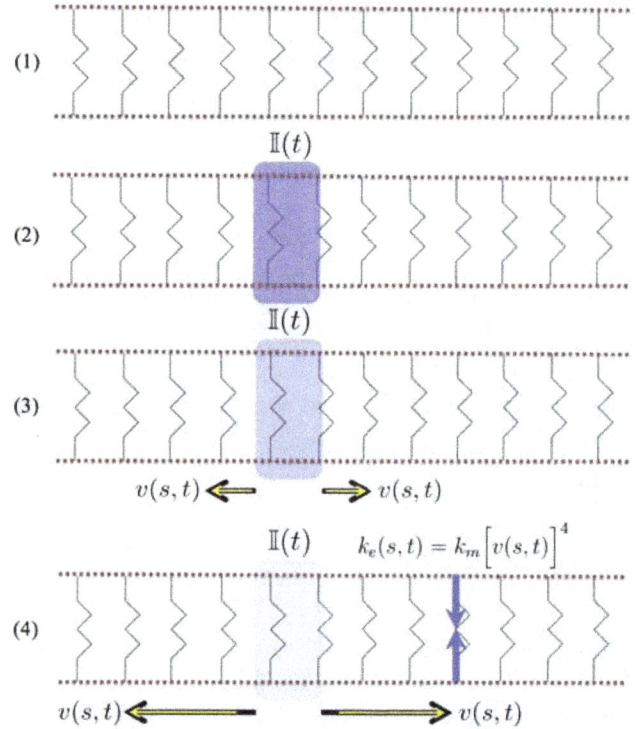

Fig. 7: Schematic of electrodynamical pumping. (1) The tube at rest; the springs connecting the top and bottom of the tube are the muscles. (2) The pacemaker initates an action potential, in which the tube will contract based on the magnitude of the signal (3)-(4) The action potential propagates along the tube, inducing contraction.

Furthermore the main electrophysiology idea behind the model is illustrated in Figure 7. In diagram 1 the flexible tube is at rest. Next 2 depicts a pacemaker initiating an input signal (current). Then that voltage (action potential) travels down the tube, while the input signal dissipates. Once the action potential reaches a muscle fiber, the tube contracts based on a non-linear relationship between spring stiffness and the magnitude of the action potential (voltage).

III. RESULTS

In this study, we conducted numerical experiments of the electro-dynamic pumping model, which encompassed fully coupled electrophysiology to pumping behavior for a heart tube, modeled as a closed racetrack geometry. We investigated various diffusivities, $\mathbb{D}$, which give rise to different pumping regimes, e.g., either a 'dynamic suction pumping-esque' or 'peristaltic-like' pumping regime. Furthermore, we explored these regimes for over 3 orders of magnitude in Wo.

A. Results of the FitzHugh-Nagumo Model

Here we present the varying action potential dynamics given via the FitzHugh-Nagumo equations, which models the electrophysiology. We explored this model for a variety of diffusive coefficients, $\mathbb{D} = \{0.1, 1.0, 10.0, 100.0\}$.

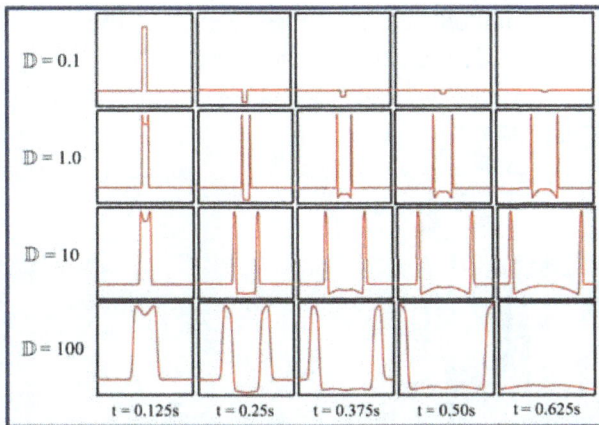

Fig. 8: Different traveling wave propagation properties arising out of the FitzHugh-Nagumo equations for varying diffusivities, $\mathbb{D} = \{0.1, 1.0, 10.0, 100.0\}$. These solutions assume a pacemaker frequency of 1 Hz and the time here corresponds to the same time, t, as in Figures 9-11.

Figure 8 illustrates the kinds of traveling action potentials that arise out of the electrophysiology model. These solutions suggest that different $\mathbb{D}$ give rise to different action potential signals. It is clear that the $\mathbb{D} = 0.1$ case resembles a signal that could be reminiscent of that of dynamic suction pumping. This is because the action potential's signal is localized to a particular region on the tube. In comparison, $\mathbb{D} = 100$ gives rise to a propagating action potential that could model a more peristaltic-like contraction. This is indicative of a coordinated peristaltic-type wave; when the

action potential propagates down the tube, it in turn causes active muscular contraction along the tube, such as those shown in Figure 6. It is clear that as the diffusivity, $\mathbb{D}$ increases, the waves propagate outwards, and with greater wave-speed. Furthermore, the wave-form itself gets wider.

The remainder of the results will be shown for full fluid-structure interaction model that incorporates these action potential signals to induce muscular contraction. In our model, the action potential only travels down the flexible portion of the tube, shown in Figure 5. The flexible portion then can active contract and relax according to the action potential signal, muscular contraction model, and material properties of the tube.

B. Results of the electro-dynamical heart tube model

In this section we present the results describing how bulk flow rates are affected by varying the diffusivity, to capture different pumping behaviors for a variety of Wo. All the simulations were run with a pacemaker frequency of 1 Hz. We note that the maximal closure of the tube was at most approximately 90% occlusion during any of the simulations.

Figures 9 and 10 illustrate the non-dimensional spatially-averaged velocity computed across a cross-section of the top of the race-track geometry vs non-dimensional time for $\mathbb{D} = 0.1$ (Figure 9) and $\mathbb{D} = 100.0$ (Figure 10). It is clear that when $\mathbb{D} = 0.1$ there is not significant bulk flow produced regardless of Wo, unlike the case when $\mathbb{D} = 100.0$, where significant bulk flow is produced over all $Wo = \{0.1, 10, 10\}$. It is also clear that the wave-form produced for $\mathbb{D} = 0.1$ undergoes many more high frequency oscillations as compared to the case for $\mathbb{D} = 100$.

Comparing corresponding Wo pumping mechanisms for a variety of $\mathbb{D} = \{0.1, 1.0, 10.0, 100.0\}$ are shown in Figure 11, where Figure 11a compares pumping regimes for $Wo = 0.1$ and Figure 11b for $Wo = 10$. It is clear that in both cases that the most bulk flow is produced when $\mathbb{D} = 100$, and some flow is produced in the cases of $\mathbb{D} = \{1, 10\}$.

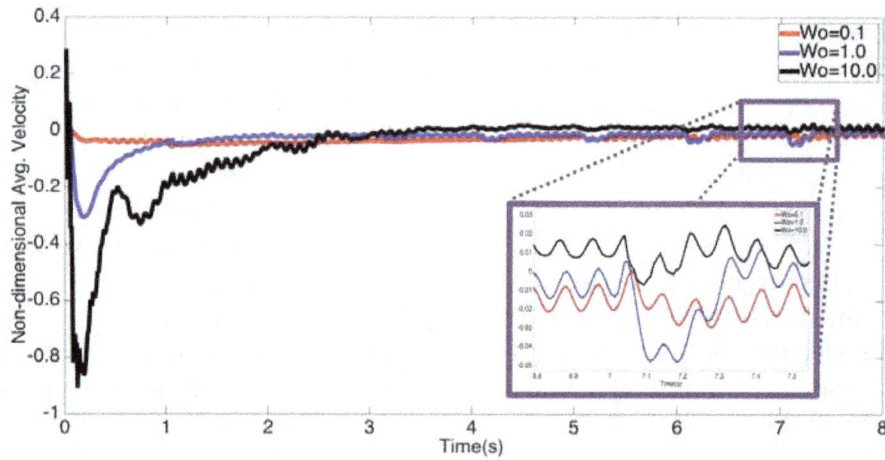

Fig. 9: The non-dimensional spatially-averaged velocity computed across a cross-section of the top of the race-track geometry vs non-dimensional time for $\mathbb{D} = 0.1$, e.g., the 'dynamic suction pumping' regime, for $Wo = \{0.1, 1.0, 10.0\}$. The zoomed in portion illustrates the resulting wave-form and the high frequency oscillations that result from this pumping regime.

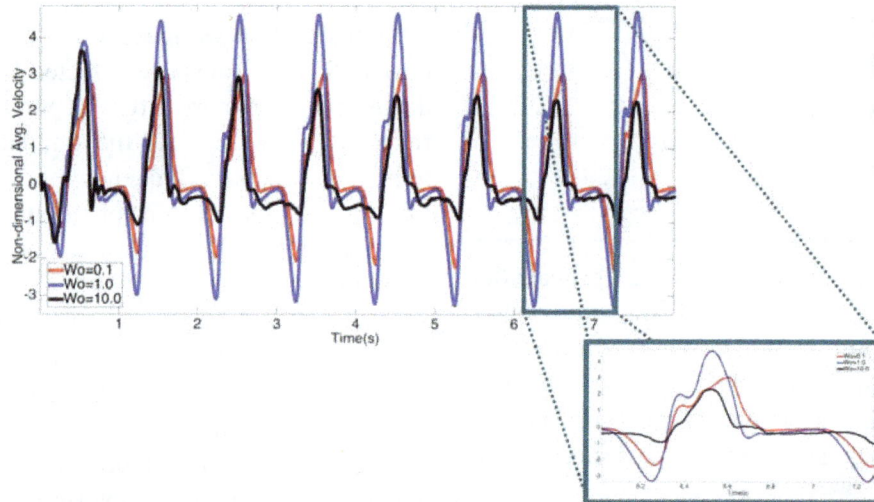

Fig. 10: The non-dimensional spatially-averaged velocity computed across a cross-section of the top of the race-track geometry vs non-dimensional time for $\mathbb{D} = 100.0$, e.g., the 'peristaltic' regime, for $Wo = \{0.1, 1.0, 10.0\}$. The zoomed in portion illustrates the resulting wave-form.

There is still backflow in the $\mathbb{D} = 100$ case and less overall backflow in the $\mathbb{D} = 10$ case.

Furthermore, the wave-form in the $\mathbb{D} = 100$ case is different between the $Wo = 0.1$ and $Wo = 10$ cases. There is a single peak for the case when $Wo = 10$ and dual peaks for $Wo = 0.1$ for the forward flow; however, in the backflow, the situation is reversed, where a dual-peak is observed for $Wo = 10$ and a single peak for $Wo = 0.1$.

In attempt to maximize bulk flow for the dynamic suction pumping-esque regime, the stretching-stiffness and bending stiffness coefficients of the tube were varied. The results are

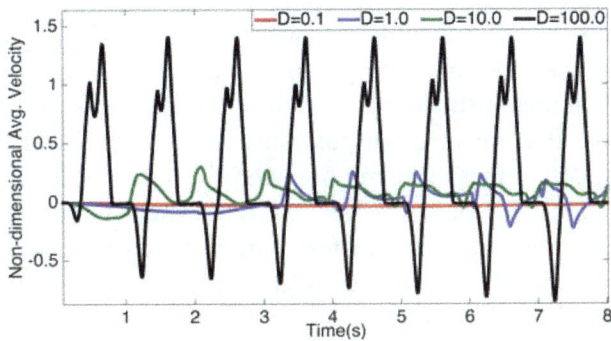

(a)

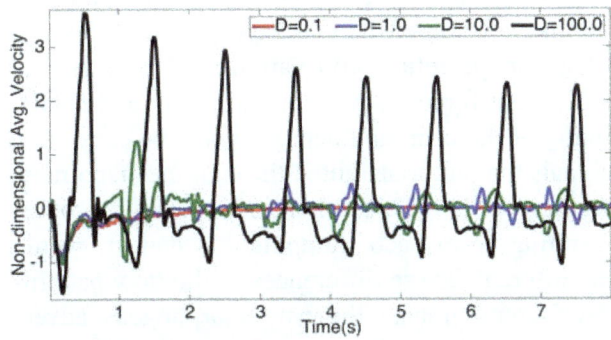

(b)

Fig. 11: A comparison of non-dimensional spatially-averaged velocity computed across a cross-section at the top of the racetrack vs non-dimensional time in the simulation for varying diffusive coefficients, $\mathbb{D} = \{0.1, 1.0, 10.0, 100.0\}$. The two plots compare different Wo, e.g., (a) $Wo = 0.1$ and (b) $Wo = 10$.

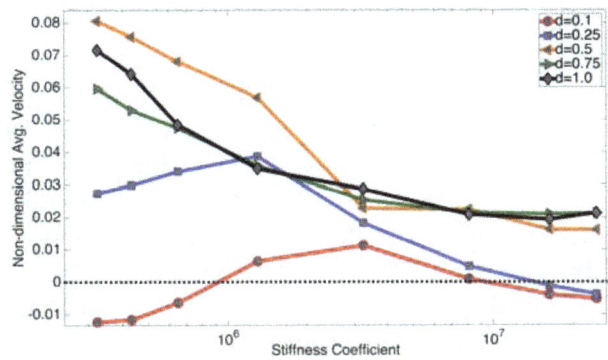

Fig. 12: A plot of non-dimensional spatially-averaged velocity computed across a cross-section at the top of the racetrack vs the non-dimensional stretching and bending stiffness co-efficients for pumping in the 'dynamic suction pumping' regime, for a variety of diffusivities, $\mathbb{D} = \{0.1, 0.25, 0.5, 0.75, 1.0\}$.

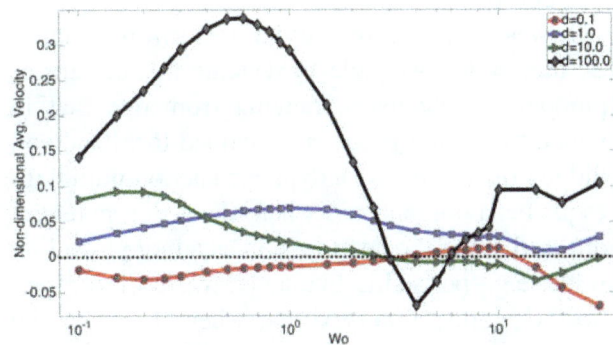

Fig. 13: A comparison of the spatially- and temporally-averaged non-dimensional velocities computed across a cross-section of the race-track vs. Wo for varying diffusivities, $\mathbb{D} = \{0.1, 1.0, 10.0, 100.0\}$.

shown in Figure 12. It is clear that as the stiffness is varied there is a non-linear relationships between flow speed (spatially- and temporally-averaged non-dimensional velocity across a cross-section of the racetrack) and stiffness. However, not a considerable amount of more bulk flow was produced from increasing these stiffness coefficients.

Lastly we compared the spatially- and temporally-averaged non-dimensional velocities across a cross-section of the racetrack against

Wo for a variety of $\mathbb{D}$. Each simulation was spatially-averaged for the same cross-sectional area and temporally-averaged across each entire simulation of $8s$. The results are shown in Figure 13. It is clear there is a non-linear relationship in average velocity and scale arising from this model of pumping in every pumping regime, given by $\mathbb{D}$. Furthermore, the highest bulk flow rates were seen in the case of $\mathbb{D} = 100$ for $Wo \sim 0.8$, which correspond to the Wo around that of tunicate tubular hearts [29], [41].

IV. DISCUSSION AND CONCLUSION

This $2D$ model coupled the propagation of action potentials, given via the FitzHugh-Nagumo equations, to the force generation and myocardial contraction, given through a non-linear spring-like muscle model, to induce pumping behavior in a flexible tube, where the fully coupled fluid-structure interaction model was solved using the immersed boundary method. This model was first described in [44], but was not used to explore a range of pumping behaviors possible due to variations on action potential diffusivity until now. We explored this effect of perturbing the diffusive coefficient of the electrophysiology model to capture different pumping regimes, including both dynamic suction pumping-esque and peristaltic-like contractile waves.

It was clear that by varying this diffusive term, $\mathbb{D}$, the model was able to recreate a spectrum of pumping mechanisms, ranging from one that in which the action potential remained localized and did not diffusive, i.e., a dynamic suction pumping-esque behavior, and one where the action potential diffused along the heart tube in as a more traveling wave, e.g., peristaltic-like active wave of contraction. Our model showed that when $\mathbb{D}$ was in the more peristaltic-like regime, i.e., $\mathbb{D} \sim 100$, more bulk flow was produced in the racetrack geometry, as compared to more negligible amounts from the dynamic suction pumping-esque regime, $\mathbb{D} \sim 0.1$. This result was consistent for the range of Wo considered.

Moreover, in all cases considered, there was a non-linear relationship between average flow rate, scale (Wo), and diffusivity (pumping behavior). More bulk flow was produced on average (both spatially and temporally), with a maximum around $Wo \sim 0.8$ than for higher Wo, up to $Wo = 30$, in the peristalic-like regime. Similar behavior, in that peristalsis produces more bulk flow than DSP, has been observed before when using prescribed pumping behavior, as in [29], [42].

However, perturbing the material properties of the tube could potentially affect bulk flow rates across all pumping regimes, given by $\mathbb{D}$. Our focus was limited to perturbing the stretching and bending stiffnesses of the tube specifically within the dynamic suction pumping-esque regime, $\mathbb{D} \sim [0.1, 1]$. Furthermore, our study only considered increasing the stiffnesses and not decreasing them. For the regime and material properties considered, we found a non-linear relationship between flow rates and stiffness.

As blood flow and the resulting hemodynamic forces are essential for proper heart development [67], it is important that the pumping model capture as much biology as possible. Each pumping regime considered here, given by the diffusivity of action potential propagation, will give rise to a different force distribution along the endothelial lining of the heart and hence impact the epigenetic signals that are transmitted through mechanotransduction [68], [69]. Furthermore the flow profiles resulting from each pumping mechanism would be different. These differences in the flow patterns itself could impact the way morphogens advect and diffuse during embryogenesis [70], [71], opening the realm to a lot more interesting biological questions to explore.

ACKNOWLEDGMENT

The authors would like to thank Steven Vogel for past conversations on scaling in various hearts as well as Lindsay Waldrop, Austin Baird, Julia Samson, and William Kier for discussions on embryonic hearts. They also wish to thank the organizers of the 2017 BIOMATH Conference in Kruger Park, South Africa. Funding for LAM was provided by the Army Research Office Staff Research Grant to Pasour, NSF DMS CAREER #1151478, NSF CBET #1511427, NSF DMS #1151478, NSF POLS #1505061, NSF IOS #1558052. Funding for N.A.B. was partially funded from an National Institutes of Health T32 grant [HL069768-14; PI, Christopher Mack] and the School of Science at The College of New Jersey.

References

[1] S. Alters, Biology: Understanding Life, Jones and Bartlett Publishers, Boston, USA, 2000.

[2] J. Xavier-Neto, R. Castro, A. Sampaio, A. Azambuja, H. Castillo, R. Cravo, M. Simoes-Costa, Parallel avenues in the evolution of hearts and pumping organs, Cell. Mol. Life Sci. 64 (2007) 719–734.

[3] C. J. Vivien, J. E. Hudson, E. R. Porrello, Evolution, comparative biology and ontogeny of vertebrate heart regeneration, Regen. Med. 1 (2016) 16012.

[4] R. L. Calabrese, C. S. Cozzens, Heart (invertebrate) from *Access Science*, https://doi.org/10.1036/1097-8542.309800 (2016).

[5] S. E. Grosskurth, D. Bhattacharya, Q. Wang, J. J. Lin, Emergence of xin demarcates a key innovation in heart evolution, PLoS One 3(8) (2008) e2857.

[6] Tunicates, https://www.coraldigest.org/index.php/Tunicates, accessed: 2017-05-30.

[7] Lancelet, https://en.wikipedia.org/wiki/Lancelet, accessed: 2017-05-30.

[8] Sea lamprey (*Petromyzon marinus*), https://nas.er.usgs.gov/queries/factsheet.aspx?SpeciesID=836, accessed: 2017-05-30.

[9] M. W. Konrad, Blood circulation in the ascidian tunicate corella inflata (corellidae), Peer J 4 (2016) e2771.

[10] N. D. Holland, T. V. Ventatesh, L. Z. Holland, D. K. Jacob, R. Bodmer, Amphink2-tin, an amphioxus homeobox gene expressed in myocardial progenitors: insights into evolution of the vertebrate heart, Dev. Biol. 255(1) (2003) 128–137.

[11] N. D. Holland, J. Y. Chen, Origin and evolution of the vertebrates: new insights from advances in molecular biology, BioEssays 23 (2001) 142–151.

[12] J. J. Smith, M. C. Keinath, The sea lamprey meiotic map improves resolution of ancient vertebrate genome duplications, Genome Res. 25(8) (2015) 1081–90.

[13] L. Percy, I. C. Potter, Description of the heart and associated blood vessels in larval lampreys, J. Zoology 208(4) (1986) 479–492.

[14] S. A. Green, M. E. Bronnera, The lamprey: A jawless vertebrate model system for examining origin of the neural crest and other vertebrate traits, Differentiation 87(1) (2015) 44–51.

[15] M. Puceat, Embryological origin of the endocardium and derived valve progenitor cells: From developmental biology to stem cell-based valve repair, Biochimica et Biophysica Acta (BBA) - Molecular Cell Research 1833(4) (2012) 917–922.

[16] H. E. Anderson, L. Christiaen, Ciona as a simple chordate model for heart development and regeneration, J. Cardiovasc. Dev. Dis. 3 (2016) 25.

[17] M. E. Kriebel, Conduction velocity and intracellular action potentials of the tunicate heart, The Journal of General Physiology 50 (1967) 2097–2107.

[18] D. J. Randall, P. S. Davie, The hearts of urochordates and cephalochordates, Comparative Anatomy and Development 1 (1980) 41–59.

[19] M. Morad, L. Cleemann, Tunicate heart as a possible model for the vertebrate heart, Fed Proc. 39(14) (1980) 3188–3194.

[20] A. Santhanakrishnan, L. A. Miller, Fluid dynamics of heart development, Cell Biochem. Biophys. 61 (2011) 1–22.

[21] T. Kenner, M. Moser, I. Tanev, K. Ono, The liebau-effect or on the optimal use of energy for the circulation of blood, Scripta Medica 73 (2000) 9–14.

[22] A. S. Forouhar, M. Liebling, A. Hickerson, A. Nasiraei-Moghaddam, H. J. Tsai, J. R. Hove, S. E. Fraser, M. E. Dickinson, M. Gharib, The embryonic vertebrate heart tube is a dynamic suction pump, Science 312 (5774) (2006) 751–753.

[23] L. D. Waldrop, L. A. Miller, Large-amplitude, short-wave peristalsis and its implications for transport, Biomechanics and Modeling in Mechanobiology (2015) 1–14.

[24] J. Manner, A. Wessel, T. M. Yelbuz, How does the tubular embryonic heart work? looking for the physical mechanism generating unidirectional blood flow in the valveless embryonic heart tube, Developmental Dynamics 239 (2010) 1035–1046.

[25] J. Meier, A novel experimental study of a valveless impedance pump for applications at lab-on-chip, microfluidic, and biomedical device size scales, Ph.D. thesis, California Institute of Technology., Pasadena, CA.

[26] D. S. Lee, H. C. Yoon, J. S. Ko, Fabrication and characterization of a bidirectional valveless peristaltic micropump and its application to a flow-type immunoanalysis, Sensors and Actuators 103 (2004) 409–415.

[27] H. T. Chang, C. Y. Lee, C. Y. Wen, Design and modeling of electromagnetic actuator in mems-based valveless impedance pump, Microsystems Technologies — Micro-and Nanosystems- Information Storage and Processing Systems 13 (2007) 1615–1622.

[28] C. Y. Lee, H. T. Chang, C. Y. Wen, A mems-based valveless impedance pump utilizing electromagnetic actuation, Journal of Micromechanics and Microengineering 18 (2008) 225–228.

[29] A. J. Baird, T. King, L. A. Miller, Numerical study of scaling effects in peristalsis and dynamic suction pumping, Biological Fluid Dynamics: Modeling, Computations, and Applications 628 (2014) 129–148.

[30] T. Bringley, S. Childress, N. Vandenberghe, J. Zhang, An experimental investigation and a simple model of a valveless pump, Physics of Fluids 20 (2008) 033602.

[31] A. Hickerson, D. Rinderknecht, M. Gharib, Experimental study of the behavior of a valveless impedance pump, Experiments in Fluids 38 (2005) 534–540.

[32] J. Ottesen, Valveless pumping in a fluid-filled closed elastic tube-system: one-dimensional theory with experimental validation, Journal of Mathematical Biology 46 (2003) 309–332.

[33] D. Auerbach, W. Moehring, M. Moser, An analytic approach to the liebau problem of valveless pumping,

Cardiovascular Engineering: An International Journal 4 (2004) 201–207.

[34] C. G. Manopoulos, D. S. Mathioulakis, S. G. Tsangaris, One-dimensional model of valveless pumping in a closed loop and a numerical solution, Physics of Fluids 18 (2006) 201–207.

[35] O. Samson, A review of valveless pumping: History, applications, and recent developments (2007). URL http://www.researchgate.net/publication/267300626_A_Review_of_Valveless_Pumping_History_Applications_and_Recent_Developments

[36] C. Babbs, Behavior of a viscoelastic valveless pump: a simple theory with experimental validation, BioMedical Engineering Online 9:42 (2010) 19832–19837.

[37] A. I. Hickerson, An experimental analysis of the characteristic behaviors of an impedance pump (ph.d. thesis), California Institute of Technology 608 (2005) 139–160.

[38] E. Jung, Two-dimensional simulations of valveless pumping using the immersed boundary method (ph.d. thesis), Courant Institute of Mathematics, New York University 608 (1999) 139–160.

[39] E. Jung, C. Peskin, 2-d simulations of valveless pumping using immersed boundary methods, SIAM Journal on Scientific Computing 23 (2001) 19–45.

[40] I. Avrahami, M. Gharib., Computational studies of resonance wave pumping in compliant tubes, Journal of Fluid Mechanics 608 (2008) 139–160.

[41] A. J. Baird, Modeling valveless pumping mechanisms (ph.d. thesis), University of North Carolina at Chapel Hill 628 (2014) 129–148.

[42] N. A. Battista, A. N. Lane, L. A. Miller, On the dynamic suction pumping of blood cells in tubular hearts, in: A. Layton, L. A. Miller (Eds.), Women in Mathematical Biology: Research Collaboration, Springer, New York, NY, 2017, Ch. 11, pp. 211–231.

[43] F. Maes, B. Chaudhry, P. V. Ransbeeck, P. Verdonck, The pumping mechanism of embryonic hearts, IFMBE Proceedings 37 (2011) 470–473.

[44] A. J. Baird, L. D. Waldrop, L. A. Miller, Neuromechanical pumping: boundary flexibility and traveling depolarization waves drive flow within valveless, tubular hearts, Japan J. Indust. Appl. Math. 32 (2015) 829–846.

[45] R. FitzHugh, Impulses and physiological states in theoretical models of nerve membrane, Biophys. J. 1(6) (1961) 445–466.

[46] C. S. Peskin, The immersed boundary method, Acta Numerica 11 (2002) 479–517.

[47] R. Mittal, C. Iaccarino, Immersed boundary methods, Annu. Rev. Fluid Mech. 37 (2005) 239–261.

[48] N. A. Battista, W. C. Strickland, L. A. Miller, Ib2d: a python and matlab implementation of the immersed boundary method, Bioinspir. Biomim. 12(3) (2017) 036003.

[49] C. Peskin, Numerical analysis of blood flow in the heart, J. Comput. Phys. 25 (1977) 220–252.

[50] S. Hieber, P. Koumoutsakos, An immersed boundary method for smoothed particle hydrodynamics of self-propelled swimmers, J. Comput. Phys. 227 (2008) 8636–8654.

[51] A. P. Hoover, L. A. Miller, A numerical study of the benefits of driving jellyfish bells at their natural frequency, J. Theor. Biol. 374 (2015) 13–25.

[52] L. A. Miller, C. S. Peskin, When vortices stick: an aerodynamic transition in tiny insect flight, J. Exp. Biol. 207 (2004) 3073–3088.

[53] L. A. Miller, C. S. Peskin, A computational fluid dynamics of clap and fling in the smallest insects, J. Exp. Biol. 208 (2009) 3076–3090.

[54] L. A. Miller, A. Santhanakrishnan, S. K. Jones, C. Hamlet, K. Mertens, L. Zhu, Reconfiguration and the reduction of vortex-induced vibrations in broad leaves, J. Exp. Biol. 215 (2012) 2716–2727.

[55] N. A. Battista, W. C. Strickland, A. Barrett, L. A. Miller, Ib2d reloaded: a more powerful python and matlab implementation of the immersed boundary method, arXiv: https://arxiv.org/abs/1707.06928.

[56] N. A. Battista, A. J. Baird, L. A. Miller, A mathematical model and matlab code for muscle-fluid-structure simulations, Integr. Comp. Biol. 55(5) (2015) 901–911.

[57] E. Tytell, C. Hsu, T. Williams, A. Cohen, L. Fauci, Interactions between internal forces, body stiffness, and fluid environment in a neuromechanical model of lamprey swimming, Proc. Natl. Acad. Sci. 107 (2010) 19832–19837.

[58] S. D. J. Y. M. S. R. Mathura, L. Sunb, Application of the immersed boundary method to fluid, structure, and electrostatics interaction in mems, Numerical Heat Transfer, Part B: Fundamentals: An International Journal of Computation and Methodology 62 (2012) 399–418.

[59] A. L. Fogelson, R. D. Guy, Immersed-boundary-type models of intravascular platelet aggregation, Comput. Methods Appl. Mech. Engrg. 197 (2008) 2087–2104.

[60] J. W. Cooley, J. W. Tukey, An algorithm for the machine calculation of complex fourier series, Math. Comput. 19 (1965) 297–301.

[61] W. H. Press, B. P. Flannery, S. A. Teukolsky, W. T. Vetterling, Fast fourier transform, Ch. 12 in Numerical Recipes in FORTRAN: The Art of Scientific Computing 2 (1992) 490–529.

[62] M. Kalk, The organization of a tunicate heart, Tissue Cell 2 (1970) 99–118.

[63] L. D. Waldrop, L. A. Miller, The role of the pericardium in the valveless, tubular heart of the tunicate, *Ciona savignyi*, J. Exp. Biol. 218 (2015) 2753–2763.

[64] M. Anderson, Electrophysiological studies on initiation and reversal of the heart beat in ciona intestinalis, J. Exp. Biol. 49 (1968) 363–385.

[65] E. M. Jucker, M. Martin-Smith, Comparative Physiology of the Heart: Current Trends, Springer, Hanover, New Hampshire, USA, 1968.

[66] A. L. Hodgkin, A. F. Huxley, Propagation of electrical signals along giant nerve fibres, Proc. Roy. Soc. Lond. B, Biol. Sci. 140 (1952) 177–183.

[67] J. R. Hove, R. W. Koster, A. S. Forouhar, G. Acevedo-

Bolton, S. E. Fraser, M. Gharib, Intracardiac fluid forces are an essential epigenetic factor for embryonic cardiogenesis, Nature 421 (6919) (2003) 172–177.

[68] A. S. French, Mechanotransduction, Annu. Rev. Physiol. 54 (1992) 135–152.

[69] M. Weckstrom, P. Tavi, Cardiac Mechanotransduction, Springer Science and Business, New York, USA, 2007.

[70] J. B. Gurdon, P. Y. Bourillot, Morphogen gradient interpretation, Nature 413 (2001) 797–803.

[71] J. L. Christian, Morphogen gradients in development: from form to function, Wiley Interdiscip Rev. Dev. Biol. 1(1) (2012) 3–15.

9

Pulsing corals: A story of scale and mixing

Julia E. Samson[1], Nicholas A. Battista[2], Shilpa Khatri[3], Laura A. Miller[1,4]
[1]Dept. of Biology, [4]Dept. of Mathematics
University of North Carolina at Chapel Hill, Chapel Hill, NC, USA
jesamson@live.unc.edu, lam9@unc.edu
[2] Dept. of Mathematics and Statistics
The College of New Jersey, Ewing, NJ, USA
battistn@tcnj.edu
[3]Applied Mathematics Unit, School of Natural Sciences
University of California Merced, Merced, CA, USA
skhatri3@ucmerced.edu

Abstract—Effective methods of fluid transport vary across scale. A commonly used dimensionless number for quantifying the effective scale of fluid transport is the frequency based Reynolds number, Re_f, which gives the ratio of inertial to viscous forces in a fluid flow. What may work well for one Re_f regime may not produce significant flows for another. These differences in scale have implications for many organisms, ranging from the mechanics of how organisms move through their fluid environment to how hearts pump at various stages in development. Some organisms, such as soft pulsing corals, actively contract their tentacles to generate mixing currents that enhance photosynthesis. Their unique morphology and the intermediate Re_f regime at which they function, where both viscous and inertial forces are significant, make them a unique model organism for understanding fluid mixing. In this paper, 3D fluid-structure interaction simulations of a pulsing soft coral are used to quantify fluid transport and describe fluid mixing across a wide range of Re_f. The results show that net transport is negligible for $Re_f < \mathcal{O}(10^1)$, and continuous upward flow is produced for $Re_f \geq \mathcal{O}(10^1)$. Sustained net transport is necessary to bring in new fluid for sampling and to remove waste. As the Re is increased well above $\mathcal{O}(10^1)$, the slow region of mixing necessary for gas exchange between the tentacles is reduced. Since corals live at Re_f between about 8 and 36, the flows they produce are defined by sustained net transport of fluid away from the coral in a continuous upward jet and a slow region of mixing between the tentacles necessary for gas exchange.

Keywords-pulsing coral; coral reefs; immersed boundary; fluid-structure interaction; computational fluid dynamics;

I. INTRODUCTION

Biological fluid transport is not only dependent upon the method of movement, but also the fluid's physical properties and the size and velocity of the organ or organism. While one mechanism for transport may work well at the macroscale, that same mechanism may not work well at the mesoscale or microscale. For example, reciprocal motion of a fish's caudal fin may not produce adequate forward propulsion if the fish is put into a considerably more viscous fluid than water. If the viscosity is high enough, the fish might not swim at all as no reciprocal fin stroke will yield any net transport of fluid. The fact that reciprocal motions do not generate net movement at small scales is famously known as the Scallop Theorem [1]. The Reynolds number, Re_f, is a dimensionless quantity that describes the ratio of inertial to viscous forces in a fluid and is used to compare fluid transport across scales. For a fluid of density ρ, dynamic viscosity μ, and some characteristic length and frequency scale L and f, respectively, a frequency-based Re_f may be defined as

$$Re_f = \frac{\rho L^2 f}{\mu}. \tag{1}$$

For a Newtonian fluid in a large domain and with a sufficiently low Re_f, it is necessary to use non-reciprocal motions to produce the net transport of fluid. One common example of a non-reciprocal motion is the use of a rotating flagellum like in many bacteria and sperm cells [2], [3]. Beyond locomotion, there are many other applications of fluid transport within biological systems such as the generation of feeding currents [4], the generation of flow for oxygen and nutrient transport [5], the internal pumping of fluids (e.g. the cardiovascular system) [6], flows generated for filtering [7], and flows for photosynthetic enhancement [8]. As is the case for locomotion, different pumping and feeding mechanisms may only be effective over some range of Re_f [9], [10].

In this paper, we quantify the flows produced by a variety of soft corals, including the genera *Xenia* and *Heteroxenia*, that actively pulse and contribute substantially to local ocean mixing, enhancing nutrient availability in reefs. Each individual polyp is made up of eight feather-like tentacles (see Figure 1) positioned at the end of an approximately 5 cm long stalk [11]. These soft corals form colonies up to 60 cm across [11], and polyps within a colony do not normally pulse in synchrony but

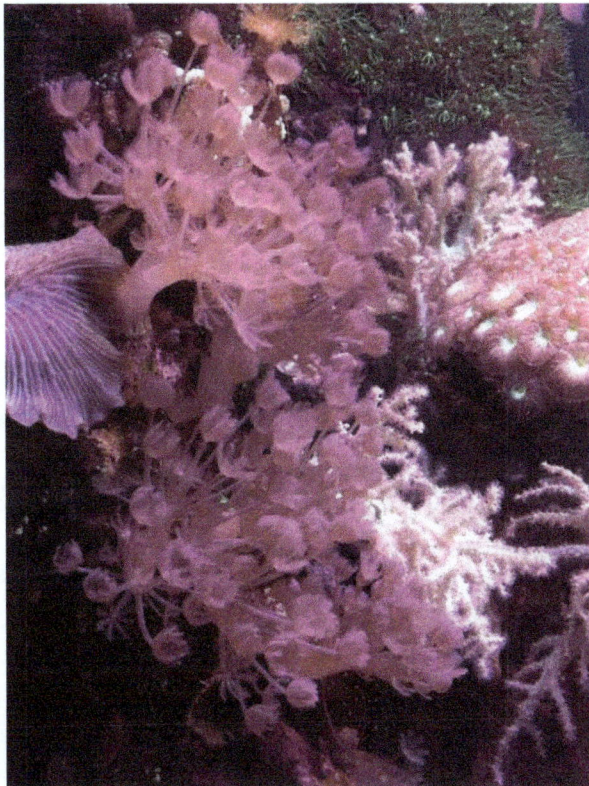

Fig. 1: Xeniid coral colonies at the Underwater Observatory, Eilat, Israel.

out of phase [12]. The pulsing motion is generated by active contraction of the muscles in the tentacles, and the expansion of the tentacles is due to passive elastic recoil. Although this behaviour is reminiscent of feeding and prey capture behaviours in other phyla like molluscs or bryozoans, past research has shown that the pulsing is linked to the removal of oxygen from the tissues [12]. This is achieved through increased mixing around the polyps and by allowing oxygen-rich water to be advected away faster than when the corals are not pulsing. Accelerating the removal of oxygen allows for the coral's symbionts to increase their photosynthetic rates, thus increasing the organism's metabolic rate.

On average, the polyp pulsing frequency is about 0.5-1 Hz, and the frequency-based Reynolds number of an individual polyp ranges from about 8 to 36 (see Section II). These corals operate at an Re_f that is much lower than most other pulsing cnidarians, including jellyfish. In particular, the pulsing soft corals operate in a much lower Re_f regime than the only other benthic cnidarian known to actively pulse to generate exchange currents, the upside-down jellyfish *Cassiopea* spp. Upside-down jellyfish host zooxanthellae in their tissues and, like corals, also benefit from their photosynthetic symbionts [13], [14]. Unlike soft corals that generate exchange currents with their tentacles, upside-down jellyfish create flow by actively contracting and relaxing their gelatinous bell. The biologically relevant Re_f for upside-down jellyfish pulsing in the benthic layer ranges from about 100 to approximately 450 (adult) [15]. As such they operate completely within the inertial range ($Re_f \gg 1$) where reciprocal motions are effective. Several experimental and computational investigations have described the fluid dynamics of upside-down jellyfish [4], [16], [17].

In this paper, we quantify the fluid dynamics of one pulsing polyp over a range of Re_f, both above and below the biologically relevant range. This fully coupled fluid-structure interaction problem is solved using the 3D immersed boundary method. We find that within the biologically relevant range, individual polyps generate a continuous upward jet using a reciprocal motion of the tentacles. This drives new fluid between the tentacles during each pulse and minimizes resampling of the same fluid volume. A slow mixing region is produced during tentacle expansion that is separated from the upward jet, which would provide sufficient time for the uptake of nutrients from the fluid and removal of waste from the tissues. Upon the next contraction, this volume of fluid is expelled and a new volume of fluid is driven between the tentacles upon the subsequent expansion. The continuous upward jet, formation of a slow mixing region during expansion, and continual flow of new fluid toward the polyp in the radial direction are not evident at $Re < 5$ when the flow becomes nearly reversible. For $Re > 40$, the magnitude of flow between the tentacles and the average vertical velocity of the upward jet is reduced.

II. METHODS

A. Coral Motion and Geometry

In this study, we use the frequency-based Reynolds number, Re_f, to describe the flows produced by the coral. The characteristic length, L_T, is set to the tentacle length and the characteristic frequency, f_{coral}, is set to the pulsation frequency. The fluid density and dynamic viscosity are set to that of sea water (see Table I).

To determine the biologically relevant range of Re_f, videos were taken of three coral colonies in the Red Sea off the coast of Eilat, Israel, and of three colonies of cultured corals in the lab. In each video, five individual polyps were tracked to determine the pulse period averaged over 20 cycles. Measurements were also taken from one tentacle on each polyp to determine the length of the tentacle. The pulsing frequency is given as a function of tentacle length in Figure 2. There was no significant correlation between pulsing frequency and size of the coral. The average Re_f was 19.64 ± 7.28 with a minimum of 8.74 and a maximum of 36.0. The average tentacle length was $(6.13 \pm 0.10) \times 10^{-3}$ m and the average pulsing frequency was 0.53 ± 0.043 Hz. For the numerical simulations performed here, we set the frequency and tentacle length to that of a typical coral where $f_{coral} = 1/1.9$ s^{-1} and $L_T = 0.0045$ m. The dynamic viscosity was varied in the simulations to study a range of Re_f, above and below that typical of soft corals. The range of Re_f studied here is 0.5, 1, 5, 10, 20, 40, and 80.

The pulsing motion of the coral was based on kinematics of five live polyps and is detailed elsewhere [18]. To summarize, the motion of the tentacles was quantified by tracking positions along a single tentacle for five pulses. Each polyp was filmed using a single Photron SA3 120K camera at either 125 or 60 frames per second in a quiescent fluid, focusing on the motion of a single tentacle that moved within the plane of focus. In each frame six approximately equispaced points were tracked along the tentacle using DLTdv5 [19]. These positions were then fit with third order polynomials. An averaged motion was constructed

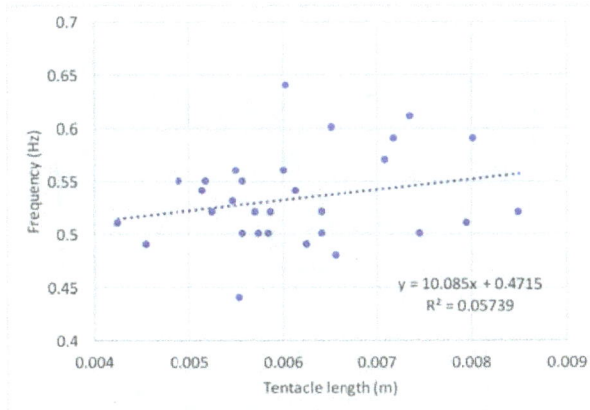

Fig. 2: Pulsing frequency vs. tentacle length for 15 corals in the field (Red Sea, Eilat, Israel) and 15 cultured corals in the lab. There is not a significant relationship between pulsing frequency and size.

Parameter	Variable	Units	Value
Domain Size	D	m	0.06
Spatial Grid Size	dx	m	$D/1024$
Lagrangian Grid Size	ds	m	$D/2048$
Time Step Size	dt	s	1.22×10^{-4}
Total Simulation Time	T	$pulses$	10
Fluid Density	ρ	kg/m^3	1000
Fluid Dynamic Viscosity	μ	$kg/(ms)$	varied
Tentacle Length	L_T	m	0.0045
Pulsing Period	P	s	1.9
Target Point Stiffness	k_{target}	$kg \cdot m/s^2$	9.0×10^{-9}

TABLE I: Numerical parameters used in the three-dimensional simulations.

by averaging the motion over the five pulses and across five polyps. The averaged motion of the tentacle was used to describe the preferred position of the immersed boundary by tethering the immersed boundary describing the tentacles to time varying target points.

The overall numerical model of the coral consisted of eight tentacles, a base, and no stem. This numerical polyp was placed in the bottom center of the computational domain (see Figure 4). Note that the presence of the stem does not significantly alter the flow and was neglected. The base of the tentacles was positioned 0.005 m above the bottom of the domain, approximating the length of the stem of the single polyp. The distance from the center of the polyp to the tip of its tentacles at full expansion was approximately 0.0045 m. The distance from the base of the polyp to the tip of the tentacles at full contraction was 0.0037 m. The length of the tentacle was determined by averaging the length measured in each frame for each polyp and then averaging over all five polyps.

The shape of each tentacle was approximated as an isosceles trapezoid with a basal width of 0.00108 m, the average width across all measured polyp tentacles. This average was found by measuring the width of the tentacle base in one frame from each video when a tentacle was parallel to the plane of focus. This distance was then used to construct the numerical tentacle. The width of the top of the tentacle was set to be one fifth of the basal width to circumvent any possible tentacle overlap when the simulated polyp is fully contracted. The average diameter of the polyp's base was measured by finding the distance between the bottom of two oppositely arranged tentacles in each frame and then averaging across all frames and all videos. This resulted in an average base diameter of 0.00106 m.

A pulsing cycle was divided into three phases as described below (see also Figure 3).

1. The coral begins with all its tentacles in an open, relaxed state. The tentacles then actively contract and the polyp closes. This takes about 28% of the pulse cycle.
2. From the contracted state, the tentacles relax back to their original expanded, resting state. The expansion phase takes about 43% of the pulse cycle.
3. The tentacles remain open and at rest for about 29% of the pulse cycle.

This process then repeats itself.

B. Numerical Method

The immersed boundary method (IB) [20] was used to solve the fully coupled fluid-structure

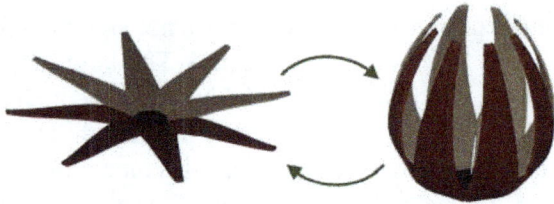

Fig. 3: A single polyp's pulsation cycle. The coral moves from its relaxed state to an actively contracted state and then relaxes back to its original, open, resting state. The two tentacle colors were chosen to differentiate the tentacles in the foreground and background.

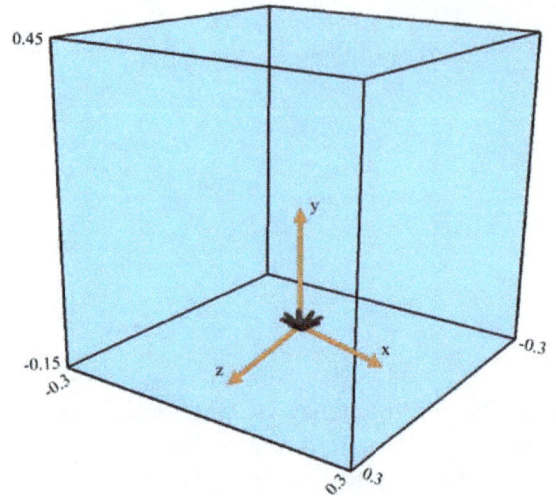

Fig. 4: The computational domain for a single coral polyp. Note that the boundaries in the x- and z-directions are periodic. The boundary conditions in the y-direction are no slip ($\mathbf{u} = 0$ at $y = -0.15$ and $y = 0.45$) .

interaction problem of a pulsing soft coral in an incompressible, viscous fluid. The IB method has been successfully applied to a variety of problems in biological fluid dynamics with an intermediate Re_f regime, i.e. $0.01 < Re_f < 1000$, including heart development [21], [22], insect flight [23], swimming [24], [25], and dating and relationships [26]. A fully parallelized implementation of the IB method with adaptive mesh refinement, IBAMR [27], was used for the simulations described here. More details on the IB method and IBAMR are found in the Appendix A.

All parameter values used in the computational model are given in Table I. A depiction of the computational domain is given in Figure 4. Note that periodic boundaries are used in the x and z directions, and no-slip conditions are used in the y-direction, corresponding to a solid boundary on the top and bottom of the domain ($\mathbf{u} = 0$ at $y = -0.15$ and $y = 0.45$). The initial conditions of the fluid are set to zero and there is no ambient flow considered. For a study including ambient flow see [28].

C. Lagrangian Coherent Structures

We computed the finite-time Lyapunov exponent (FTLE) to determine Lagrangian coherent structures (LCSs) [29], [30] using Visit 2.12.3 [31]. Within flow fields, LCSs can reveal particle transport patterns that are of potential biological importance, such as in particle capture, predator-prey interactions [32], [33], and locomotion [34].

In essence, LCSs provide a method to untangle the overall dynamics of the system in a simplified framework. Trajectories were computed using an instantaneous snapshot of the 3D vector field, and the FTLEs were computed on a regular 128^3 grid using a forward Dormand-Prince (Runge-Kutta) integrator with a relative tolerance of 0.001, an absolute tolerance of 0.0001, a maximum advection time of 0.1s, and a maximum number of steps of 1000.

III. RESULTS

Figures 5-7 show snapshots of the velocity and vorticity generated during the fourth pulsation cycle for three different numerical simulations corresponding to $Re_f = 0.5$, 10, and 80. The velocity vectors point in the direction of flow, the length of the vectors correspond to the magnitude of the flow, and the colormap corresponds to the value of the vorticity taken in the z-direction (out of plane). Both vorticity and fluid velocity were taken on a 2D plane passing through the central axis of the polyp. The tentacles are shown in pink in 3D. The snapshots taken correspond to 5%,

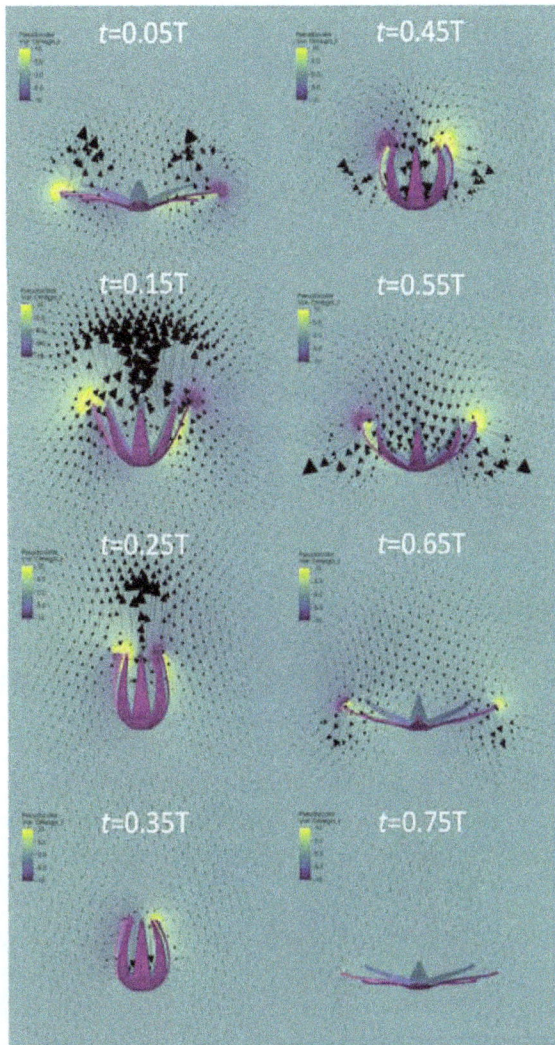

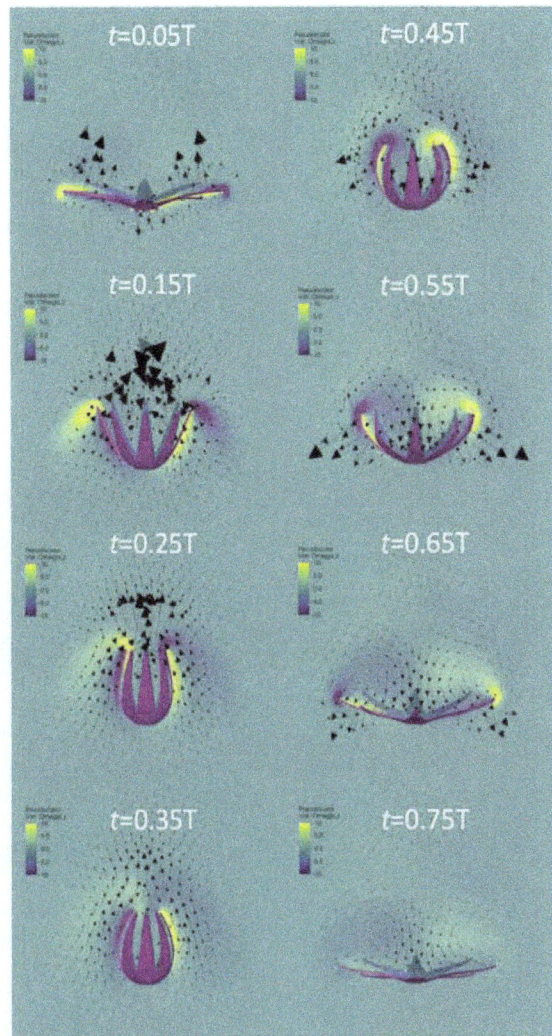

Fig. 5: The z-component of vorticity and the velocity vector field taken on a 2D plane through the central axis of the coral at $Re_f = 0.5$. This Re_f corresponds to a smaller scale than would be observed in nature. The colormap shows the value of ω_z, the arrows point in the direction of flow, and the length of the vectors correspond to the magnitude of the flow. Shapshots are taken during the fourth pulse at times that are 5%, 15%, 25%, 35%, 45%, 55%, 65%, and 75% through the cycle.

Fig. 6: The z-component of vorticity and the velocity vector field taken on a 2D plane through the central axis of the coral at $Re_f = 10$. This Re_f corresponds to a typical coral polyp.The colormap shows the value of ω_z, the arrows point in the direction of flow, and the length of the vectors correspond to the magnitude of the flow. Snapshots are taken during the fourth pulse at times that are 5%, 15%, 25%, 35%, 45%, 55%, 65%, and 75% through the cycle.

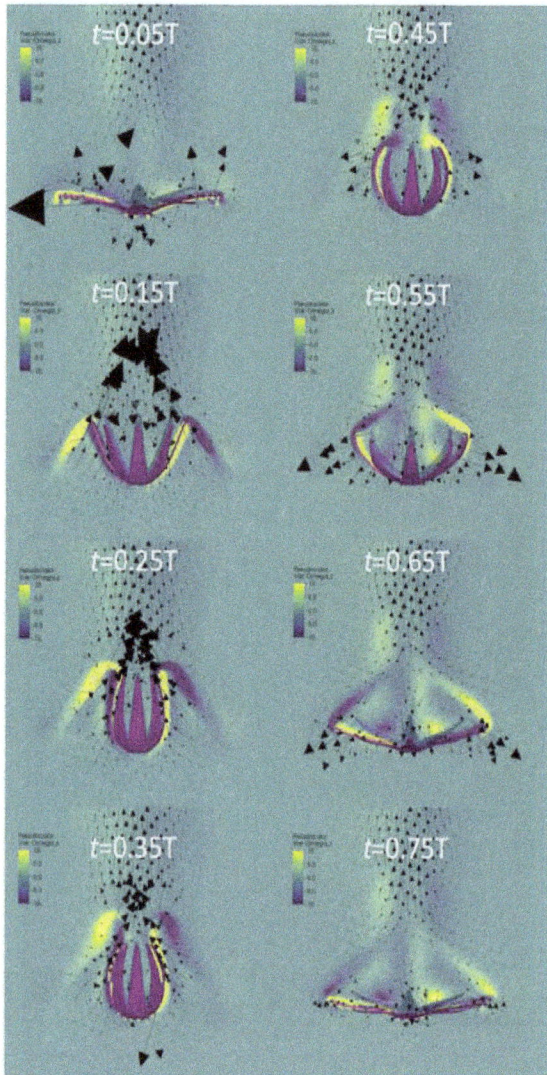

Fig. 7: The z-component of vorticity and the velocity vector field taken on a 2D plane through the central axis of the coral at $Re_f = 80$. This Re_f corresponds to a very large, fast pulsing coral polyp. The colormap shows the value of ω_z, the arrows point in the direction of flow, and the length of the vectors correspond to the magnitude of the flow. Snapshots are taken during the fourth pulse at times that are 5%, 15%, 25%, 35%, 45%, 55%, 65%, and 75% through the cycle.

15%, 25%, 35%, 45%, 55%, 65%, and 75% of the pulsing cycle such that the first three frames show the contraction phase, the next four frames show the expansion phase, and the last frame shows the polyp at rest.

During contraction, regardless of Re_f, there is a clear upwards jet. In addition, vorticity is generated at the tips of the tentacles. At the beginning of expansion ($t = 0.35T$), oppositely spinning vortices are formed at the tips of each tentacle. At higher Re_f, particularly $Re_f = 80$, the vortices formed during contraction separate from the tentacle tips and are advected upwards. The motion of these vortices helps to maintain a strong upward jet above the polyp. At the lower Re_f, (e.g. $Re_f = 0.5$), these vortices quickly dissipate. The direction of flow above the coral also reversed such that fluid is pulled downward between the tentacles. For $Re_f < 1$, the flow is nearly reversible, that is, any fluid pushed away from the polyp during contraction is pulled back during expansion. At intermediate Re_f (e.g. $Re_f = 10$), an upward jet is observed above the polyp during expansion, and fluid below this jet mixes between the tentacles.

During the resting phase (last frame), the fluid comes to rest in the lower Re_f cases. Although the strength of the upwards jet in the $Re_f = 80$ case is greatest, the magnitude of the flow between the tentacles produced by vortices generated during expansion are greater in the $Re_f = 0.5$ and 10 cases. We find strong mixing between the tentacles for $Re_f \leq 30$; this mixing decreases for $Re_f > 30$. This indicates that, near the biologically relevant Re_f, the morphology and motion of the tentacles allow for greater mixing close to the polyp itself.

To compare the relative strength of the upward jets generated by coral polyps across scales, we averaged the y-component of the velocity (in the vertical direction) within a box that was drawn from the tips of the tentacles during full contraction to one tentacle length above that point ($-0.0063m < Y < -0.0018m$). The width of the box was set equal to the diameter of the fully

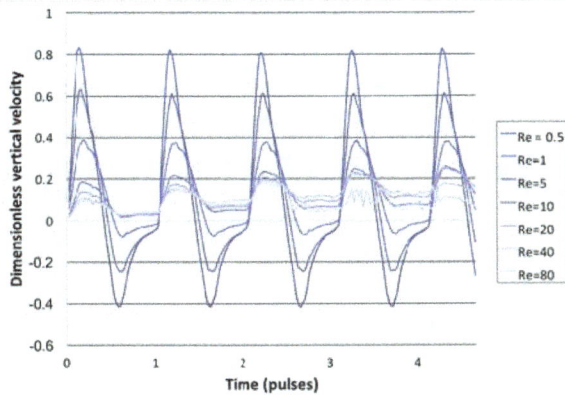

Fig. 8: The spatially averaged dimensionless vertical flow upwards over the polyp (u_y) versus time for five pulse cycles. $Re_f = 0.5$, 1, 5, 10, 20, 40, and 80 are shown. Flow velocity is nondimensionalized using the tentacle length and pulse duration.

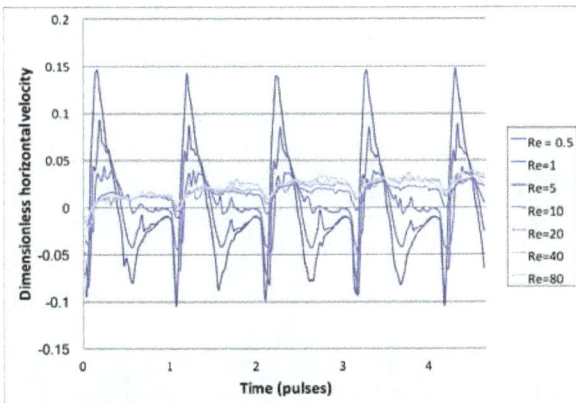

Fig. 9: The spatially averaged dimensionless horizontal flow towards the polyp (u_x) over time during five pulse cycles. $Re_f = 0.5$, 1, 5, 10, 20, 40, and 80 are shown. Velocity is given as tentacle lengths per pulse.

expanded polyp ($-0.0045m < X, Z < 0.0045m$). The average vertical velocity versus time for five pulses is shown in Figure 8 for $Re_f = 0.5$, 1, 5, 10, 20, 40, and 80. Note that the velocities are nondimensionalized by the tentacle length and pulse duration such that $U' = U/\frac{L_T}{P} = U/\frac{0.0045}{1.9}$.

Each Re_f investigated showed a peak average velocity in the upward jet that corresponds to the

end of the contraction phase. Moreover, the largest maximal peak in average velocity corresponds to the lowest $Re_f = 0.5$ case, while the lowest peak corresponds to the highest case, $Re_f = 80$. This is partially due to the fact that we average over a relatively large box. Additionally, the region of motion is larger at lower Re_f due to the relatively large boundary layers (recall that Re_f is lowered by increasing only dynamic viscosity). Immediately following contraction, as the polyp begins to expand, the average velocity drops for each Re_f. In the cases for $Re_f < 5$ there is significant backflow, where the average velocity becomes negative, reaches a minimum, and then slowly approaches zero. Around $Re_f \geq 10$ the average vertical flow decreases during tentacle expansion; however, the net average flow remains upwards. This is significant as the continuous upward jet allows new fluid to be brought to the polyp throughout the pulsing cycle.

While the transition to continuous upward flow occurs at $Re_f = 10$, for $10 \leq Re_f \leq 30$, we have also seen that the tentacle morphology allows for greater mixing near the polyp itself. This suggests that the polyp may be able to enhance its nutrient uptake or waste removal. Note that since the $Re_f = 80$ case has a continuous upward jet but little mixing near the polyp, wastes as well as nutrients would continuously be expelled away from the polyp, leaving less possibility for nutrient absorption. The opposite occurs for the case of $Re_f < 10$, where there is more mixing near the polyp, but the resulting flows are unable to remove wastes away from the polyp.

To compare the relative strength of the flow towards the polyp, we averaged the x-component of the velocity (in the horizontal direction) within a box that was drawn from the tips of the tentacles during full expansion to one tentacle length to the left of that point ($-0.009m < X < -0.0045m$), and in the z-direction, the box was drawn along the diameter of the polyp fully expanded ($-0.0045m < Z < 0.0045m$). In the vertical direction, the box was drawn from the polyp base to the top of the fully contracted tentacle ($-0.01m < Y < -0.0063m$). The average horizontal dimensionless velocity (tentacle lengths per pulse) versus time for five pulses is given in Figure 9 for $Re_f = 0.5, 1, 5, 10, 20, 40$, and 80.

For all cases of Re_f considered as the polyp begins to contract, the average flow is away from the polyp during the first 5% of the pulsation period, with the highest average velocities corresponding to the lowest Re_f, $Re_f = 0.5$. The lowest average velocity corresponds to the highest Re_f, $Re_f = 80$. The initial negative values are due to the whip-like motion of the tentacles at the beginning of contraction. Highest average velocities are seen at the lowest Re_f due to the relatively larger boundary layers. After the initial contraction motion, the average velocities become positive, indicating bulk flow towards the polyp. For all Re_f, the average velocity increases until the contraction phase is over. The highest peak average velocity, again, corresponds to the lowest Re_f, $Re_f = 0.5$; however, for $Re_f \geq 10$, their associated peaks of average velocity are almost equivalent. Moreover, for $Re_f \geq 10$, the average velocity remains towards the polyp and almost constant during the expansion and relaxation phases. At the start of the next contraction phase, the average velocity dips, once again within the first $\sim 5\%$ of the pulsation cycle. In contrast, for $Re_f \leq 5$, once the expansion phase begins, the average velocity decreases. For $Re_f \leq 1$, the average velocity decreases, reaches a minimum, and then approaches zero. In the case of $Re_f = 5$, during expansion, the average velocity monotonically decreases toward zero before the start of the next pulsation cycle.

Figure 10 shows temporally and spatially averaged flows as a function of Re_f. The vertical flow above the coral from Figure 8 is temporally averaged during the fourth pulse and plotted in Figure 10A. Figure 10B illustrates the horizontal flow in Figure 9 temporally averaged over the fourth pulse as a function of the Re_f. Both graphs highlight two flow phenomena that depend on Re_f. As the Re_f is lowered, the tentacles entrain a larger volume of fluid. This in turn leads to larger spatially averaged velocities due to the wider jet. Also as the

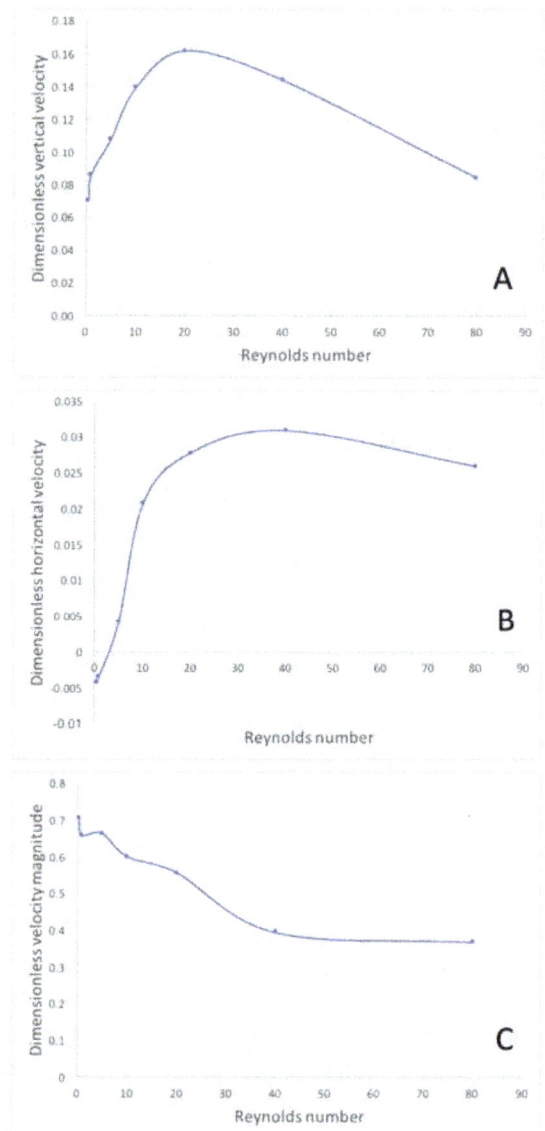

Fig. 10: Temporally and spatially averaged vertical flow above the polyp (A), horizontal flow in the x–direction towards the polyp (B), and velocity magnitude between the tentacles (C) as a function of Re_f. Note that the velocities are nondimensionalized by the tentacle length and pulse duration.

Re_f is decreased, the flow becomes increasingly reversible: the flow moves up and away from the polyp during contraction and back towards the polyp during expansion. Net volumetric flow is maximized for Re_f between about 20 and 30. Reduction in net flow is observed for $Re_f \approx 1$ and lower because the flow becomes reversible. The net flow is reduced as Re_f increases above 30 because the width of the upwards jet decreases.

As a coarse metric of the amount of mixing near the polyp, the magnitude of the velocity of the flow between the tentacles, was spatially and temporally averaged during the last pulse in a volume defined by $-0.001m < X < 0.001m$, $-0.009m < Y < -0.001m$, and $-0.001m < Z < 0.001m$. This averaged flow as a function of Re_f is shown in Figure 10C. The magnitude of flow generally decreases for increasing Re_f, suggesting that more of the fluid is directed into a narrow upward jet as the polyps grow larger. On the other hand, strong flow is generated between the tentacles at Re_f below the biologically relevant range.

A. Lagrangian Coherent Structures

Figure 11 shows contours of the logarithm of the finite-time Lyapunov exponents (FTLE) which illustrate the instantaneous Lagrangian coherent structures (LCS). The contours are shown in a 2D slice through the central axis during the fourth pulsing cycle for Re_f =0.5, 20, and 80. Note that the LCS were calculated using the entire 3D flow field. Small values of the FTLE highlight regions where flow is attractive, and large values of the FTLE indicate areas in which the flow is repelling [30]. In the case of the polyp, LCSs can be used to highlight regions of fluid that the polyp may sample or that may pass by without interacting with it.

In the biologically relevant case (B) and at higher Re_f (C), we see that fluid is pulled towards the polyp and pushed into the upward jet during the contraction phase ($t = 0.073T$ and $t = 0.17T$). The FTLE values are small between the tentacles during contraction, indicating that this fluid is pushed upward and into the vertical jet

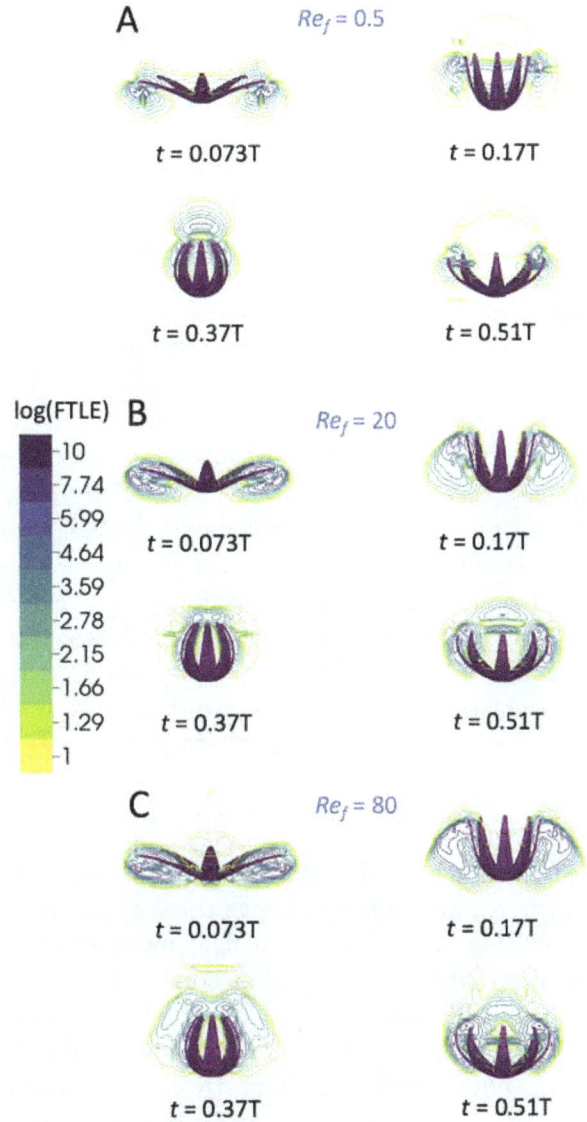

Fig. 11: Contour plot of the finite time Lyapunov exponents (FTLE) illustrating the instantaneous Lagrangian coherent structures during a single polyp's pulsing cycle for (A) $Re_f = 0.5$, (B) $Re_f = 20$ and (C) $Re_f = 80$, using a logarithmic scale.

during this phase. The large FTLE values near the tentacles show that fluid is repelling around the tentacles and the starting vortices. Comparison with the viscous dominated case at $Re_f = 0.5$ (A) shows a region of larger FTLE values between the tentacles. This indicates that the fluid near the bottom of the polyp does not mix as well with the upward jet and is not fully expelled during contraction.

During expansion ($t = 0.37T$ and $t = 0.51T$), large FTLE values directly above the polyp and between the tentacles indicate a region of mixing that is separated from the upward jet in the biologically relevant case ($Re_f = 20$). We also see larger FTLE values in the higher Re_f case (C), but now a more complicated pattern between the tentacles indicating separated mixing regions. For the viscous dominated case (A), the FTLE values are low once the tentacles have partially expanded ($t = 0.51T$). This indicates that the upward jet and the mixing region between the tentacles is no longer separated, and indeed fluid is pulled from above the polyp and into the region between the tentacles. At this Re_f, a new volume of fluid would not be sampled during each pulse.

IV. CONCLUSION

The results of this paper highlight important Re_f transitions in the exchange currents generated by pulsing soft coral. From field measurements, we determined the Re_f of a coral polyp to be 19.64 ± 7.82 with a range of about 8 to 36. In this regime, the flow around the coral polyp is defined by a continuous upward jet, nearly continual radial flow towards the polyp, a slow region of mixing between the tentacles during expansion, and the ejection of the fluid volume into the upward jet during contraction. This pattern implies that a new volume of fluid is brought to the polyp during each polyp cycle that is slow mixed around the tentacles, allowing time for the removal of oxygen from the tissues. Note that the continuous upward jet is significant since, at these scales, the polyp is able to remove waste up and away from the coral colony.

For $Re_f \leq 5$ (below the biologically relevant range), significant backflow is observed during the pulsing cycle. This would result in resampling of the same fluid and reduce waste removal and nutrient exchange. For $Re_f \geq 40$ (above the biologically relevant range), the continuous upward jet becomes narrower, reducing the net transport of fluid away from the coral. The magnitude of flow between the tentacles is also reduced, which could result in less nutrient absorption and exchange.

Spatially and temporally averaged horizontal flow towards the polyp and vertical flow above the polyp show that mass transfer is enhanced across the biologically relevant range of $8 < Re_f < 36$. Spatially and temporally averaged velocity magnitude between the tentacles show that there is less transport near the tentacle base at higher Re_f. Our limited sample of live polyps is insufficient, however to show that an active polyp may not be found at either higher or lower Re_f. Accordingly, it would be interesting to extensively search for the smallest and largest pulsing corals, calculate their effective Re_f, and determine whether or not their pulsing behavior is adapted to push the behavior into more viscous or inertial dominated regimes.

ACKNOWLEDGMENT

The authors would like to thank Uri Shavit and Roi Holzman for introducing us to pulsing soft corals and for their assistance in the field, and the organizers of the 2017 BIOMATH meeting at Kruger Park, South Africa. The authors would also like to acknowledge funding from NSF PHY grant #1505061 (to S.K.) and #1504777 (to L.A.M.), NSF DMS grant #1151478 (to L.A.M.), and NSF DMS grant #1127914 (to the Statistical and Applied Mathematical Sciences Institute). Travel support for J.E.S. was obtained from the Company of Biologists, and J.E.S. was supported by an HHMI International Student Research Fellowship and the Women Diver's Hall of Fame.

APPENDIX

A three-dimensional formulation of the immersed boundary method is discussed here. For a full review of the immersed boundary method, please see Peskin [20].

A. Governing Equations of IB

The governing equations for an incompressible, viscous fluid motion are given below:

$$\rho\left[\frac{\partial \mathbf{u}}{\partial t}(\mathbf{x}, t) + \mathbf{u}(\mathbf{x}, t) \cdot \nabla \mathbf{u}(\mathbf{x}, t)\right]$$
$$= \nabla p(\mathbf{x}, t) + \mu \Delta \mathbf{u}(\mathbf{x}, t) + \mathbf{F}(\mathbf{x}, t), \quad (2)$$

$$\nabla \cdot \mathbf{u}(\mathbf{x}, t) = 0, \quad (3)$$

where $\mathbf{u}(\mathbf{x}, t)$ is the fluid velocity, $p(\mathbf{x}, t)$ is the pressure, $\mathbf{F}(\mathbf{x}, t)$ is the force per unit area applied to the fluid by the immersed boundary, ρ and μ are the fluid's density and dynamic viscosity, respectively. The independent variables are the time t and the position $\mathbf{x}$. The variables $\mathbf{u}, p,$ and $\mathbf{F}$ are all written in an Eulerian frame on the fixed Cartesian mesh, $\mathbf{x}$.

The interaction equations, which handle the communication between the Eulerian (fluid) grid and Lagrangian (boundary) grid are written as the following two integral equations:

$$\mathbf{F}(\mathbf{x}, t) = \int \mathbf{f}(s, t) \delta\left(\mathbf{x} - \mathbf{X}(s, t)\right) dq \quad (4)$$

$$\mathbf{U}(\mathbf{X}(s, t)) = \int \mathbf{u}(\mathbf{x}, t) \delta\left(\mathbf{x} - \mathbf{X}(s, t)\right) d\mathbf{x} \quad (5)$$

where $\mathbf{f}(s, t)$ is the force per unit length applied by the boundary to the fluid as a function of Lagrangian position, s, and time, t, $\delta(\mathbf{x})$ is a three-dimensional delta function, and $\mathbf{X}(s, t)$ gives the Cartesian coordinates at time t of the material point labeled by the Lagrangian parameter, s. The Lagrangian forcing term, $\mathbf{f}(s, t)$, gives the deformation forces along the boundary at the Lagrangian parameter, s. Equation (4) applies this force from the immersed boundary to the fluid through the external forcing term in Equation (2). Equation (5) moves the boundary at the local fluid velocity. This enforces the no-slip condition. Each integral transformation uses a three-dimensional Dirac delta function kernel, δ, to convert Lagrangian variables to Eulerian variables and vice versa.

The way deformation forces are computed, e.g., the forcing term, $\mathbf{f}(s, t)$, in the integrand of Equation (4), is specific to the application. To prescribe the motion of the coral boundary, the boundary points are tethered to target points, which can be moved in a prescribed fashion. The prescribed motion of the boundary itself comes through a penalty term, tethering the Lagrangian points to the target points. The equation describing this model is

$$\mathbf{f}(s, t) = k_{targ}\left(\mathbf{Y}(s, t) - \mathbf{X}(s, t)\right), \quad (6)$$

where k_{targ} is a stiffness coefficient and $\mathbf{Y}(s, t)$ is the prescribed position of the target boundary. Note that $\mathbf{Y}(s, t)$ is a function of both the Lagrangian parameter, s, and time, t. Details on other forcing terms can be found in [26], [35].

The delta functions in these Eqs.(4-5) are the heart of the IB. In approximating these integral transformations, the following discretized and regularized delta functions, $\delta_h(\mathbf{x})$ [20], are used,

$$\delta_h(\mathbf{x}) = \frac{1}{h^3}\phi\left(\frac{x}{h}\right)\phi\left(\frac{y}{h}\right)\phi\left(\frac{z}{h}\right), \quad (7)$$

where $\phi(r)$ is defined as

$$\phi(r) = \begin{cases} \frac{1}{8}(3 - 2|r| + \sqrt{1 + 4|r| - 4r^2}), & 0 \leq |r| < 1, \\ \frac{1}{8}(5 - 2|r| + \sqrt{-7 + 12|r| - 4r^2}), & 1 \leq |r| < 2, \\ 0, & 2 \leq |r|. \end{cases}$$

B. Numerical Algorithm

As stated in the main text, we impose periodic and no slip boundary conditions on the rectangular domain . To solve Equations (2), (3),(4) and (5) we need to update the velocity, pressure, position of the boundary, and force acting on the boundary at time $n + 1$ using data from time n. The IB does this in the following steps [20], with an additional step (4b) for IBAMR [36], [27]:

Step 1: Find the force density, $\mathbf{F}^n$ on the immersed boundary, from the current boundary configuration, $\mathbf{X}^n$.

Step 2: Use Equation (4) to spread this boundary force from the Lagrangian boundary mesh to the Eulerian fluid lattice points.

Step 3: Solve the Navier-Stokes equations, Equations (2) and (3), on the Eulerian grid. Upon doing so, we are updating $\mathbf{u}^{n+1}$ and p^{n+1} from $\mathbf{u}^n$, p^n, and $\mathbf{f}^n$. Note that a staggered grid projection scheme is used to perform this update.

Step 4: (4a) Update the material positions, $\mathbf{X}^{n+1}$, using the local fluid velocities, $\mathbf{U}^{n+1}$, computed from $\mathbf{u}^{n+1}$ and Equation (5). (4b) If on a selected time-step for adaptive mesh refinement, refine the Eulerian grid in areas of the domain that contain the immersed structure or where the vorticity exceeds a predetermined threshold, .

We note that Step 4b is from the IBAMR implementation of IB. IBAMR is an IB framework written in C++ that provides discretization and solver infrastructure for partial differential equations on block-structured locally refined Eulerian grids [37], [38] and on Lagrangian meshes. Adaptive mesh refinement (AMR) achieves higher accuracy between the Lagrangian and Eulerian mesh by increasing grid resolution in areas of the domain where the vorticity exceeds a certain threshold and in areas of the domain that contain an immersed boundary. AMR improves the computational efficiency by decreasing grid resolution in areas that do not necessitate high resolution.

The Eulerian grid was locally refined near both the immersed boundaries and regions of vorticity where $|\omega| > 0.50$. This Cartesian grid was structured as a hierarchy of four nested grid levels where the finest resolved grid was assigned a resolution of $dx = D/1024$, see Table I. A 1:4 spatial step size ratio was used between each successive grid refinements. The Lagrangian spatial step resolution was chosen to be twice the resolution of the finest Eulerian grid, with $ds = D/2048$.

References

[1] E. Purcell, Life at low reynolds number, Am. J. Phys. 45 (1977) 3–11.

[2] R. H. Dillon, L. J. Fauci, X. Yang, Sperm motility and multiciliary beating: An integrative mechanical model, Computers & Mathematics with Applications 52(5) (2006) 749–758.

[3] S. D. Olson, L. J. Fauci, S. S. Suarez, Mathematical modeling of calcium signaling during sperm hyperactivation, Mol. Hum. Reprod. 17(8) (2011) 500–510.

[4] C. Hamlet, L. A. Miller, T. Rodriguez, A. Santhanakrishnan, The fluid dynamics of feeding in the upside-down jellyfish, The IMA Volumes on Mathematics and its Applications: Natural Locomotion in Fluids and Surfaces 155 (2012) 35–51.

[5] G. A. Truskey, F. Yuan, D. F. Katz, Transport Phenomena in Biological Systems, 2nd Edition, Pearson Prentice Hall, Upper Saddle River, NJ, 2004.

[6] C. S. Peskin, D. M. McQueen, Fluid dynamics of the heart and its valves, in: F. R. Adler, M. A. Lewis, J. C. Dalton (Eds.), Case Studies in Mathematical Modeling: Ecology, Physiology, and Cell Biology, Prentice-Hall, New Jersey, 1996, Ch. 14, pp. 309–338.

[7] A. Y. Cheer, M. A. Koehl, Fluid flow through filtering appendages of insects, Math. Med. and Biol.: A Journal of the IMA 4(3) (1987) 185–199.

[8] O. H. Shapiro, V. I. Fernandex, M. Garren, J. S. Guasto, F. P. Debaillon-Vesque, E. Kramarsky-Winter, A. Vardi, R. Stocker, Vortical ciliary flows actively enhance mass transport in reef corals, PNAS 111(37) (2014) 13391–13396.

[9] A. Baird, T. King, L. A. Miller, Numerical study of scaling effects in peristalsis and dynamic suction pumping, Biol. Fluid Dyn. Model. Comput. Appl. (2014) 129148.

[10] R. Holzman, V. China, S. Yaniv, M. Zilka, Hydrodynamic constraints of suction feeding in low reynolds numbers, and the critical period of larval fishes, Integrative and Comparative Biology 55 (2015) 4861.

[11] E. Lieske, R. F. Myers, Coral Reef Guide: Red Sea, 2nd Edition, Harper Collins, NY, NY, 2004.

[12] M. Kremien, U. Shavit, T. Mass, A. Genin, Benefit of pulsation in soft corals, PNAS 110(22) (2013) 8978–8983.

[13] E. H. Kaplan, R. T. Peterson, S. L. Kaplan, A Field Guide to Southeastern and Caribbean Seashores: Cape Hatteras to the Gulf Coast, Florida, and the Caribbean, 2nd Edition, Houghton Mifflin Harcourt, Boston, MA, 1988.

[14] W. K. Fitt, K. Costley, The role of temperature in survival of the polyp stage of the tropical rhizostome jellyfish cassiopea xamachana, J. Exp. Marine Biol. Ecol. 222 (1998) 79–91.

[15] C. Hamlet, Mathematical modeling, immersed boundary simulation, and experimental validation of the fluid flow around the upside-down jellyfish *Cassiopea xamachana*, Ph.D. thesis, University of North Carolina at Chapel Hill, Chapel Hill, NC (2011).

[16] C. Hamlet, L. A. Miller, Feeding currents of the upside-down jellyfish in the presence of background flow, Bull. Math. Bio. 74(11) (2012) 2547–2569.

[17] C. Hamlet, L. A. Miller, Effects of grouping behavior, pulse timing and organism size on fluidflow around the upside-down jellyfish, *Cassiopea sp.*, Biological Fluid Dynamics: Modeling,Computation, and Applications, Contemporary Mathematics, American Mathematical Society 628 (2014) 173–187.

[18] J. E. Samson, D. Ray, U. Shavit, R. Holzman, L. A. Miller, S. Khatri, Pulsing corals are efficient mesoscale mixers.

[19] T. L. Hedrick, Software techniques for two- and three-

dimensional kinematic measurements of biological and biomimetic systems, Bioinspiration and Biomimetics 3 (3) (2008) 034001.

[20] C. S. Peskin, The immersed boundary method, Acta Numerica 11 (2002) 479–517.

[21] N. A. Battista, A. N. Lane, L. A. Miller, On the dynamic suction pumping of blood cells in tubular hearts, in: A. Layton, L. A. Miller (Eds.), Women in Mathematical Biology: Research Collaboration, Springer, New York, NY, 2017, Ch. 11, pp. 211–231.

[22] N. A. Battista, A. N. Lane, J. Liu, L. A. Miller, Fluid dynamics of heart development: Effects of trabeculae and hematocrit, Math. Med. Biol.doi:10.1093/imammb/dqx018.

[23] S. K. Jones, R. Laurenza, T. L. Hedrick, B. E. Griffith, L. A. Miller, Lift- vs. drag-based for vertical force production in the smallest flying insects, J. Theor. Biol. 384 (2015) 105–120.

[24] A. P. Hoover, L. A. Miller, A numerical study of the benefits of driving jellyfish bells at their natural frequency, J. Theor. Biol. 374 (2015) 13–25.

[25] A. P. Hoover, B. E. Griffith, L. A. Miller, Quantifying performance in the medusan mechanospace with an actively swimming three-dimensional jellyfish model, J. Fluid. Mech. 813 (2017) 1112–1155.

[26] N. A. Battista, W. C. Strickland, L. A. Miller, Ib2d: a python and matlab implementation of the immersed boundary method, Bioinspir. Biomim. 12(3) (2017) 036003.

[27] B. E. Griffith, An adaptive and distributed-memory parallel implementation of the immersed boundary (ib) method (2014) [cited October 21, 2014].
URL https://github.com/IBAMR/IBAMR

[28] N. A. Battista, J. E. Samson, S. Khatri, L. A. Miller, Under the sea: Pulsing corals in ambient flow, Mathematical Methods and Models in Biosciences, Biomath Forum, Sofia (2017) 22–35 url: http://www.biomathforum.org/biomath/index.php/texts/article/view/1107.

[29] G. Haller, Lagrangian coherent structures from approximate velocity data, Phys. Fluids 14 (2002) 1851–1861.

[30] S. C. Shadden, F. Lekien, J. E. Marsden, Definition and properties of lagrangian coherent structures from finite-time lyapunov exponents in two-dimensional aperiodic flows, Physica D 212 (2005) 271.

[31] H. Childs, E. Brugger, B. Whitlock, J. Meredith, S. Ahern, D. Pugmire, K. Biagas, M. Miller, C. Harrison, G. H. Weber, H. Krishnan, T. Fogal, A. Sanderson, C. Garth, E. W. Bethel, D. Camp, O. Rübel, M. Durant, J. M. Favre, P. Navrátil, VisIt: An End-User Tool For Visualizing and Analyzing Very Large Data, in: High Performance Visualization–Enabling Extreme-Scale Scientific Insight, 2012, pp. 357–372.

[32] T. Sapsis, J. Peng, G. Haller, Instabilities on prey dynamics in jellyfish feeding, Bulletin of Mathematical Biology 73 (8) (2011) 1841–1856. doi:10.1007/s11538-010-9594-4.
URL https://doi.org/10.1007/s11538-010-9594-4

[33] J. Peng, J. O. Dabiri, Transport of inertial particles by lagrangian coherent structures: application to predator-prey interaction in jellyfish feeding, Journal of Fluid Mechanics 623 (2009) 75–84.

[34] M. M. Wilson, J. Peng, J. O. Dabiri, J. D. Eldredge, Lagrangian coherent structures in low reynolds number swimming, J. Phys.: Condens. Matter 21 (20) (2009) 204105.

[35] N. A. Battista, W. C. Strickland, A. Barrett, L. A. Miller, Ib2d reloaded: a more powerful python and matlab implementation of the immersed boundary method, arXiv: https://arxiv.org/abs/1707.06928.

[36] B. E. Griffith, Simulating the blood-muscle-vale mechanics of the heart by an adaptive and parallel version of the immsersed boundary method, Ph.D. thesis, Courant Institute of Mathematics, New York University, New York, NY (2005).

[37] M. J. Berger, J. Oliger, Adaptive mesh refinement for hyperbolic partial-differential equations, J. Comput. Phys. 53 (3) (1984) 484–512.

[38] M. J. Berger, P. Colella, Local adaptive mesh refinement for shock hydrodynamics, J. Comput. Phys. 82 (1) (1989) 64–84.

SUPPLEMENTARY MATERIAL LINKED TO THE ONLINE VERSION

1. Movie of velocity and vorticity of flow field around pulsing coral at $Re_f = 0.5$.
2. Movie of velocity and vorticity of flow field around pulsing coral at $Re_f = 10$.
3. Movie of velocity and vorticity of flow field around pulsing coral at $Re_f = 80$.

10

Operator splitting and discontinuous Galerkin methods for advection-reaction-diffusion problem: Application to plant root growth

Emilie Peynaud
CIRAD, UMR AMAP, Yaoundé, Cameroun
AMAP, University of Montpellier, CIRAD, CNRS, INRA, IRD, Montpellier, France
University of Yaoundé 1, National Advanced School of Engineering, Yaoundé, Cameroon
emilie.peynaud@cirad.fr

Abstract—Motivated by the need of developing numerical tools for the simulation of plant root growth, this article deals with the numerical resolution of the C-Root model. This model describes the dynamics of plant root apices in the soil and it consists in a time dependent advection-reaction-diffusion equation whose unique unknown is the density of apices. The work is focused on the implementation and validation of a suitable numerical method for the resolution of the C-Root model on unstructured meshes. The model is solved using Discontinuous Galerkin (DG) finite elements combined with an operator splitting technique. After a brief presentation of the numerical method, the implementation of the algorithm is validated in a simple test case, for which an analytic expression of the solution is known. Then, the issue of the positivity preservation is discussed. Finally, the DG-splitting algorithm is applied to a more realistic root system and the results are discussed.

Keywords-Time dependent advection-reaction-diffusion; Operator splitting; Discontinuous Galerkin method; Plant root growth simulation;

I. INTRODUCTION

The article is devoted to the numerical modeling of plant root growth. This work has been originally motivated by the need of developing numerical tools for the simulation of plant growth dynamics. Due to the difficulty of doing non-destructive observations of the underground part of plants (that allow to do long term studies of the dynamics of tree roots for example), mathematical models are achieving an essential role. Several theoretical and numerical challenges arise in the field of the simulation of the dynamics of plant roots [48], [47], [38], [2], [39]. The mathematical description of plant root is not trivial, due to the presence of many interactions arising in the rhizosphere and also due to the diversity of plant root types. Mathematical models based on the use of partial differential equations are useful tools to simulate the evolution of root densities in space and time [43], [44], [44], [45], [46], [41], [40], [1]. This formalism facilitates the coupling with physical models such as water and nutrient transports [42], [43], [44], [41], [49]. And the computational time for the simulation of such models is not dependent on the number of roots which is useful for applications at large scale. The C-Root model [1] is a generic model of the dynamics of root density growth. This model takes only one unknown which is related to root densities such as the density of apices, root length density or biomass density. It has only three parameters. The model is said to be generic in the sense that it can apply to a wide variety of root system types. The model consists in a single time-dependent advection-reaction-diffusion equation, and one of the challenge is to numerically solve the equation. In [1] and [2] the authors solved the problem with the finite difference method on Cartesian mesh grids combined with an operator splitting technique. Unfortunately, Cartesian mesh grids do not allow easily to mesh complex soil geometries. From the theoretical and computational point of view, Cartesian grids also lead to difficulties for a rigorous study and validation of the model. That is why this article focus on the development and implementation of a suitable numerical method for the resolution of the C-Root model on triangular mesh grids, that allow to mesh complex geometries. However, one of the main difficulties in the C-root model is that the advection and diffusion terms are not always of the same order of magnitude. It depends on the phase of the root system development [2]. As a result, the properties of the equation may vary along the simulation: it can be either close to a hyperbolic problem or close to a parabolic problem.

In a previous work [3], the use of the Discontinuous Galerkin method has been implemented and validated. Indeed, the usual choice of the classical Lagrange finite element method suffers from a lack of stability when the advection term is dominant [4]. For this reason, we implemented a discontinuous Galerkin (DG) method for both the advection and diffusion terms. All the three operators where solved simultaneously using the same time approximation scheme (θ-scheme).

However, as explained in [6], for multi-biophysic problems it is not efficient to use the same numerical scheme for the different operators of the system. For example, we may want to use the Euler explicit scheme for the advection term and an Euler implicit scheme for the diffusion. The operator splitting technique [7], [8] is a well known alternative for the resolution of equations having a multi-biophysic behaviour that allows the use of different time schemes for each operator of the equation. The idea of the splitting technique is to split the problem into smaller and simpler parts of the problem so that each part can be solved by an efficient and suitable time scheme. This methods has been used for a wide range of applications dealing with the advection-reaction-diffusion equation [9]. Operator splitting techniques have been extensively used in combination with finite difference methods [10], [2], finite volume methods [11], [12] but also with Continuous Galerkin methods [13], [14], [15], [16], [17]. To the best of my knowledge, only very few articles deal with the use of the operator splitting technique in combination with the discontinuous Galerkin approximation [18], [19], [20], [21]. In this paper, we present a new application of the operator splitting technique combined with discontinuous finite elements.

The paper is structured as follows. In section II, the C-root growth model [1], [2], [3] is briefly described. An analysis is also provided, where I showed the existence and uniqueness of a positive real solution. In section III, the splitting operator technique is introduced and applied to the C-Root model, combined with the use of discontinuous Galerkin approximations. In section IV, the algoritm is implemented and validated using a simple test case for which an analytic expression of the solution is known. As an application, I provide simulations of the development of eucalyptus roots in section V. Finally, the paper ends with a conclusion and further improvements.

II. THE MODEL

A. Modelling root growth with PDE: the C-Root model

The C-Root model [1] was developed to simulate the growth of dense root networks, usually composed of fine roots, with negligible secondary thickening. As presented in [1], the unknown variable u is the number of apices per unit volume, but it can also stand for the density of fine root biomass. The soil is considered as a subdomain of $\mathbb{R}^d$ (with $d = 1, 2$ or 3). It is assumed that Ω has smooth boundaries (Lipschitz boundaries) denoted $\partial\Omega$. The C-Root model combines advection, diffusion and reaction, which aggregate the main biological processes involved in root growth, such as primary growth, ramification and root death. The reaction operator gives the quantity of apices (or root biomass) produced in time, whereas advection and diffusion operators spatially distribute the whole apices (or biomass) in the domain.

The reaction operator describes the evolution in time of the root biomass in a given domain. In the C-Root model it is a linear term characterized by the scalar parameter ρ which is the growth rate of the root system. The diffusion corresponds to the spread of the root biomass over space. It is described by the parameter σ which is a $d \times d$ matrix that characterizes the growth of the root biomass in any direction exploiting free space in the soil. The advection corresponds to the displacement of the root biomass in a direction and velocity given by $\mathbf{v}$ which is a vector in $\mathbb{R}^d$.

On the boundaries of Ω, what happens for the quantity being transported is different depending if the growth makes the roots to come inside Ω or to go outside of Ω. If $\mathbf{v}$ is going inside Ω (at the inlet boundary) the root biomass u will enter the domain and increase. On edges where $\mathbf{v}$ is going out of the domain (outlet boundary) the root biomass u is going to be pushed out of Ω. Since this phenomena is oriented (causality) and the behaviour of the solution is different on inlet and outlet boundaries, we need to specify in the model these parts of the boundaries. Mathematically, it is required to define the inlet boundary with respect to $\mathbf{v}$ as

$$\partial\Omega^- = \{\mathbf{x} \in \partial\Omega : (\mathbf{v} \cdot \mathbf{n})(\mathbf{x}) < 0\}. \quad (1)$$

The outlet boundary Ω^+ is given by $\partial\Omega^+ = \partial\Omega \backslash \partial\Omega^-$. The dynamics of the root system is studied between the time t_0 and t_1 with $0 \leq t_0 < t_1$. The problem reads as follow: find u such that

$$\begin{cases} \partial_t u + \mathbf{v} \cdot \nabla u - \nabla \cdot (\sigma \nabla u) + \rho u = 0 \\ \qquad\qquad\qquad\qquad \text{in }]t_0, t_1[\times \Omega \\ u(t_0) = u_0 \text{ at } \{t_0\} \times \Omega \\ \mathbf{n} \cdot \sigma \nabla u = g \text{ on }]t_0, t_1[\times \partial\Omega \\ (\mathbf{n} \cdot \mathbf{v})u = g_{in} \text{ on }]t_0, t_1[\times \partial\Omega^- \end{cases} \quad (2)$$

where $g \in L^2(\partial\Omega)$ and $g_{in} \in L^2(\partial\Omega^-)$ are given. And u_0 is the given initial solution.

Problem (2) is known as the time dependent advection- reaction-diffusion problem and belongs to the class of parabolic partial differential equations. This equation is a model problem that often occurs in fluid mechanics but also in many other applications in life sciences (see for instance [22], [23], [24]).

Depending on the boundary conditions, the problem has different meanings. To simplify the presentation we only consider the Neumann boundary condition combined with an inlet boundary condition at the inlet of the domain. The Neumann condition specifies the value of the normal derivative of the solution at the boundary of the domain. The inlet boundary condition specifies the quantity of u convected by $\mathbf{v}$ that enters in the domain.

B. The weak problem

Since the goal is to solve the problem on unstructured meshes, the spatial operators are approximated using finite element methods. Within this framework, it is classical to write the problem in a variational form. Let us first introduce some functional spaces [50].

- The space $H^1(\Omega)$ defined such that $H^1(\Omega) = \{v \in L^2(\Omega) : \nabla v \in L^2(\Omega)\}$ is a Hilbert space when equipped with the norm $\| \cdot \|_{1,\Omega}$. We recall that $\forall v \in H^1(\Omega)$, $\|v\|_{1,\Omega} = (v, v)_{1,\Omega}$

and the scalar product $(\cdot, \cdot)_{1,\Omega}$ is defined by $\forall v \in H^1(\Omega)$,

$$(u, v)_{1,\Omega} = \int_\Omega uv \, dx + \int_\Omega \nabla u \cdot \nabla v \, dx.$$

- We denote $L^2(]t_0, t_1[, H)$ the space of H-valued functions whose norm in H is in $L^2(]t_0, t_1[)$. This space is a Hilbert space for the norm

$$\|u\|_{L^2(]t_0,t_1[,H)} = \left(\int_{t_0}^{t_1} \|u(t)\|_H^2 \right)^{1/2}.$$

- Let $B_0 \subset B_1$ be two reflexive Hilbert spaces with continuous embedding, we denote $\mathcal{W}(B_0, B_1)$ the space of functions $v :$ $]t_0, t_1[\longrightarrow B_0$ such that $v \in L^2(]t_0, t_1[, B_0)$ and $d_t v \in L^2(]t_0, t_1[, B_1)$. Equipped with the norm

$$\begin{aligned} \|u\|_{\mathcal{W}(B_0,B_1)} &= \|u\|_{L^2(]t_0,t_1[,B_0)} \\ &\quad + \|d_t u\|_{L^2(]t_0,t_1[,B_1)}, \end{aligned}$$

the space $\mathcal{W}(B_0, B_1)$ is a Hilbert space [25].

Using the previous functional spaces, I now define the problem in the following weak form: Find u in W such that $\forall v \in H$

$$\langle d_t u(t), v(t) \rangle_{H', H} + a(t, u, v) = \ell(t, v) \text{ a.e. } t \in]t_0, t_1[$$
$$u(t_0) = u_0, \tag{3}$$

where $W = \mathcal{W}(H^1(\Omega), (H^1(\Omega))')$ and $H = H^1(\Omega)$ and

$$\ell(t, v) = \int_{\partial\Omega} g(t) v \, d\gamma \tag{4}$$

$$a(t, u, v) = a_A(t, u, v) + a_D(t, u, v) + a_R(t, u, v) \tag{5}$$

with

$$a_A(t, u, v) = \int_\Omega v(\mathbf{v}(t, \mathbf{x}) \cdot \nabla u) \, dx, \tag{6}$$

$$a_D(t, u, v) = \int_\Omega \nabla v \cdot \sigma(t, \mathbf{x}) \cdot \nabla u \, dx, \tag{7}$$

$$a_R(t, u, v) = \int_\Omega \rho(t) uv \, dx. \tag{8}$$

One can prove that problems (3) and (2) are equivalent almost everywhere in $]t_0, t_1[\times\Omega$. Let us

assume that there is a constant $\sigma_0 > 0$ such that

$$\forall \xi \in \mathbb{R}^d, \ \sum_{i,j=1}^{d} \sigma_{ij} \xi_i \xi_j \geq \sigma_0 \|\xi\|_d^2 \text{ a.e. in } \Omega. \tag{9}$$

In addition, I assume that

$$\inf_{\mathbf{x}\in\Omega} \left(\sigma - \frac{1}{2}(\nabla \cdot \mathbf{v}) \right) > 0 \text{ and } \inf_{\mathbf{x}\in\partial\Omega} (\mathbf{v} \cdot \mathbf{n}) \geq 0. \tag{10}$$

Under assumption (9) and (10), one can prove that the problem is well-posed for sufficiently smooth $\mathbf{v}$, σ and ρ (see for instance [25]).

C. The positivity preserving property of the solution

In the framework of our applications to the simulation of root biomass densities one of the crucial property of the problem is the preservation of the positity of the solution along time. For a positive initial solution u_0, the solution of (3) stays positive.

Proposition II-C.1. Let $u_0 \in L^2(\Omega)$ and $f \in L^2(]t_0, t_1[, L^2(\Omega))$. We consider u the solution of (3) in W. We assume that $u_0(\mathbf{x}) \geq 0$ a.e. in Ω and $g(t, \mathbf{x}) \geq 0$ a.e. in $]t_0, t_1[\times\partial\Omega$. Then $u(t, \mathbf{x}) \geq 0$ a.e in $]t_0, t_1[\times\Omega$.

Proof: I follow [25]. See also [26], [27]. We consider the function u^- defined by

$$u^- = \frac{1}{2}(|u| - u).$$

Let us note that

$$u^- = \begin{cases} 0 \text{ a.e in }]t_0, t_1[\times\Omega, & \text{if } u \geq 0 \text{ a.e in }]t_0, t_1[\times\Omega, \\ -u \text{ a.e in }]t_0, t_1[\times\Omega, & \text{if } u < 0 \text{ a.e in }]t_0, t_1[\times\Omega. \end{cases}$$

That is we have

$$u^- \geq 0 \text{ a.e. in }]t_0, t_1[\times\Omega. \tag{11}$$

We verify that u^- is an admissible test function in W. Using the following obvious equations

$$(\nabla|u|)^2 = (\nabla u)^2$$
$$\nabla|u| \cdot \nabla|u| = \nabla u \cdot \nabla u$$
$$u\nabla|u| = |u|\nabla u$$
$$u\nabla u = |u|\nabla|u|$$

that are valid a.e in $]t_0, t_1[\times\Omega$ we can verify that

$$a(t, u^-, u^-) = -a(t, u, u^-).$$

By adding the same quantity on both sides of the equation we get

$$\langle d_t u^-, u^-\rangle + a(t, u^-, u^-) = \langle d_t u^-, u^-\rangle - a(t, u, u^-).$$

Since u satisfy (3) we have

$$\langle d_t u^-, u^-\rangle + a(t, u^-, u^-) = \langle d_t u^-, u^-\rangle + \langle d_t u, u^-\rangle - \ell(t, u^-).$$

One can notice that $\langle d_t u^-, u^-\rangle + \langle d_t u, u^-\rangle = 0$. Then we have

$$\frac{1}{2}d_t\|u^-\|_{0,\Omega}^2 + a(t, u^-, u^-) = -\ell(t, u^-) \leq 0,$$

with $g(t, \mathbf{x}) \geq 0$ a.e in $]t_0, t_1[\times\partial\Omega$. Now from the coercivity of the bilinear form a we obtain

$$\frac{1}{2}d_t\|u^-\|_{0,\Omega}^2 + c\|u^-\|_{0,\Omega}^2$$
$$\leq \frac{1}{2}d_t\|u^-\|_{0,\Omega}^2 + a(t; u^-, u^-) \leq 0,$$

where c is a strictly positive constant. The estimate is then

$$\frac{1}{2}d_t\|u^-\|_{0,\Omega}^2 \leq -c\|u^-\|_{0,\Omega}^2.$$

By the Gronwall lemma we have that

$$\forall t \in [t_0, t_1] \times \Omega, \|u^-(t)\|_{0,\Omega}^2 \leq e^{-2ct}\|u^-(0)\|_{0,\Omega}^2.$$

Since $c > 0$ and $t \geq t_0 \geq 0$, we have that $e^{-2ct} \leq 1$, so we obtain

$$\forall t \in [t_0, t_1] \times \Omega, \|u^-(t)\|_{0,\Omega}^2 \leq \|u^-(0)\|_{0,\Omega}^2.$$

Since $u(0) = u_0 \geq 0$ a.e in Ω we have $u^-(0) = 0$ a.e in Ω. So we deduce that

$$\forall t \in [t_0, t_1] \times \Omega, \|u^-(t)\|_{0,\Omega}^2 \leq 0.$$

But from the definition of u^- we have $u^- \geq 0$ a.e in $]t_0, t_1[\times\Omega$. So we deduce that $\|u^-(t)\|_{0,\Omega}^2 = 0$ and thus $u^-(t) = 0$ a.e in $]t_0, t_1[\times\Omega$. It means that $u \geq 0$ a.e in $]t_0, t_1[\times\Omega$ by definition of u^-.

III. APPROXIMATION OF THE MODEL

A. The operator splitting technique

Here we focus on the implementation of the operator splitting technique. The time interval $[t_0, t_1]$ is divided in N subspaces of size δt such that $[t_0, t_1] = \cup_{n=1,N}]t_n, t_{n+1}[$ with $\cap_{n=1,N}]t_n, t_{n+1}[= \emptyset$. At each iteration step we solve the following problems

- Find $u_A \in H$ such that $\forall v \in H$, for a.e $t \in]t_n, t_{n+1}[$,

$$\langle d_t u_A(t), v(t)\rangle_{H',H} + a_A(t, u_A, v) = 0$$
$$u_A(t_n) = u(t_n).$$

- Find $u_D \in H$ such that $\forall v \in H$, for a.e $t \in]t_n, t_{n+1}[$,

$$\langle d_t u_D(t), v(t)\rangle_{H',H} + a_D(t, u_D, v) = \ell(t, v)$$
$$u_D(t_n) = u_A(t_{n+1}).$$

- Find $u_R \in H$ such that $\forall v \in H$, for a.e $t \in]t_n, t_{n+1}[$,

$$\langle d_t u_R(t), v(t)\rangle_{H',H} + a_R(t, u_R, v) = 0$$
$$u_R(t_n) = u_D(t_{n+1}).$$

- Set $u(t_{n+1}) = u_R(t_{n+1})$.

The bilinear forms $a_A(t, u, v)$, $a_D(t, u, v)$ and $a_R(t, u, v)$ are respectively given by (6), (7) and (8). And $\ell(t, v)$ is the linear form (4). If the operators are commutative, then the splitting error vanishes. Otherwise, if the operators are not commutative, then the splitting error does not vanish and a second order splitting would be required (see [6]). In the following, I present the different schemes related to each operator.

B. The advection step: DG upwind scheme

The advection step consists in solving the following transport problem : Find u such that $\forall v \in H$, for a.e $t \in]t_n, t_{n+1}[$,

$$\langle d_t u(t), v(t)\rangle_{H',H} + a_A(t, u, v) = 0 \qquad (12)$$
$$u_A(t_n) = u_R(t_n) \qquad (13)$$

where $a_A(t, u, v)$ is the bilinear form (6). For the space approximation of this problem, we implemented the DG upwind method presented below.

Let $\mathcal{T}_h$ be a regular family of decomposition in triangles of the domain Ω such that

$$\Omega = \bigcup_{i=1}^{N} \bar{K}_i \text{ and } K_i \cap K_j = \emptyset, \forall i \neq j.$$

The h subscript in $\mathcal{T}_h$ denotes the size of the mesh cells and it is defined by

$$h = \max_{K \in \mathcal{T}_h} h_K$$

where h_K is the diameter of the element K. Let $\mathcal{E}_h$ be the set of edges of the elements of $\mathcal{T}_h$. Among the elements of $\mathcal{E}_h$ we denote by $\mathcal{E}_h^b$ the set of edges belonging to $\partial \Omega$. The sets $\mathcal{E}_h^{b,-}$ and $\mathcal{E}_h^{b,+}$ are the sets of edges belonging to $\partial\Omega^-$ and $\partial\Omega^+$ respectively. And $\mathcal{E}_h^i$ is the set of interior edges. Let us consider an element of $\mathcal{E}_h^i$. We denote by T^+ and T^- the two mesh elements sharing the edge e so that $e = \partial T^+ \cap \partial T^-$ where the minus and plus superscripts depend on the direction of the advection vector. By convention we suppose that $\mathbf{v}$ goes from T^- to T^+ that is $\mathbf{v} \cdot \mathbf{n}_e^+ < 0$ and $\mathbf{v} \cdot \mathbf{n}_e^- > 0$ where $\mathbf{n}_e^+$ (resp. $\mathbf{n}_e^-$) is the outward normal vector of e in T^+ (resp. T^-). When it is not necessary to distinguish the orientation of the normal vectors $\mathbf{n}_e^+$ and $\mathbf{n}_e^-$ we denote by $\mathbf{n}$ the unitary normal of e.

Let us consider the advection problem on each element K_i of the domain : for all K_i, $i = 1, N$ we look for u the solution of the equation (12) defined on K_i. Similarly to the problem defined on all the domain Ω, we look for a solution u that is in $L^2(K_i)$ and such that ∇u is in $L^2(K_i)$ for all K_i in $\mathcal{T}_h$. Let us introduce the following broken Sobolev space:

$$H^1(\mathcal{T}_h) = \{ v \in L^2(\Omega) : \nabla v \in L^2(K_i)$$
$$\text{and } v \in H^{1/2+\varepsilon}(K_i), \forall K_i \in \mathcal{T}_h \}$$

with ε a positive real number. The trace of the functions of $H^1(\mathcal{T}_h)$ are meaningful on $e \subset K_i$, $\forall K_i \in \mathcal{T}_h$. The functions v of $H^1(\mathcal{T}_h)$ have two traces along the edges e. We denote v_e^+ the trace of v along e on the side of triangle T^+ and v_e^- the trace of v along e on the side of T^-. On edges

that are subsets of $\partial\Omega$ the trace is unique and we can note

$$v_e^+ = v \text{ if } e \in \mathcal{E}_h^{b,-} \text{ and } v_e^- = v \text{ if } e \in \mathcal{E}_h^{b,+},$$

and by convention, we set

$$v_e^- = 0 \text{ if } e \in \mathcal{E}_h^{b,-} \text{ and } v_e^+ = 0 \text{ if } e \in \mathcal{E}_h^{b,+}.$$

The jump of functions of $H^1(\mathcal{T}_h)$ across the internal edge e is defined by:

$$[\![v]\!] = v_e^+ - v_e^-, \forall e \in \mathcal{E}_h^i.$$

For edges belonging to the boundary of Ω we take

$$[\![v]\!] = v_e, \forall e \in \mathcal{E}_h^{b,-} \text{ and } [\![v]\!] = -v_e, \forall e \in \mathcal{E}_h^{b,+},$$

with v_e the trace of v along e. The mean value of u on e is defined by

$$\{\!\{v\}\!\} = \frac{1}{2}(v_e^+ + v_e^-), \forall e \in \mathcal{E}_h^i.$$

Besides for edges on the boundaries we take

$$\{\!\{v\}\!\} = v_e, \forall e \in \mathcal{E}_h^b.$$

Let us denote by $\mathcal{X}$ the functional space defined such that

$$\mathcal{X} = \{ v :]t_0, t_1[\longrightarrow H^1(\mathcal{T}_h) :$$
$$v \in L^2(]t_0, t_1[, H^1(\mathcal{T}_h));$$
$$\text{and } d_t v \in L^2(]t_0, t_1[, H^1(\mathcal{T}_h)') \}.$$

This space is a Hilbert space equipped with the norm

$$\|v\|_{\mathcal{X}} = \|v\|_{L^2(]t_0,t_1[, H^1(\mathcal{T}_h))} + \|v\|_{L^2(]t_0,t_1[, H^1(\mathcal{T}_h)')}.$$

The DG variational formulation of the advection step written on the broken Sobolev space takes the following form: Find u in $\mathcal{X}$ such that for a.e $t \in]t_0, t_1[$, $\forall v \in H^1(\mathcal{T}_h)$

$$\langle d_t u(t), v(t) \rangle_{H^1(\mathcal{T}_h)', H^1(\mathcal{T}_h)} + a_h^{up}(t; u, v) = \ell_h^{up}(t; v),$$

$$u(t_0) = u_0,$$

where the form $a_h^{up}(t; u, v)$ is the approximation of the advection term. It consists in the upwind formulation of the DG method [28]. It reads:

$$a_h^{up}(t; u, v) = \sum_{K \in \mathcal{T}_h} \int_K u(\rho v - \mathbf{v} \cdot \nabla v) dx$$
$$- \sum_{e \in \mathcal{E}_h^{b,\pm} \cup \mathcal{E}_h^{b,+}} \int_e |\mathbf{v} \cdot \mathbf{n}_e^+| u_e^- [\![v]\!] ds. \quad (14)$$

The approximated linear form of the the right hand side reads

$$\ell_h^{up}(t; v) = - \sum_{e \in \mathcal{E}_h^{b,-}} \int_e (\mathbf{v} \cdot \mathbf{n}_e^+) g_{in} v_e^+ ds.$$

The DG-formulation (14) is consistent and stable, see for example [32]. The discontinuous Galerkin method consists in searching the solution in the approximation space $\mathcal{X}_h$ defined such that

$$\mathcal{X}_h = \left\{ v :]t_0, t_1[\longrightarrow W_h^k; v \in L^2(]t_0, t_1[, W_h^k);$$

$$\text{and } d_t v \in L^2(]t_0, t_1[, (W_h^k)')) \right\},$$

where W_h^k is given by

$$W_h^k = \left\{ v_h \in L^2(\Omega); \forall K \in \mathcal{T}_h, v_h|_K \in \mathbb{P}^k \right\}.$$

Let us note that the functions of W_h^k can be discontinuous from one element of the mesh to the other. Let us note that W_h^k is embedded in $H^1(\mathcal{T}_h)$ so that $\mathcal{X}_h \subset \mathcal{X}$. This problem can be written in a matrix form. Let us denote $(\lambda_i)_{i=1,n}$ the basis of the finite dimensional subspace W_h^k where $n = \dim(W_h^k)$. In this basis the approximated solution takes the form:

$$u_h(t, x, y) = \sum_{i=1}^n \xi_i(t) \lambda_i(x, y),$$

where the $\xi_i(t)$ are the degrees of freedom. Let us define X the vector of degrees of freedom:

$$X(t) = (\xi_1(t), \ldots, \xi_n(t))^T.$$

The approximated problem then reduces to find $X(t) \in [\mathcal{C}^2(0, T)]^n$ such that

$$\mathbf{M} \frac{dX(t)}{dt} + \mathbf{A}^{up}(t) X(t) = L_h^{up}(t)$$

$$\mathbf{M} X(0) = \mathbf{M} X_0$$

where $\mathbf{M}$ and $\mathbf{A}^{up}(t)$ are two matrices defined such that

$$\mathbf{M} = (M_{i,j})_{i,j} \text{ and } M_{i,j} = \sum_{T \in \mathcal{T}_h} \int_K \lambda_i \lambda_j dx,$$

$$(15)$$

$$\mathbf{A}^{up} = \left(A_{i,j}^{up} \right)_{i,j} \text{ and } A_{i,j}^{up} = a_h^{up}(t; , u, v), \quad (16)$$

and $L_h^{up}(t)$ is the vector of size n defined such that $\left(L_h^{up}(t) \right)_i = \ell_h^{up}(t; \lambda_i)$ for $i = 1, n$. The problem reduces to a linear system of ordinary differential equations. The time approximation is based on a finite difference scheme.

At each iteration step we solve the following problem: Find $X^{N+1} \in \mathbb{R}^n$ such that

$$\frac{1}{\delta t} \mathbf{M} \left(X^{N+1} - X^N \right)$$

$$+ (1 - \theta) \mathbf{A}^{up} X^N + \theta \mathbf{A}^{up} X^{N+1} \quad (17)$$

$$= (1 - \theta) L_h^{up,N} + \theta L_h^{up,N+1}$$

and $\mathbf{M} X^0 = \mathbf{M} X_0,$

where θ is a real parameter taken in $[0, 1]$. For $\theta = 0$, we have the explicit Euler schema. For $\theta = 1$, it is the implicit Euler schema. For $\theta = 1/2$, it is the Crank-Nicolson schema.

C. The diffusion step

The diffusion step consists in solving the following problem : Find u such that $\forall v \in H$, for a.e $t \in]t_n, t_{n+1}[$,

$$\langle d_t u(t), v(t) \rangle_{H', H} + a_D(t; u, v) = \ell(t; v)$$

$$u(t_n) = u_A(t_n)$$

where $a_D(t; u, v)$ is the bilinear form (7) and $\ell(t; v)$ is the linear form (4). In the setting introduced before, the DG variational formulation of the diffusion step written in the broken Sobolev space takes the following form: Find u in $\mathcal{X}$ such that $\forall v \in H^1(\mathcal{T}_h)$, for a.e. $t \in]t_0, t_1[$

$$\langle d_t u(t), v \rangle_{H^1(\mathcal{T}_h)', H^1(\mathcal{T}_h)} + a_h^{ip}(t; u, v) = \ell_h^{ip}(t; v)$$

$$u(t_0) = u_0.$$

The form $a_h^{ip}(t; u, v)$ is the approximation of the diffusion term. It consists in the interior penalty formulation (IP) that reads

$$a_h^{ip}(t; u, v) = \sum_{K \in \mathcal{T}_h} \int_K \sigma \nabla u \cdot \nabla v \, dx$$

$$- \sum_{e \in \mathcal{E}_h^i} \int_e \{\sigma \nabla u\} \cdot \mathbf{n}_e^+ [\![v]\!] \, ds$$

$$+ \sum_{e \in \mathcal{E}_h^i} \int_e \{\sigma \nabla v\} \cdot \mathbf{n}_e^+ [\![u]\!] \, ds$$

$$+ \sum_{e \in \mathcal{E}_h^i} \frac{\eta}{h_e} \int_e [\![u]\!][\![v]\!] \, ds,$$

where η is a positive penalization factor. The linear form $\ell_h^{ip}(t; v)$ is given by $\ell_h^{ip}(t; v) = \sum_{e \in \mathcal{E}_h^b} \int_e gv \, ds$. This formulation was introduced in [31] and is known as the non-symmetric interior penalty (NSIP) formulation, see [30], [32]. In matrix form the problem reduces to find $X(t) \in [\mathcal{C}^2(0, T)]^n$ such that

$$\mathbf{M} \frac{dX(t)}{dt} + \mathbf{A}^{ip}(t)X(t) = L_h^{ip}(t)$$

$$\mathbf{M}X(0) = \mathbf{M}X_0$$

where $\mathbf{M}$ is defined by (15) and $\mathbf{A}^{ip}$ is defined such that

$$\mathbf{A}^{ip} = \left(A_{i,j}^{ip} \right)_{i,j} \quad \text{and} \quad A_{i,j}^{up} = a_h^{ip}(t;, u, v).$$

The vector $L_h^{ip}(t)$ is such that $\left(L_h^{ip}(t) \right)_i = \ell_h^{ip}(t; \lambda_i)$ for $i = 1, n$. Similarly to the advection step, the time approximation of the problem is based on a finite difference scheme of the form (17).

D. The reaction step

The reaction step consists in solving the following problem : Find u such that $\forall v \in H$, for a.e. $t \in]t_n, t_{n+1}[$

$$\langle d_t u(t), v(t) \rangle_{H', H} + a_R(t; u, v) = 0$$

$$u(t_n) = u_D(t_n)$$

where $a_R(t; u, v)$ is the bilinear form (8). This problem takes the following matrix form find

$X(t) \in [\mathcal{C}^2(0, T)]^n$ such that

$$\frac{dX(t)}{dt} + \rho X(t) = 0$$

$$X(0) = X_0$$

where we recall that ρ is a constant real parameter. This problem can be solved by an exact scheme (a kind of schemes that provide exact solutions, i.e. a solution equal to the analytical solution). At each iteration we find X^{N+1} such that

$$\frac{1}{\Phi(\delta t)} \left(X^{N+1} - X^N \right) = -\rho X^N$$

with $\Phi(\delta t) = \frac{1}{\rho}(1 - \exp(-\rho \delta t))$. This scheme is unconditionally stable, meaning that we can choose the time step independently from the space step. It is also positively stable, meaning that if $X^N \geq 0$ so is X^{N+1}.

IV. VALIDATION OF THE SPLITTING ALGORITHM WITH A SIMPLE TEST CASE

Problem (3) has been already solved using discontinuous Galerkin elements (DG) [3]. Advection and Diffusion operators were solved simultaneously using the Crank-Nicolson scheme providing stable results. However, even for simple test cases some simulations did not always provide positive numerical solutions. One reason is that the same time approximation scheme is not necessarily suitable for both the advection and for the diffusion. That is why a new operator splitting algorithm has been implemented with a different time scheme for each operator.

The goal of this section is to validate the implementation of the code. To this end I compare the convergence of the approximation with and without the splitting technique. I briefly explore the question of the positivity of the approximated solution.

A. Description of the simple test-case

First let me introduce a simplified test-case for the validation of the splitting algorithm. Set $L > 0$, and $\Omega =]-L; L[^2$. Let $\mathbf{v} = (v_1, v_2) \in \mathbb{R}^2$ and

$d \in \mathbb{R}$ be a constant and $0 \leq t_0 \leq t_1$. Find u such that

$$\frac{\partial u}{\partial t} + \mathbf{v} \cdot \nabla u + \rho u = d\Delta u \text{ in }]t_0, t_1[\times\Omega,$$

$$u(x, y, 0) = u_0(x, y) \text{ on } \{t_0\} \times \Omega, \qquad (18)$$

$$\mathbf{n} \cdot \nabla u = g \text{ on }]t_0, t_1[\times\partial\Omega.$$

$$\mathbf{n} \cdot \mathbf{v} u = g_{in} \text{ on }]t_0, t_1[\times\partial\Omega^-.$$

The initial condition and the boundary condition are chosen such that the solution of problem (18) is explicitly given by $\forall(x, y, t) \in \Omega\times]t_0, t_1[$

$$u(x, y, t) = c_0 \left(\frac{a^2}{a^2 + td}\right) \kappa(x, y, t)e^{-\rho t}.$$

with

$$\kappa(x, y, t)$$
$$= c_0 \exp\left(-\frac{(x - x_0 - tv_1)^2 + (y - y_0 - tv_2)^2}{4(a^2 + td)}\right)$$

where $c_0 > 0$, $a > 0$, x_0 and y_0 are real parameters and v_1 and v_2 are the two components of $\mathbf{v}$. Notice that $u(x, y, t) > 0$ for all (x, y, t) in $\Omega\times]t_0, t_1[$.

B. Numerical validation and convergence

To validate the implementation of the splitting technique, I ran the previous test case with different mesh sizes and time steps and I computed the global L^2-errors such that

$$e_h = \left(\delta t \sum_{k=1}^{N} \|u(t_k) - u_h(t_k)\|_{0,\Omega}^2\right)^{1/2}$$

where $t_k = t_0 + k\delta t$, with $k \in \mathbb{N}_*^+$ and $t_N = t_1$.

The flexibility of the splitting technique allows to choose different time schemes for each operator. I consider a θ-scheme with $\theta = 0$ (explicit Euler), $\theta = 1$ (implicit Euler), and $\theta = \frac{1}{2}$ (Crank-Nicolson) for both the advection step and the diffusion step, and I consider an exact scheme for the reaction step. For the simulations I took the parameters such that $\mathbf{v} = (0.1, 0)^T$, $\sigma = 0.01$ and $\rho = -1$. The triangular meshes used for the simulations are identified by h which is the size of the biggest triangle of the mesh. Table I, page 9, gives the number of triangles and the number of nodes of each mesh used for the simulations. Choosing

h ($\approx$)	number of triangles	number of nodes
2.63×10^{-1}	68	45
1.31×10^{-1}	272	157
6.57×10^{-2}	1 088	585
3.29×10^{-2}	4 352	2 257
1.64×10^{-2}	17 408	8 865
8.22×10^{-3}	69 632	35 137
4.11×10^{-3}	278 528	139 905

TABLE I
TRIANGULAR MESHES USED FOR THE SIMULATIONS.

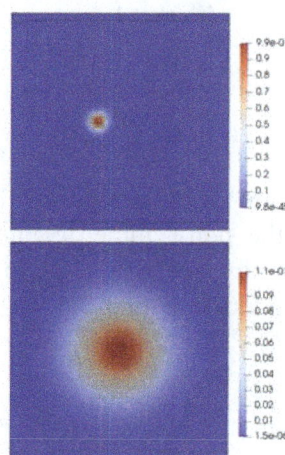

Fig. 1. Solution of the validation test case at $t = t_0$ (left) and $t = t_1$ (right) computed using the DG method with p^1-finite elements and the Euler implicit scheme ($\theta = 1$) and the operator splitting technique with $h \approx 8.2 \times 10^{-3}$ and $\delta t = 10^{-2}$.

$L = 1/2$, the simulations are performed between $t_0 = 0$ and $t_1 = 1$ for different values of the time step δt. Fig. 1, page 9, shows the solution at $t = t_0$ and $t = t_1$. The code is implemented in Fortran 90 and it is run under a 64-bit Linux operating system on a 8-core processor Intel®Core™i7-7820HQ at a frequency of 2.9GHz and with 32 GB of RAM. The sparse matrices resulting from the finite element approximation are inverted using a solver provided by the library MUMPS [51], [52].

According to Fig. 2, page 10, all the three temporal schemes provide results with approximately the same level of accuracy with a spatial convergence rate of 2 computed with the global L^2-

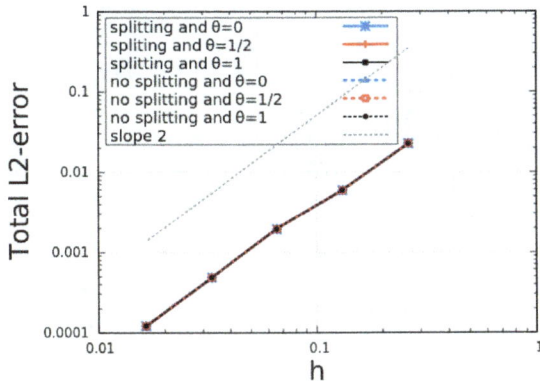

Fig. 2. Convergence of the solution with respect to the mesh size: plot of the total L^2-error computed between $t = 0$ and $t = 1$ with and without the splitting technique for the explicit Euler scheme ($\theta = 0$), the implicit Euler scheme ($\theta = 1$) and the Crank-Nicolson scheme ($\theta = 1/2$) for $\delta t = 5 \times 10^{-5}$.

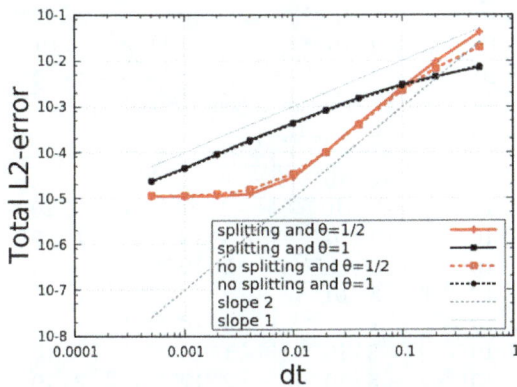

Fig. 3. Convergence of the solution with respect to the time step: plot of the total L^2-error computed between $t = 0$ and $t = 1$ with and without the splitting technique for the implicit Euler scheme ($\theta = 1$) and the Crank-Nicolson scheme ($\theta = 1/2$) for $h = 4.1 \times 10^{-3}$.

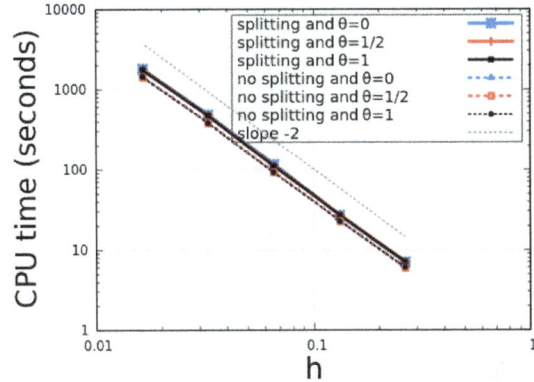

Fig. 4. Validation of the test case: plot of the CPU time against the mesh size (h) for the computations performed with a processor Intel®Core™i7-7820HQ at 2.9 GHz and RAM 32 GB, between $t = 0$ and $t = 1$ with and without the splitting technique for the explicit Euler scheme ($\theta = 0$), the implicit Euler scheme ($\theta = 1$) and the Crank-Nicolson scheme ($\theta = 1/2$) for $\delta t = 5 \times 10^{-5}$.

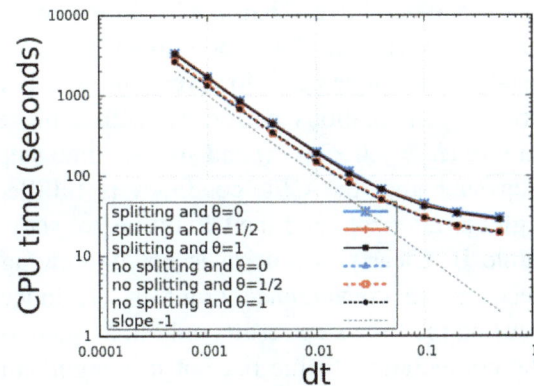

Fig. 5. Validation of the test case: plot of the CPU time against the time step (δt) for the computations performed with a processor Intel®Core™i7-7820HQ at 2.9 GHz and RAM 32 GB, between $t = 0$ and $t = 1$ with and without the splitting technique for the explicit Euler scheme ($\theta = 0$), the implicit Euler scheme ($\theta = 1$) and the Crank-Nicolson scheme ($\theta = 1/2$) for $h \approx 4.1 \times 10^{-3}$ and δt ranging from 5×10^{-1} to 5×10^{-4}. Note that the computations performed here with $\theta = 0$ gave unstable results.

norm. The same order of convergence is obtained when the problem is solved without the splitting technique.

Figure 3 on page 10 shows that the Crank-Nicolson scheme ($\theta = 1/2$) converges in δt^2 while the Euler Implicit scheme ($\theta = 1$) converges in δt, with and without the splitting technique. The convergence rate in time has to be computed with a really refined mesh grid (here $h \approx 4.1 \times 10^{-3}$).

$\delt t$	global L^2-errors	min dof	t_+	CPU time
$1 \cdot 10^{-3}$	unstable	unstable	-	95 s
$2 \cdot 10^{-4}$	$1.24 \cdot 10^{-4}$	$-1 \cdot 10^{-4}$	0.221	475 s
$1 \cdot 10^{-4}$	$1.22 \cdot 10^{-4}$	$-1 \cdot 10^{-4}$	0.218	838 s
$5 \cdot 10^{-5}$	$1.21 \cdot 10^{-4}$	$-1 \cdot 10^{-4}$	0.216	1672 s
$2.5 \cdot 10^{-5}$	$1.20 \cdot 10^{-4}$	$-9 \cdot 10^{-5}$	0.215	3376 s
$1 \cdot 10^{-5}$	$1.20 \cdot 10^{-4}$	$-9 \cdot 10^{-5}$	0.215	10183 s

TABLE II

COMPUTATIONS PERFORMED WITH THE SPLITTING TECHNIQUE AND THE EXPLICIT EULER SCHEME ($\theta = 0$) WITH $h \approx 1.6 \cdot 10^{-2}$ (INTEL®CORE™I7-7820HQ AT 2.9 GHz, RAM 32 GB).

It results an additional cost in term of CPU time, since it behaves like $1/h^2$, as shown on figure 4 page 10. For bigger values of h the plot of the errors gave convergence order in time less than 1 and 2 for the implicit Euler scheme and the Crank-Nicolson scheme respectively. As expected, the explicit Euler scheme is conditionally stable, such that, when the CFL condition is fulfilled, the computational time becomes prohibitive. Indeed, it behaves like $1/\delta t$, as shown on figure 5. For instance, the computation with $h \approx 4.1 \times 10^{-3}$ and $\delta t = 10^{-5}$ takes more than 30 hours with the device specified above. That is why, in the rest of the paper, we will only focus on implicit Euler and Crank-Nicolson schemes. However I present here additional computations performed with a bigger mesh size ($h \approx 1.6 \times 10^{-2}$) and smaller time steps chosen such that the CFL condition is fulfilled. The global L^2-errors and the CPU time are shown on table II. Clearly, the mesh is not fine enough to recover the convergence order in δt, indeed decreasing the time step results only in an increase of the computational time but not in a significant decrease of the errors.

C. Some comments on the positivity

1) Positivity of the full problem: Table III on page 11 and table V on page 12 give the minimum values of the degrees of freedom (dof) obtained during the simulations performed respectively with and without the splitting technique. The minimum value of the dof is defined such that $\min_{t_k}(\min_{i=1,n} X_i^k)$ where X_i^k is the ith dof at time t_k. This quantity gives an idea about the

δt	$h \approx 1.6 \cdot 10^{-2}$	$h \approx 8.2 \cdot 10^{-3}$	$h \approx 4.1 \cdot 10^{-3}$
$5 \cdot 10^{-1}$	$-7 \cdot 10^{-1}$	$-7 \cdot 10^{-1}$	$-7 \cdot 10^{-1}$
$2 \cdot 10^{-1}$	$-2 \cdot 10^{-1}$	$-2 \cdot 10^{-1}$	$-2 \cdot 10^{-1}$
$1 \cdot 10^{-1}$	$-8 \cdot 10^{-4}$	$-1 \cdot 10^{-3}$	$-2 \cdot 10^{-3}$
$4 \cdot 10^{-2}$	$-2 \cdot 10^{-4}$	$-6 \cdot 10^{-4}$	$-1 \cdot 10^{-3}$
$2 \cdot 10^{-2}$	$-3 \cdot 10^{-13}$	$-2 \cdot 10^{-4}$	$-6 \cdot 10^{-4}$
$1 \cdot 10^{-2}$	$-6 \cdot 10^{-5}$	$-2 \cdot 10^{-20}$	$-1 \cdot 10^{-4}$
$4 \cdot 10^{-3}$	$-3 \cdot 10^{-4}$	$4 \cdot 10^{-65}$	$5 \cdot 10^{-66}$
$2 \cdot 10^{-3}$	$-3 \cdot 10^{-4}$	$-9 \cdot 10^{-11}$	$5 \cdot 10^{-88}$
$1 \cdot 10^{-3}$	$-2 \cdot 10^{-4}$	$-8 \cdot 10^{-10}$	$8 \cdot 10^{-114}$
$1 \cdot 10^{-4}$	$-9 \cdot 10^{-5}$	$-1 \cdot 10^{-11}$	$-4 \cdot 10^{-35}$

Crank-Nicolson scheme ($\theta = 1/2$)

δt	$h \approx 1.6 \cdot 10^{-2}$	$h \approx 8.2 \cdot 10^{-3}$	$h \approx 4.1 \cdot 10^{-3}$
$5 \cdot 10^{-1}$	$-5 \cdot 10^{-8}$	$-9 \cdot 10^{-9}$	$-2 \cdot 10^{-9}$
$2 \cdot 10^{-1}$	$4 \cdot 10^{-9}$	$1 \cdot 10^{-9}$	$4 \cdot 10^{-10}$
$1 \cdot 10^{-1}$	$3 \cdot 10^{-11}$	$3 \cdot 10^{-11}$	$1 \cdot 10^{-11}$
$4 \cdot 10^{-2}$	$6 \cdot 10^{-17}$	$5 \cdot 10^{-17}$	$5 \cdot 10^{-17}$
$2 \cdot 10^{-2}$	$3 \cdot 10^{-23}$	$2 \cdot 10^{-23}$	$2 \cdot 10^{-23}$
$1 \cdot 10^{-2}$	$1 \cdot 10^{-31}$	$4 \cdot 10^{-32}$	$3 \cdot 10^{-32}$
$4 \cdot 10^{-3}$	$-3 \cdot 10^{-5}$	$1 \cdot 10^{-48}$	$6 \cdot 10^{-49}$
$2 \cdot 10^{-3}$	$-5 \cdot 10^{-5}$	$2 \cdot 10^{-65}$	$2 \cdot 10^{-66}$
$1 \cdot 10^{-3}$	$-7 \cdot 10^{-5}$	$-1 \cdot 10^{-13}$	$2 \cdot 10^{-88}$
$1 \cdot 10^{-4}$	$-9 \cdot 10^{-5}$	$-7 \cdot 10^{-12}$	$-3 \cdot 10^{-43}$

Implicit Euler scheme ($\theta = 1$)

TABLE III

MINIMUM VALUE OF THE DOF ($\min_{i,k} X_i^k$) COMPUTED WITH THE SPLITTING ALGORITHM WITH THE CRANK-NICOLSON SCHEME (TOP) AND THE IMPLICIT EULER SCHEME (BOTTOM).

stability and the positivity preserving behaviour of the schemes. Tables III and V clearly show that the schemes are not always positivity preserving. In case where the approximated solution is not positive for all $t > t_0$ I also check if it becomes non-negative for larger time ie. if there is $t_+ > t_0$

δt	$h \approx 1.6 \cdot 10^{-2}$	$h \approx 8.2 \cdot 10^{-3}$	$h \approx 4.1 \cdot 10^{-3}$
$5 \cdot 10^{-1}$	-	-	-
$2 \cdot 10^{-1}$	-	-	-
$1 \cdot 10^{-1}$	-	-	-
$4 \cdot 10^{-2}$	-	-	-
$2 \cdot 10^{-2}$	0.24	-	-
$1 \cdot 10^{-2}$	0.07	0.10	-
$4 \cdot 10^{-3}$	0.172	t_0	t_0
$2 \cdot 10^{-3}$	0.202	0.044	t_0
$1 \cdot 10^{-3}$	0.210	0.079	t_0
$1 \cdot 10^{-4}$	0.2140	0.0939	0.0297

Crank-Nicolson scheme ($\theta = 1/2$)

δt	$h \approx 1.6 \cdot 10^{-2}$	$h \approx 8.2 \cdot 10^{-3}$	$h \approx 4.1 \cdot 10^{-3}$
$5 \cdot 10^{-1}$	1	1	1
$2 \cdot 10^{-1}$	t_0	t_0	t_0
$1 \cdot 10^{-1}$	t_0	t_0	t_0
$4 \cdot 10^{-2}$	t_0	t_0	t_0
$2 \cdot 10^{-2}$	t_0	t_0	t_0
$1 \cdot 10^{-2}$	t_0	t_0	t_0
$4 \cdot 10^{-3}$	0.072	t_0	t_0
$2 \cdot 10^{-3}$	0.132	t_0	t_0
$1 \cdot 10^{-3}$	0.173	0.029	t_0
$1 \cdot 10^{-4}$	0.2102	0.0874	0.0217

Implicit Euler scheme ($\theta = 1$)

TABLE IV

POSITIVITY THRESHOLD (t_+) COMPUTED WITH THE SPLITTING ALGORITHM AND THE CRANK-NICOLSON SCHEME (TOP) AND THE IMPLICIT EULER SCHEME (BOTTOM).

δt	$h \approx 1.6 \cdot 10^{-2}$	$h \approx 8.2 \cdot 10^{-3}$	$h \approx 4.1 \cdot 10^{-3}$
$5 \cdot 10^{-1}$	$-4 \cdot 10^{-1}$	$-4 \cdot 10^{-1}$	$-4 \cdot 10^{-1}$
$2 \cdot 10^{-1}$	$-1 \cdot 10^{-1}$	$-1 \cdot 10^{-1}$	$-1 \cdot 10^{-1}$
$1 \cdot 10^{-1}$	$1 \cdot 10^{-13}$	$1 \cdot 10^{-13}$	$1 \cdot 10^{-13}$
$4 \cdot 10^{-2}$	$1 \cdot 10^{-21}$	$1 \cdot 10^{-21}$	$9 \cdot 10^{-22}$
$2 \cdot 10^{-2}$	$3 \cdot 10^{-30}$	$1 \cdot 10^{-30}$	$1 \cdot 10^{-30}$
$1 \cdot 10^{-2}$	$-6 \cdot 10^{-5}$	$1 \cdot 10^{-42}$	$9 \cdot 10^{-43}$
$4 \cdot 10^{-3}$	$-3 \cdot 10^{-4}$	$3 \cdot 10^{-64}$	$5 \cdot 10^{-65}$
$2 \cdot 10^{-3}$	$-3 \cdot 10^{-4}$	$-8 \cdot 10^{-11}$	$3 \cdot 10^{-87}$
$1 \cdot 10^{-3}$	$-2 \cdot 10^{-4}$	$-7 \cdot 10^{-10}$	$3 \cdot 10^{-113}$
$1 \cdot 10^{-4}$	$-9 \cdot 10^{-5}$	$-1 \cdot 10^{-11}$	$-8 \cdot 10^{-35}$

Crank-Nicolson scheme ($\theta = 1/2$)

δt	$h \approx 1.6 \cdot 10^{-2}$	$h \approx 8.2 \cdot 10^{-3}$	$h \approx 4.1 \cdot 10^{-3}$
$5 \cdot 10^{-1}$	$2 \cdot 10^{-5}$	$2 \cdot 10^{-5}$	$2 \cdot 10^{-5}$
$2 \cdot 10^{-1}$	$1 \cdot 10^{-7}$	$1 \cdot 10^{-7}$	$1 \cdot 10^{-7}$
$1 \cdot 10^{-1}$	$6 \cdot 10^{-10}$	$5 \cdot 10^{-10}$	$5 \cdot 10^{-10}$
$4 \cdot 10^{-2}$	$1 \cdot 10^{-15}$	$1 \cdot 10^{-15}$	$1 \cdot 10^{-15}$
$2 \cdot 10^{-2}$	$6 \cdot 10^{-22}$	$5 \cdot 10^{-22}$	$5 \cdot 10^{-22}$
$1 \cdot 10^{-2}$	$1 \cdot 10^{-30}$	$7 \cdot 10^{-31}$	$6 \cdot 10^{-31}$
$4 \cdot 10^{-3}$	$-2 \cdot 10^{-5}$	$2 \cdot 10^{-47}$	$8 \cdot 10^{-48}$
$2 \cdot 10^{-3}$	$-4 \cdot 10^{-5}$	$2 \cdot 10^{-64}$	$2 \cdot 10^{-65}$
$1 \cdot 10^{-3}$	$-7 \cdot 10^{-5}$	$-3 \cdot 10^{-13}$	$2 \cdot 10^{-87}$
$1 \cdot 10^{-4}$	$-9 \cdot 10^{-5}$	$-7 \cdot 10^{-12}$	$-1 \cdot 10^{-42}$

Implicit Euler scheme ($\theta = 1$)

TABLE V

MINIMUM VALUE OF THE DOF ($\min_{i,k} X_i^k$) COMPUTED WITHOUT THE SPLITTING ALGORITHM AND THE CRANK-NICOLSON SCHEME (TOP) AND IMPLICIT EULER SCHEME (BOTTOM).

such that $X_i^k \geq 0, \forall i = 1, n$ for all $t_k > t_+ > t_0$. The smallest such t_+, if it exists, is referred as the positivity threshold, as defined in [36]. Table IV on page 12 and table VI on page 13 give the positivity thresholds computed with and without the splitting technique respectively.

For the Crank-Nicolson scheme ($\theta = 1/2$) and the implicit Euler scheme ($\theta = 1$) the positivity is obtained under a specific condition on the time step and the mesh size. For a given mesh size, the time step δt must be bounded from above, but also from below to guarantee that the solution stays positive all along the simulation. In the case of the splitting technique those bounds are more restrictive than in the case of the resolution of the full problem without splitting. Those bounds are also more restrictive in the case of the Crank-Nicolson ($\theta = 1/2$) scheme than in the case of the

implicit Euler scheme ($\theta = 1$). Refining the mesh results in less restrictions on the time step but also lead to additional computational time.

With the Crank-Nicolson scheme ($\theta = 1/2$), for a given mesh size, if δt is too big, there is no threshold of positivity in $t_k \in]t_0, t_1]$ and the computed solution is not non-negative all along the simulation. For $\theta = 1/2$ and $\theta = 1$, still with a given mesh size, if δt is too small, the simulations showed that there is a threshold of positivity t_+ such that the approximated solution becomes non-negative for $t_k \geq t_+$. The thresholds of positivity slightly depend on the time step and tend to increase when the time step δt is decreased. The computations clearly showed that the positivity thresholds diminish with the mesh size h (see for example [36]).

Altogether, the positivity of the approximated

δt	$h\approx 1.6\cdot 10^{-2}$	$h\approx 8.2\cdot 10^{-3}$	$h\approx 4.1\cdot 10^{-3}$
$5\cdot 10^{-1}$	-	1	1
$2\cdot 10^{-1}$	0.8	0.8	0.8
$1\cdot 10^{-1}$	t_0	t_0	t_0
$4\cdot 10^{-2}$	t_0	t_0	t_0
$2\cdot 10^{-2}$	t_0	t_0	t_0
$1\cdot 10^{-2}$	0.08	t_0	t_0
$4\cdot 10^{-3}$	0.188	t_0	t_0
$2\cdot 10^{-3}$	0.208	0.050	t_0
$1\cdot 10^{-3}$	0.213	0.083	t_0
$1\cdot 10^{-4}$	0.2143	0.0942	0.0300

Crank-Nicolson scheme ($\theta = 1/2$)

δt	$h\approx 1.6\cdot 10^{-2}$	$h\approx 8.2\cdot 10^{-3}$	$h\approx 4.1\cdot 10^{-3}$
$5\cdot 10^{-1}$	t_0	t_0	t_0
$2\cdot 10^{-1}$	t_0	t_0	t_0
$1\cdot 10^{-1}$	t_0	t_0	t_0
$4\cdot 10^{-2}$	t_0	t_0	t_0
$2\cdot 10^{-2}$	t_0	t_0	t_0
$1\cdot 10^{-2}$	t_0	t_0	t_0
$4\cdot 10^{-3}$	0.0720	t_0	t_0
$2\cdot 10^{-3}$	0.1380	t_0	t_0
$1\cdot 10^{-3}$	0.1760	0.0290	t_0
$1\cdot 10^{-4}$	0.2104	0.0877	0.0221

Implicit Euler scheme ($\theta = 1$)

TABLE VI

POSITIVITY THRESHOLD (t_+) COMPUTED WITHOUT THE SPLITTING ALGORITHM AND THE CRANK-NICOLSON SCHEME (TOP) AND THE IMPLICIT EULER SCHEME (BOTTOM).

δt	$h\approx 1.6\cdot 10^{-2}$	$h\approx 8.2\cdot 10^{-3}$	$h\approx 4.1\cdot 10^{-3}$
$5\cdot 10^{-1}$	$-5\cdot 10^{-1}$	$-5\cdot 10^{-1}$	$-5\cdot 10^{-1}$
$2\cdot 10^{-1}$	$-2\cdot 10^{-1}$	$-2\cdot 10^{-1}$	$-2\cdot 10^{-1}$
$1\cdot 10^{-1}$	$4\cdot 10^{-15}$	$4\cdot 10^{-15}$	$4\cdot 10^{-15}$
$4\cdot 10^{-2}$	$6\cdot 10^{-23}$	$5\cdot 10^{-23}$	$4\cdot 10^{-23}$
$2\cdot 10^{-2}$	$2\cdot 10^{-31}$	$8\cdot 10^{-32}$	$6\cdot 10^{32}$
$1\cdot 10^{-2}$	$-4\cdot 10^{-5}$	$1\cdot 10^{-43}$	$6\cdot 10^{43}$
$4\cdot 10^{-3}$	$-3\cdot 10^{-4}$	$4\cdot 10^{-65}$	$5\cdot 10^{-66}$
$2\cdot 10^{-3}$	$-3\cdot 10^{-4}$	$-5\cdot 10^{-11}$	$5\cdot 10^{-88}$
$1\cdot 10^{-3}$	$-2\cdot 10^{-4}$	$-6\cdot 10^{-10}$	$8\cdot 10^{-114}$
$1\cdot 10^{-4}$	$-9\cdot 10^{-5}$	$-1\cdot 10^{-11}$	$-1\cdot 10^{-34}$

Crank-Nicolson scheme ($\theta = 1/2$)

δt	$h\approx 1.6\cdot 10^{-2}$	$h\approx 8.2\cdot 10^{-3}$	$h\approx 4.1\cdot 10^{-3}$
$5\cdot 10^{-1}$	$4\cdot 10^{-6}$	$4\cdot 10^{-6}$	$4\cdot 10^{-5}$
$2\cdot 10^{-1}$	$2\cdot 10^{-8}$	$2\cdot 10^{-8}$	$2\cdot 10^{-8}$
$1\cdot 10^{-1}$	$2\cdot 10^{-11}$	$2\cdot 10^{-11}$	$2\cdot 10^{-11}$
$4\cdot 10^{-2}$	$5\cdot 10^{-17}$	$5\cdot 10^{-17}$	$5\cdot 10^{-17}$
$2\cdot 10^{-2}$	$3\cdot 10^{-23}$	$2\cdot 10^{-23}$	$2\cdot 10^{-23}$
$1\cdot 10^{-2}$	$1\cdot 10^{-31}$	$4\cdot 10^{-32}$	$3\cdot 10^{-32}$
$4\cdot 10^{-3}$	$-2\cdot 10^{-5}$	$1\cdot 10^{-48}$	$6\cdot 10^{-49}$
$2\cdot 10^{-3}$	$-4\cdot 10^{-5}$	$2\cdot 10^{-65}$	$2\cdot 10^{-66}$
$1\cdot 10^{-3}$	$-7\cdot 10^{-5}$	$-3\cdot 10^{-13}$	$2\cdot 10^{-88}$
$1\cdot 10^{-4}$	$-9\cdot 10^{-5}$	$-7\cdot 10^{-12}$	$-2\cdot 10^{-42}$

Implicit Euler scheme ($\theta = 1$)

TABLE VII

MINIMUM VALUE OF THE DOF ($\min_{i,k} X_i^k$) COMPUTED FOR THE PURE DIFFUSION PROBLEM WITH THE CRANK-NICOLSON SCHEME (TOP) AND THE IMPLICIT EULER SCHEME (BOTTOM).

solution is obtained at the expense of the computational cost, but for a given mesh size h computations performed with too small time step can also lead to a loss of positivity for small t_k. In [36] (and references therein), Thomée showed that threshold values of $t_k > 0$ may exist such that $X(t) > 0$ when $t > t_k$.

At this stage, one may wonder how each term of the splitting behaves in terms of positivity preservation. The reaction term is approximated using an exact scheme, so obviously the positivity of the solution is preserved. What about the diffusion and the advection term ?

2) Positivity of the pure diffusion problem:
Here I set $\mathbf{v} = (0,0)$ and $\rho = 0$, while keeping all others parameters to the same values as previously. Table VII clearly shows that the Crank-Nicolson scheme ($\theta = 1/2$) is positivity preserving under a CFL-like condition with upper and lower bounds, like in the previous test. The implicit Euler scheme ($\theta = 1$) seems to be more favorable, since it preserves the positivity even for big values of the time step. For both the Crank-Nicolson ($\theta = 1/2$) and implicit Euler ($\theta = 1$) schemes, the approximated solution suffers from a loss of positivity for small values of t_k when the time step is too small. According to table VIII, there are positivity thresholds, like in [36] which indeed deals with the heat equation.

3) Positivity of the pure advection problem:
Here I set $\sigma = 0$ and $\rho = 0$, while keeping all others parameters to the same values as in the first test. Table IX shows that none of the computations performed gave a non negative solutions, even though the minimum value of the dof can be really close to zero for small mesh sizes. Besides, I did

δt	$h \approx 1.6 \cdot 10^{-2}$	$h \approx 8.2 \cdot 10^{-3}$	$h \approx 4.1 \cdot 10^{-3}$
$5 \cdot 10^{-1}$	-	-	-
$2 \cdot 10^{-1}$	0.8	0.8	0.8
$1 \cdot 10^{-1}$	t_0	t_0	t_0
$4 \cdot 10^{-2}$	t_0	t_0	t_0
$2 \cdot 10^{-2}$	t_0	t_0	t_0
$1 \cdot 10^{-2}$	0.1	t_0	t_0
$4 \cdot 10^{-3}$	0.2	t_0	t_0
$2 \cdot 10^{-3}$	0.216	0.054	t_0
$1 \cdot 10^{-3}$	0.219	0.085	t_0
$1 \cdot 10^{-4}$	0.2203	0.0956	0.0303

Crank-Nicolson scheme ($\theta = 1/2$)

δt	$h \approx 1.6 \cdot 10^{-2}$	$h \approx 8.2 \cdot 10^{-3}$	$h \approx 4.1 \cdot 10^{-3}$
$5 \cdot 10^{-1}$	t_0	t_0	t_0
$2 \cdot 10^{-1}$	t_0	t_0	t_0
$1 \cdot 10^{-1}$	t_0	t_0	t_0
$4 \cdot 10^{-2}$	t_0	t_0	t_0
$2 \cdot 10^{-2}$	t_0	t_0	t_0
$1 \cdot 10^{-2}$	t_0	t_0	t_0
$4 \cdot 10^{-3}$	0.084	t_0	t_0
$2 \cdot 10^{-3}$	0.148	t_0	t_0
$1 \cdot 10^{-3}$	0.184	0.032	t_0
$1 \cdot 10^{-4}$	0.2165	0.0893	0.0224

Implicit Euler scheme ($\theta = 1$)

TABLE VIII

POSITIVITY THRESHOLD (t_+) COMPUTED FOR THE PURE DIFFUSION PROBLEM WITH THE CRANK-NICOLSON SCHEME (TOP) AND THE IMPLICIT EULER SCHEME (BOTTOM).

δt	$h \approx 1.6 \cdot 10^{-2}$	$h \approx 8.2 \cdot 10^{-3}$	$h \approx 4.1 \cdot 10^{-3}$
$5 \cdot 10^{-1}$	$-2 \cdot 10^{-1}$	$-2 \cdot 10^{-1}$	$-2 \cdot 10^{-1}$
$2 \cdot 10^{-1}$	$-4 \cdot 10^{-2}$	$-4 \cdot 10^{-2}$	$-4 \cdot 10^{-2}$
$1 \cdot 10^{-1}$	$-4 \cdot 10^{-3}$	$-2 \cdot 10^{-3}$	$-1 \cdot 10^{-3}$
$4 \cdot 10^{-2}$	$-1 \cdot 10^{-3}$	$-4 \cdot 10^{-7}$	$-5 \cdot 10^{-7}$
$2 \cdot 10^{-2}$	$-1 \cdot 10^{-3}$	$-6 \cdot 10^{-8}$	$-2 \cdot 10^{-13}$
$1 \cdot 10^{-2}$	$-1 \cdot 10^{-3}$	$-4 \cdot 10^{-8}$	$-3 \cdot 10^{-28}$
$4 \cdot 10^{-3}$	$-1 \cdot 10^{-3}$	$-3 \cdot 10^{-8}$	$-1 \cdot 10^{-28}$
$2 \cdot 10^{-3}$	$-1 \cdot 10^{-3}$	$-3 \cdot 10^{-8}$	$-1 \cdot 10^{-28}$
$1 \cdot 10^{-3}$	$-1 \cdot 10^{-3}$	$-3 \cdot 10^{-8}$	$-1 \cdot 10^{-28}$
$1 \cdot 10^{-4}$	$-1 \cdot 10^{-3}$	$-3 \cdot 10^{-8}$	$-1 \cdot 10^{-28}$

Crank-Nicolson scheme ($\theta = 1/2$)

δt	$h \approx 1.6 \cdot 10^{-2}$	$h \approx 8.2 \cdot 10^{-3}$	$h \approx 4.1 \cdot 10^{-3}$
$5 \cdot 10^{-1}$	$-1 \cdot 10^{-4}$	$-1 \cdot 10^{-10}$	$-8 \cdot 10^{-34}$
$2 \cdot 10^{-1}$	$-2 \cdot 10^{-4}$	$-6 \cdot 10^{-10}$	$-3 \cdot 10^{-32}$
$1 \cdot 10^{-1}$	$-4 \cdot 10^{-4}$	$-1 \cdot 10^{-9}$	$-2 \cdot 10^{-31}$
$4 \cdot 10^{-2}$	$-6 \cdot 10^{-4}$	$-4 \cdot 10^{-9}$	$-9 \cdot 10^{-31}$
$2 \cdot 10^{-2}$	$-8 \cdot 10^{-4}$	$-7 \cdot 10^{-9}$	$-2 \cdot 10^{-30}$
$1 \cdot 10^{-2}$	$-9 \cdot 10^{-4}$	$-1 \cdot 10^{-8}$	$-5 \cdot 10^{-30}$
$4 \cdot 10^{-3}$	$-1 \cdot 10^{-3}$	$-1 \cdot 10^{-8}$	$-8 \cdot 10^{-30}$
$2 \cdot 10^{-3}$	$-1 \cdot 10^{-3}$	$-2 \cdot 10^{-8}$	$-2 \cdot 10^{-29}$
$1 \cdot 10^{-3}$	$-1 \cdot 10^{-3}$	$-2 \cdot 10^{-8}$	$-4 \cdot 10^{-29}$
$1 \cdot 10^{-4}$	$-1 \cdot 10^{-3}$	$-3 \cdot 10^{-8}$	$-9 \cdot 10^{-29}$

Implicit Euler scheme ($\theta = 1$)

TABLE IX

MINIMUM VALUE OF THE DOF ($\min_{i,k} X_i^k$) COMPUTED FOR THE PURE ADVECTION PROBLEM WITH THE CRANK-NICOLSON SCHEME (TOP) AND THE IMPLICIT EULER SCHEME (BOTTOM).

not observe any positivity threshold. The approximated solution stays non positive all along the simulation. However I run additional simulations with even smaller mesh size ($h \approx 2.0 \times 10^{-3}$ and $\delta t = 10^{-4}$). This time the computed solution was positive at the beginning of the simulation (before $t_- = 1.9 \times 10^{-3}$), pointing the existence of a threshold of negativity, to finally reaching a negative minimum values of dof (around -10^{-44}). Unfortunately, this threshold of negativity is really small compared to the ending time of the computation ($t_1 = 1$), while the computational time was reaching more than 14 hours (Intel®Core™i7-7820HQ at 2.9 GHz, RAM 32 GB) for both the Crank-Nicolson and the implicit Euler schemes.

In fact it is well known that for the advection term the solution can be polluted by overshoot and undershoot oscillations near a discontinuity or a sharp layer, see [34], [33], [35], [30]. For low order accurate spacial approximations one can prove the positivity preserving property of the scheme [33]. But for high order schemes slopes limiters are often required to guarantee the positivity of the approximated solution. When slope limiters are used, explicit time schemes seem to be suitable for the advection [6]. However, in the next section we will only privilege a numerical scheme that is unconditionally stable, i.e. the Crank-Nicolson scheme, that is a two-order scheme.

V. APPLICATION TO THE SIMULATION OF ROOT SYSTEM GROWTH

In this section, I apply the previous DG-splitting approach to solve numerically the C-Root model. First, I detail the parameters used for the simulations, then, I present and validate the results of the simulations.

A. *The C-Root parameters for Eucalyptus root growth*

The parameters and operators' coefficients are chosen based on the previous calibration done in [2]. The diffusion coefficient, σ, is build using the following Gaussian function

$$f_{\alpha,\mu}(x,y) = \frac{\alpha}{\sqrt{2\pi}} \exp\left(-\frac{(r(x,y)-\mu)^2}{2}\right)$$

where $r(x,y) = \sqrt{(x-x_0)^2 + (y-y_0)^2}$ and $(x_0, y_0) \in \Omega =]-L, L[$. The function $f_{\alpha,\mu}(x,y)$ depends on two real and positive parameters: α, related to the maximum amplitude of $f_{\alpha,\mu}$, and μ, the distance from (x_0, y_0) to the point where the function $f_{\alpha,\mu}$ reaches its maximum.

The diffusion tensor is taken such that

$$\sigma(x,y) = f_{\alpha_d,\mu_d}(x,y) \begin{pmatrix} 1 & 0 \\ 0 & 1 \end{pmatrix},$$

for all $(x,y) \in \Omega$, and $\alpha_d, \mu_d \in \mathbb{R}^+$ are given parameters. The advection vector is taken such that $\mathbf{v}(x,y) = (0, -v_0)^T$, for all $(x,y) \in \Omega$, with v_0 a positive constant. The reaction term is constant in space and splited into two contributions: β_r and μ_r, the branching and mortality rates, respectively. That is

$$\rho = \beta_r - \mu_r \in \mathbb{R}.$$

The branching rate, β_r, is estimated from biological knowledge: it is equal to zero before 9 months and equal to $1/3$ after, since no roots die before 9 month. However, for the following simulations we will not distinguish the contribution of β_r and μ_r, so that the reaction term will only be described by the parameter ρ.

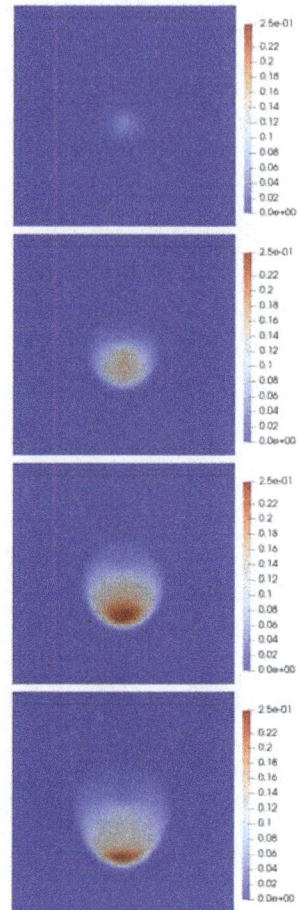

Fig. 6. Density of apices computed at $t = 6$, $t = 12$, $t = 18$ and $t = 24$ months (from the left to the right and from the top to the bottom).

B. *Some simulations*

For the simulation the initial solution is chosen equal to the following function:

$$u_0(x,y) = A\left[\frac{\exp(b(1-x))}{(\exp(-b(1-x)) + \exp(b(1-x)))}\right.$$
$$\left. - \frac{\exp(b(-1-x))}{(\exp(-b(-1-x)) + \exp(b(-1-x)))}\right]$$
$$\times\left[\frac{\exp(b(1-y))}{(\exp(-b(1-y)) + \exp(b(1-y)))}\right.$$
$$\left. - \frac{\exp(b(-1-y))}{(\exp(-b(-1-y)) + \exp(b(-1-y)))}\right]$$

with $A = 2 \cdot 10^{-4}$ and $b = 1$. The parameters' values μ_r, α_d, μ_d are estimated using the code

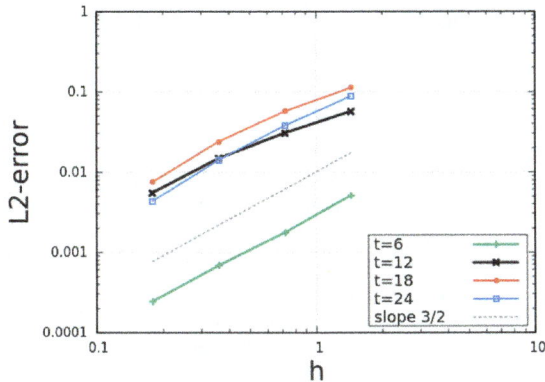

h ($\approx$)	$\delta t = 10^{-1}$	$\delta t = 10^{-2}$	$\delta t = 10^{-3}$
1.44	1 s.	7 s.	66 s.
7.18×10^{-1}	16 s.	46 s.	6 min.
3.59×10^{-1}	70 s.	3.5 min.	27 min.
1.79×10^{-1}	7 min.	17 min.	2 h.
8.97×10^{-2}	50 min.	2h30	9 h.

TABLE X

COMPUTATIONAL TIMES FOR THE SIMULATIONS OF A ROOT SYSTEM GROWTH PERFORMED (WITH THE PROCESSOR INTEL®CORE™I7-7820HQ AT 2.9 GHZ, RAM 32 GB) BETWEEN $t = 1$ AND $t = 24$ MONTHS WITH THE DG-SPLITTING ALGORITHM AND THE CRANK-NICOLSON SCHEME ($\theta = 1/2$).

Fig. 7. L^2-error with respect to the solution obtained with the mesh of size $h \approx 9.87 \times 10^{-2}$ and $\delta t = 10^{-3}$ computed at $t = 6$, $t = 12$, $t = 18$ and $t = 24$ months and plotted against the mesh size.

described in [2]. I run the simulations from $t_0 = 1$ to $t_1 = 24$ months, with $L = 13$. The simulations are performed for different values of the mesh size. Fig. 6, page 15, shows the solution computed at four different stages of the root system development. One can notice the diffusion of the apices in the soil and also the transport of the apices from the top to the bottom of the soil layer. Since there is no analytic solution, the convergence of the computation is evaluated by measuring the L^2-errors with respect to the approximated solution computed with the finest mesh ($h \approx 8.97 \times 10^{-2}$) and with $\delta t = 10^{-3}$. The curves of the errors against the mesh size are plotted on figure 7 and clearly show that the DG-splitting algorithm converges with a convergence rate of almost two. However, one can note that the mesh sizes and the time steps chosen for the simulations presented here might not be small enough. The positivity of the solution is not preserved at all times and the full convergence might not be acheived. Unfortunatly, refining the mesh sizes and the time steps can lead to prohibitive computational time as shown on table X. On top of that simulation of root system growth can last for a long period of time, particularly for trees. Finally, this application shows promising results for future simulations of

the root system growth, provided that the computational cost is not limiting. Further simulations requiring much more computational power has to be done to check if the convergence is acheived. This application also point out the difficulties related to the rigorous simulation validation in realistic test-cases of root system growth.

VI. CONCLUSION

In this work, a discontinuous Galerkin approximation method based on unstructured mesh combined with operator splitting has been described, implemented and tested, to solve an advection-diffusion-reaction equation used to model the growth of root systems. The code has been validated in a simple test case for which an analytic expression of the solution is known. The computations showed that the method convergences with a convergence rate of two in space with P^1-finite elements. A convergence rate of one and two in time were obtained for respectively the implicit Euler scheme and the Crank-Nicolson scheme both with and without the splitting technique. The computations of those convergence rates required the use of fine mesh grids. For the explicit Euler scheme, such fine mesh computations were not performed since they require really small time steps to fulfill the CFL condition, resulting in huge additional computational cost. Indeed the computational time of the DG-splitting algorithm behaves like $1/\delta t$ and $1/h^2$ where δt and h are respectively the time step and the mesh size.

Similarly, the positivity of the approximated solution is obtained at the expense of the computational time since it requires meshes of small size and small time steps. In fact, there is a CFL-like condition for positivity that has to be fulfilled to guarantee the positivity of the approximated solution. But for a given mesh size computations performed with too small time step can also lead to a loss of positivity at the beginning of the computation [36]. In that cases, the computations showed that there is a positivity threshold in time after which the solution becomes positive. This positivity threshold clearly appeared to diminish with the mesh size. This behavior is specific to the diffusion term. For the advection term, the computations also showed that the positivity of the solution can be preserved, but only at the beginning of the simulation and it required a really small mesh size and time step leading to huge computational time. Further studies in terms of numerical analysis has to be done in that direction.

I also performed a more realistic simulation of root system growth. The computations showed that the algorithm converged but additional simulations with smaller time steps and mesh sizes might be performed to recover the full convergence order and positivity. Validation of the computation, but above all the computational time appeared to be the major limitations of the root growth simulation based on the C-Root model, particularly when it comes to deal with trees for which the life span is rather a long period of time. Further improvements on the numerical method has to be done so that the scheme preserves the positivity of the approximated solution under acceptable CFL conditions in terms of computational time. However, our work shows promising results for the simulation of the C-Root model which appears to be an appropriate methodology for future improvements, like root-soil coupling or nonlinear terms arising to handle competition phenomena.

Acknowledgment

The author would like to thank Y. Dumont (CIRAD, University of Pretoria) for discussions and valuable comments about the numerical schemes.

References

[1] A. Bonneu, Y. Dumont, H. Rey, C. Jourdan and T. Fourcaud, A minimal continuous model for simulating growth and development of plant root systems, Plant and Soil, Springer, 2012, 354, 211-22. https://doi.org/10.1007/s11104-011-1057-7.

[2] E. Tillier and A. Bonneu, Operator splitting for solving C-Root, a minimalist and continuous model of root system growth, Plant Growth Modeling, Simulation, Visualization and Applications (PMA), 2012 IEEE Fourth International Symposium on, 2012, 396-402.https://doi.org/10.1109/PMA.2012.6524863.

[3] E. Peynaud, T. Fourcaud and Y. Dumont Numerical resolution of the C-Root model using Discontinuous Galerkin methods on unstructured meshes: application to the simulation of root system growth, 2016 IEEE International Conference on Functional-Structural Plant Growth Modeling, Simulation, Visualization and Applications (FSPMA), IEEE, FSPMA. Qingdao: IEEE, 2016, 158-166. https://doi.org/10.1109/FSPMA.2016.7818302.

[4] H.-G. Roos, M. Stynes and L. Tobiska, Robust numerical methods for singularly perturbed differential equations: convection-diffusion-reaction and flow problems Springer Science & Business Media, 2008, 24. https://doi.org/10.1007/978-3-540-34467-4.

[5] X. Zhang and CW. Shu, Maximum-principle-satisfying and positivity-preserving high-order schemes for conservation laws: survey and new developments. Proceedings of the Royal Society of London A: Mathematical, Physical and Engineering Sciences, 2011, 467, 2752-2776. https://doi.org/10.1098/rspa.2011.0153.

[6] W. Hundsdorfer and J. G. Verwer, Numerical solution of time-dependent advection-diffusion-reaction equations. Springer Science & Business Media, 2013, 33. DOI 10.1007/978-3-662-09017-6

[7] G. Strang, On the construction and comparison of difference schemes. SIAM Journal on Numerical Analysis, SIAM. 5, 506-517. 1968. https://doi.org/10.1137/0705041.

[8] N. Janenko, The method of fractional steps. Springer. 1971. https://doi.org/10.1007/978-3-642-65108-3.

[9] A. Chertock and A. Kurganov, On splitting-based numerical methods for convection-diffusion equations Numerical methods for balance laws, Aracne Editrice Srl Rome, 2010, 24, 303-343.

[10] D. Lanser and J.G. Verwer, Analysis of operator splitting for advection-diffusion-reaction problems from air pollution modelling, Journal of computational and applied mathematics, Elsevier, 111, 1-2, 1999, 201-216. https://doi.org/10.1016/S0377-0427(99)00143-0.

[11] J. Kačur, B. Malengier, M. Remešíková, Convergence of an operator splitting method on a bounded domain for a convection-diffusion-reaction system, Journal of

Mathematical Analysis and Applications. 348, 894-914, 2008. https://doi.org/10.1016/j.jmaa.2008.08.017.

[12] M. Remešíková, Numerical solution of two-dimensional convection-diffusion-adsorption problems using an operator splitting scheme, Applied mathematics and computation, 184, 116-130, 2007. https://doi.org/10.1016/j.amc.2005.06.018.

[13] J.C. Chrispell, V. Ervin V and E. Jenkins, A fractional step θ-method for convection-diffusion problems, Journal of Mathematical Analysis and Applications, 333, 204-218, 2007.https://doi.org/10.1016/j.jmaa.2006.11.059.

[14] S. Ganesan and L. Tobiska, Operator-splitting finite element algorithms for computations of high-dimensional parabolic problems, Applied Mathematics and Computation, Elsevier, 219, 2013. 6182-6196. https://doi.org/10.1016/j.amc.2012.12.027.

[15] M. Wheeler and C. Dawson, An operator-splitting method for advection-diffusion-reaction problems, MAFELAP Proceedings, 6, 463-482, 1987.

[16] R. Anguelov, C. Dufourd and Y. Dumont, Simulations and parameter estimation of a trap-insect model using a finite element approach, Mathematics and Computers in Simulation, 2017, 133, 47-75. https://doi.org/10.1016/j.matcom.2015.06.014.

[17] C. Dufourd and Y. Dumont, Impact of environmental factors on mosquito dispersal in the prospect of sterile insect technique control, Computers & Mathematics with Applications, Elsevier, 2013, 66, 1695-1715. https://doi.org/10.1016/j.camwa.2013.03.024.

[18] V. Girault, B. Rivière and M.F. Wheeler, A splitting method using discontinuous Galerkin for the transient incompressible Navier-Stokes equations, ESAIM: Mathematical Modelling and Numerical Analysis, EDP Sciences,39, 1115-1147, 2005. https://doi.org/10.1051/m2an:2005048.

[19] J. Zhu, Y.T. Zhang, S.A. Newman and M. Alber, Application of discontinuous Galerkin methods for reaction-diffusion systems in developmental biology, Journal of Scientific Computing, Springer, 2009, 40, 391-418. https://doi.org/10.1007/s10915-008-9218-4.

[20] N. Ahmed, G. Matthies and L. Tobiska, Finite element methods of an operator splitting applied to population balance equations Journal of Computational and Applied Mathematics, Elsevier. 236, 1604-1621, 2011. https://doi.org/10.1016/j.cam.2011.09.025.

[21] R. Zhang, J. Zhu, A.F Loula and X. Yu, Operator splitting combined with positivity-preserving discontinuous Galerkin method for the chemotaxis model, Journal of Computational and Applied Mathematics. 302, 312 - 326, 2016. https://doi.org/10.1016/j.cam.2016.02.018.

[22] L. Edelstein-Keshet, Mathematical models in biology, SIAM, 46, 1988. ISBN 0-89871-554-7.

[23] B. Perthame, Transport equations in biology. Springer Science & Business Media, 2006. https://doi.org/10.1007/978-3-7643-7842-4.

[24] B. Perthame, Parabolic equations in biology. Growth, reaction, mouvement and diffusion, Springer, 2015. https://doi.org/10.1007/978-3-319-19500-1_1

[25] A. Ern, J.L Guermond, Theory and practice of finite elements, Springer, 159, 2004. https://doi.org/10.1007/978-1-4757-4355-5.

[26] V. Volpert, Elliptic Partial Differential Equations: Volume 2: Reaction-Diffusion Equations, Springer, 104, 2014. https://doi.org/10.1007/978-3-0348-0813-2.

[27] D. Kuzmin, A guide to numerical methods for transport equations, Friedrich-Alexander-Universitt, Erlangen-Nrnberg, 2010.

[28] F. Brezzi, L. D. Marini and E. Süli, Discontinuous Galerkin methods for first-order hyperbolic problems, Mathematical models and methods in applied sciences, World Scientific, 2004, 14, 1893-1903. https://doi.org/10.1142/S0218202504003866.

[29] W. H. Reed and T. R. Hill, Triangular mesh methods for the neutron transport equation, Los Alamos Scientific Lab., N. Mex.(USA), report LA-UR-73-479, 1973.

[30] B. Rivière, Discontinuous Galerkin methods for solving elliptic and parabolic equations: theory and implementation, Society for Industrial and Applied Mathematics, 2008. https://doi.org/10.1137/1.9780898717440

[31] J. Oden, I. Babuŝka and C. E. Baumann, A discontinuous hp-finite element method for diffusion problems, Journal of computational physics, Elsevier 146, 491-519, 1998. https://doi.org/10.1006/jcph.1998.6032.

[32] D. A. Di Pietro and A. Ern, Mathematical aspects of discontinuous Galerkin methods Springer, 69, 2011. https://doi.org/10.1007/978-3-642-22980-0.

[33] X. Zhang and CW. Shu CW. Maximum-principle-satisfying and positivity-preserving high-order schemes for conservation laws: survey and new developments, Proceedings of the Royal Society of London A: Mathematical, Physical and Engineering Sciences, 2011, 467, 2752-2776. https://doi.org/10.1098/rspa.2011.0153.

[34] B. Cockburn and CW. Shu, Runge-Kutta discontinuous Galerkin methods for convection-dominated problems. Journal of scientific computing, Springer, 2001, 16, 173-261. https://doi.org/10.1023/A:1012873910884.

[35] JS. Hesthaven and T. Warburton, Nodal discontinuous Galerkin methods: algorithms, analysis, and applications. Springer, 2007, 54. https://doi.org/10.1007/978-0-387-72067-8.

[36] V. Thomée, On positivity preservation in some finite element methods for the heat equation. International Conference on Numerical Methods and Applications, 2014, 13-24. https://doi.org/10.1007/978-3-319-15585-2_2.

[37] J. Zhu, YT. Zhang, SA. Newman and M. Alber, Application of discontinuous Galerkin methods for reaction-diffusion systems in developmental biology. Journal of Scientific Computing, Springer, 2009, 40, 391-418. https://doi.org/10.1007/s10915-008-9218-4.

[38] J.-F. Barczi, H. Rey, S. Griffon and C. Jourdan, DigR: a generic model and its open source simulation software to mimic three-dimensional root-system architecture

diversity. Annals of Botany, 2018, 121, 5, 1089-1104, https://doi.org/10.1093/aob/mcy018.

[39] L. X. Dupuy, M. Vignes, An algorithm for the simulation of the growth of root systems on deformable domains. Journal of Theoretical Biology, 2012, 310, 164-174. https://doi.org/10.1016/j.jtbi.2012.06.025.

[40] L. Dupuy, P. J. Gregory, A. G. Bengough, Root growth models: towards a new generation of continuous approaches. Journal of experimental botany, Soc Experiment Biol, 2010. https://doi.org/10.1093/jxb/erp389.

[41] P. Bastian, A. Chavarria-Krauser, C. Engwer, W. Jäger, S. Marnach, M. Ptashnyk, Modelling in vitro growth of dense root networks Journal of theoretical biology, Elsevier, 2008, 254, 99-109. https://doi.org/10.1016/j.jtbi.2008.04.014.

[42] T. Roose, A. Schnepf, Mathematical models of plant-soil interaction Philosophical Transactions of the Royal Society of London A: Mathematical, Physical and Engineering Sciences, The Royal Society, 2008, 366, 4597-4611, https://doi.org/10.1098/rsta.2008.0198.

[43] S. G. Adiku, R. D. Braddock, C. W. Rose, 1996, Simulating root growth dynamics, Environmental Software 11 : 99-103. https://doi.org/10.1016/S0266-9838(96)00041-X.

[44] H. Hayhoe, 1981, Analysis of a diffusion model for plant root growth and an application to plant soil-water uptake, Soil Science 131 : 334-343.

[45] M. Heinen, A. Mollier, P. De Willigen, 2003 Growth of a root system described as diffusion numerical model and application, Plant and Soil 252 : 251-265. https://doi.org/10.1023/A:1024749022761.

[46] V. R. Reddy, Ya. A. Pachepsky, 2001, Testing a convective dispersive model of two dimensional root growth and proliferation in a greenhouse experiment with mare plants, Annals of Botany 87 : 759-768. https://doi.org/10.1006/anbo.2001.1409.

[47] P. -H. Tournier, F. Hecht, M. Comte, Finite element model of soil water and nutrient transport with root uptake: explicit geometry and unstructured adaptive meshing. Transp. Porous Media 106 (2), 487504 (2015). https://doi.org/10.1007/s11242-014-0411-7.

[48] M. Comte, Analysis and Simulation of a Model of Phosphorus Uptake by Plant Roots in Current Research in Nonlinear Analysis: In Honor of Haim Brezis and Louis Nirenberg, Rassias, T. M. (Ed.), Springer International Publishing, 2018, 85-97. https://doi.org/10.1007/978-3-319-89800-1_4.

[49] F. Gérard, Cé Blitz-Frayret, P. Hinsinger, L. Pagès, Modelling the interactions between root system architecture, root functions and reactive transport processes in soil Plant and Soil, 2017, 413, 161-180. https://doi.org/10.1007/s11104-016-3092-x.

[50] H. Brezis, Functional analysis, Sobolev spaces and partial differential equations. Springer Science & Business Media, 2010. https://doi.org/10.1007/978-0-387-70914-7.

[51] P. R. Amestoy, I. S. Duff, J. Koster and J.-Y. L'Excellent, A fully asynchronous multifrontal solver using distributed dynamic scheduling, SIAM Journal of Matrix Analysis and Applications, Vol 23, No 1, pp 15-41 (2001). https://doi.org/10.1137/S0895479899358194.

[52] P. R. Amestoy, A. Guermouche, J.-Y. L'Excellent and S. Pralet, Hybrid scheduling for the parallel solution of linear systems. Parallel Computing Vol 32 (2), pp 136-156 (2006). https://doi.org/10.1016/j.parco.2005.07.004.

11

Mechanotransduction caused by a point force in the extracellular space

Bradley J. Roth
Department of Physics, Oakland University
Rochester, MI, USA
roth@oakland.edu

Abstract—The mechanical bidomain model is a mathematical description of biological tissue that focuses on mechanotransduction. The model's fundamental hypothesis is that differences between the intracellular and extracellular displacements activate integrins, causing a cascade of biological effects. This paper presents analytical solutions of the bidomain equations for an extracellular point force. The intra- and extracellular spaces are incompressible, isotropic, and coupled. The expressions for the intra- and extracellular displacements each contain three terms: a monodomain term that is identical in the two spaces, and two bidomain terms, one of which decays exponentially. Near the origin the intracellular displacement remains finite and the extracellular displacement diverges. Far from the origin the monodomain displacement decays in inverse proportion to the distance, the strain decays as the distance squared, and the difference between the intra- and extracellular displacements decays as the distance cubed. These predictions could be tested by applying a force to a magnetic nanoparticle embedded in the extracellular matrix and recording the mechanotransduction response.

Keywords-analytical solution; extracellular matrix; integrin; intracellular cytoskeleton; mathemat-ical model; mechanotransduction; mechanical bidomain model; point source.

I. INTRODUCTION

Mechanotransduction is the process by which biological tissues grow and remodel in response to mechanical signals. One cause of mechanotransduction might be a cascade of biological responses triggered by activation of integrin molecules in the cell membrane [2], [3], [16]. A force acting on the extracellular matrix is transmitted to the cytoskeleton via these integrins, thereby coupling the intra- and extracellular spaces. Much research on mechanotransduction is qualitative, but to predict quantitatively how tissue responds to applied forces we need a mathematical model [12]. Many studies in mechanobiology analyze individual cells and molecules, but to describe tissues and organs we require a macroscopic model that averages over the cellular and molecular scales. Yet, this macroscopic model must predict the activation of integrin molecules.

One mathematical model that describes mechan-otransduction is the mechanical bidomain model

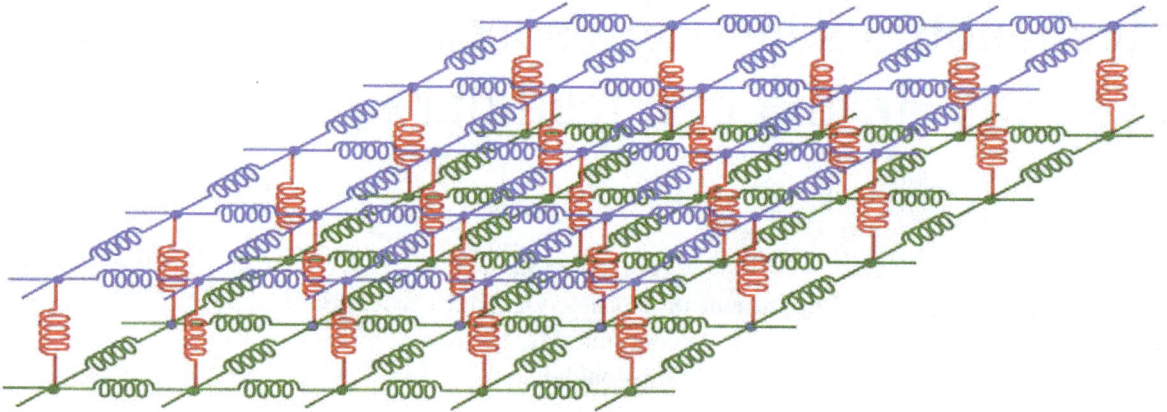

Fig. 1. A schematic illustration of the mechanical bidomain model. The green springs represent the intracellular cytoskeleton, the blue the extracellular matrix, and the red the integrins. The figure illustrates a two-dimensional version of the model, but this article analyzes a three-dimensional version.

[11], [15]. It predicts displacements of the intra- and extracellular spaces individually. The difference between the intra- and extracellular displacements results in a force on the integrins that couple the two spaces. A schematic illustration of the model is shown in Figure 1. One of the most important properties of a mathematical model is how it responds to a point source. Often complicated responses can be expressed as a convolution of the point source response, so knowing how tissue responds to a point force provides insight into its general behavior.

In this paper, I derive analytical expressions describing how the mechanical bidomain model responds to a point source in the extracellular space. Experimentally, this could be approximated by, for instance, applying a magnetic force on a superparamagnetic nanoparticle [7], [8]. Magnetic tweezers [5] have been used to exert forces on single cells or individual molecules. The technique, however, could be applied to intact tissue where a nanoparticle is embedded in the extracellular matrix. When a force is exerted by the nanoparticle it pulls on the matrix, which stretches the integrins embedded in the membranes of nearby cells, triggering mechanotransduction [9].

II. METHODS

I assume the intra- and extracellular spaces are incompressible and isotropic, and their strains are small and linear. Incompressibility implies that the intracellular displacement $\mathbf{u}$ and the extracellular displacement $\mathbf{w}$ are both divergenceless. I use spherical coordinates (r, θ, ϕ) with the force applied at the origin and acting along the z axis ($\theta = 0$). By symmetry there are no displacements or derivatives in the ϕ direction. In that case $\mathbf{u}$ and the intracellular strain ϵ_i are related by [10]

$$\epsilon_{irr} = \frac{\partial u_r}{\partial r}, \tag{1}$$

$$\epsilon_{i\theta\theta} = \frac{1}{r}\frac{\partial u_\theta}{\partial \theta} + \frac{u_r}{r}, \tag{2}$$

$$\epsilon_{i\phi\phi} = \frac{u_\theta}{r}\cot\theta + \frac{u_r}{r}, \tag{3}$$

$$\epsilon_{ir\theta} = \frac{1}{2}\left(\frac{1}{r}\frac{\partial u_r}{\partial \theta} + \frac{\partial u_\theta}{\partial r} - \frac{u_\theta}{r}\right), \tag{4}$$

with analogous relationships in the extracellular space. The intracellular stress τ_i and the intracellular strain are related by

$$\tau_{irr} = -p + 2\nu\epsilon_{irr}, \tag{5}$$

$$\tau_{i\theta\theta} = -p + 2\nu\epsilon_{i\theta\theta}, \qquad (6)$$

$$\tau_{i\phi\phi} = -p + 2\nu\epsilon_{i\phi\phi}, \qquad (7)$$

$$\tau_{ir\theta} = 2\nu\epsilon_{ir\theta}, \qquad (8)$$

where p is the intracellular pressure and ν is the intracellular shear modulus. Similar stress-strain relationships exist for the extracellular pressure q and extracellular shear modulus μ. The equations of mechanical equilibrium are [10], [15]

$$-\frac{\partial p}{\partial r} + 2\nu\left[\frac{\partial \epsilon_{irr}}{\partial r} + \frac{1}{r}\frac{\partial \epsilon_{ir\theta}}{\partial \theta}\right.$$
$$\left. + \frac{1}{r}\left(2\epsilon_{irr} - \epsilon_{i\theta\theta} - \epsilon_{i\phi\phi} + \cot\theta\,\epsilon_{ir\theta}\right)\right]$$
$$= K\left(u_r - w_r\right), \qquad (9)$$

$$-\frac{1}{r}\frac{\partial p}{\partial \theta} + 2\nu\left[\frac{\partial \epsilon_{ir\theta}}{\partial r} + \frac{1}{r}\frac{\partial \epsilon_{i\theta\theta}}{\partial \theta}\right.$$
$$\left. + \frac{1}{r}\left(\left(\epsilon_{i\theta\theta} - \epsilon_{i\phi\phi}\right)\cot\theta + 3\epsilon_{ir\theta}\right)\right]$$
$$= K\left(u_\theta - w_\theta\right), \qquad (10)$$

$$-\frac{\partial q}{\partial r} + 2\mu\left[\frac{\partial \epsilon_{err}}{\partial r} + \frac{1}{r}\frac{\partial \epsilon_{er\theta}}{\partial \theta}\right.$$
$$\left. + \frac{1}{r}\left(2\epsilon_{err} - \epsilon_{e\theta\theta} - \epsilon_{e\phi\phi} + \cot\theta\,\epsilon_{er\theta}\right)\right]$$
$$+ F\delta\left(r\right)\cos\theta$$
$$= -K\left(u_r - w_r\right), \qquad (11)$$

$$-\frac{1}{r}\frac{\partial q}{\partial \theta} + 2\mu\left[\frac{\partial \epsilon_{er\theta}}{\partial r} + \frac{1}{r}\frac{\partial \epsilon_{e\theta\theta}}{\partial \theta}\right.$$
$$\left. + \frac{1}{r}\left(\left(\epsilon_{e\theta\theta} - \epsilon_{e\phi\phi}\right)\cot\theta + 3\epsilon_{er\theta}\right)\right]$$
$$- F\delta\left(r\right)\sin\theta$$
$$= -K\left(u_\theta - w_\theta\right), \qquad (12)$$

where K is the integrin spring constant coupling the two spaces, F is the force applied to the extracellular space, and $\delta(r)$ is the delta function. I assume that the displacements and pressures go to zero at large r.

To picture the problem physically, imagine that in Figure 1 a point in the extracellular matrix (one of the blue dots) is pulled to the right by an attached nanoparticle. This force would displace the extracellular matrix (blue springs), which would stretch the integrins coupling the two spaces (red springs). The integrins would then pull on the cytoskeleton, causing the intracellular space to be displaced.

III. RESULTS

Equations 9-12 were solved using the method of undetermined coefficients. The solution is

$$u_r = \frac{F}{8\pi\left(\nu + \mu\right)}\cos\theta$$
$$\left\{\frac{2}{r} - \frac{4\sigma^2}{r^3} + 4\left[\frac{\sigma^2}{r^3} + \frac{\sigma}{r^2}\right]e^{-\frac{r}{\sigma}}\right\}, \qquad (13)$$

$$u_\theta = \frac{F}{8\pi\left(\nu + \mu\right)}\sin\theta$$
$$\left\{-\frac{1}{r} - \frac{2\sigma^2}{r^3} + 2\left[\frac{\sigma^2}{r^3} + \frac{\sigma}{r^2} + \frac{1}{r}\right]e^{-\frac{r}{\sigma}}\right\}, \qquad (14)$$

$$w_r = \frac{F}{8\pi\left(\nu + \mu\right)}\cos\theta$$
$$\left\{\frac{2}{r} + \frac{\nu}{\mu}\frac{4\sigma^2}{r^3} - 4\frac{\nu}{\mu}\left[\frac{\sigma^2}{r^3} + \frac{\sigma}{r^2}\right]e^{-\frac{r}{\sigma}}\right\}, \qquad (15)$$

$$w_\theta = \frac{F}{8\pi\left(\nu + \mu\right)}\sin\theta$$
$$\left\{-\frac{1}{r} + \frac{\nu}{\mu}\frac{2\sigma^2}{r^3} - 2\frac{\nu}{\mu}\left[\frac{\sigma^2}{r^3} + \frac{\sigma}{r^2} + \frac{1}{r}\right]e^{-\frac{r}{\sigma}}\right\}, \qquad (16)$$

$$p = 0, \qquad (17)$$

$$q = \frac{F}{4\pi}\frac{\cos\theta}{r^2}. \qquad (18)$$

Each expression for the displacement contains a monodomain term (first term in the brace) that is the same in the intra- and extracellular spaces, and two bidomain terms that are different in the two spaces (one is $-\nu/\mu$ times the other). The first bidomain term is proportional to σ^2, where $\sigma = \sqrt{\frac{\nu\mu}{K(\nu+\mu)}}$ is a length constant characteristic of the mechanical bidomain model [15]. The exponential in the second bidomain term decays with length constant σ.

The displacements (Eqs. 13-16) have interesting properties as r goes to zero. If you expand the exponential as a Taylor series, you will find that the terms in the expression for the intracellular displacement that are singular at the origin cancel and it remains finite there. The extracellular displacement, however, diverges at the origin as $1/r$ as expected for a delta function source in the extracellular space. At large distances ($r \gg \sigma$) bidomain terms decay more rapidly than monodomain terms.

The fundamental hypothesis of the mechanical bidomain model is that mechanotransduction depends on the difference $\mathbf{u} - \mathbf{w}$ [15]. The monodomain terms are the same in the two spaces and do not contribute to $\mathbf{u} - \mathbf{w}$; only the bidomain terms generate the displacement difference that drives mechanotransduction,

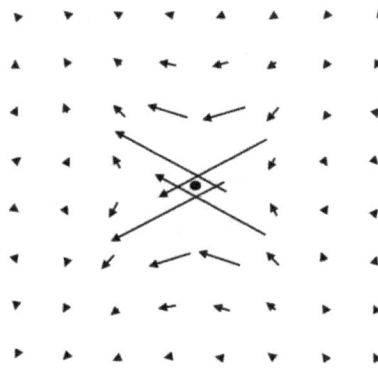

$$u_r - w_r = \frac{F}{8\pi\mu} \cos\theta \left\{ -\frac{4\sigma^2}{r^3} + 4\left[\frac{\sigma^2}{r^3} + \frac{\sigma}{r^2}\right] e^{-\frac{r}{\sigma}} \right\},$$

$$u_\theta - w_\theta = \frac{F}{8\pi\mu} \sin\theta \left\{ -\frac{2\sigma^2}{r^3} + 2\left[\frac{\sigma^2}{r^3} + \frac{\sigma}{r^2} + \frac{1}{r}\right] e^{-\frac{r}{\sigma}} \right\}.$$

For $r \gg \sigma$ the exponentials are negligible and the difference in displacements falls as $1/r^3$.

Figure 2 shows the extracellular displacement, $\mathbf{w}$, the intracellular displacement, $\mathbf{u}$, and their difference, $\mathbf{u} - \mathbf{w}$, in the plane corresponding to a constant angle ϕ. Near the source, $\mathbf{u}$ $\mathbf{w}$ resembles $-\mathbf{w}$. Far from the source, $\mathbf{u} - \mathbf{w}$ is small compared to $\mathbf{u}$ and $\mathbf{w}$ individually.

Extracellular displacement, w

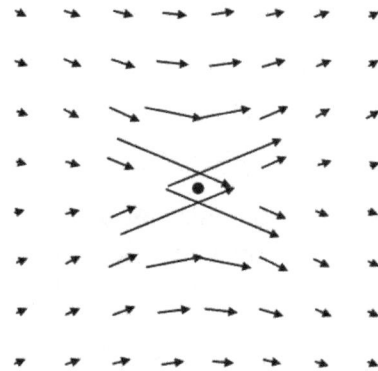

Intracellular displacement, u

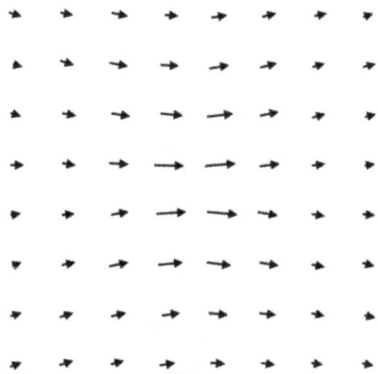

u - w

σ

Fig. 2.   The extarcellular displacement, $\mathbf{w}$, the intracellular displacement, $\mathbf{u}$, and their difference, $\mathbf{u}$-$\mathbf{w}$. The calculation assumes $\nu = \mu$. The black dot indicates the position of the point source, corresponding to an applied force F acting to the right.

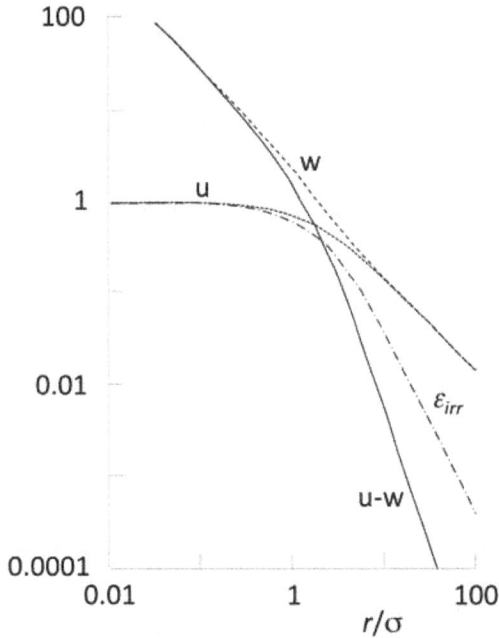

Fig. 3. u_r, w_r, u_r - w_r, and ϵ_{irr} as functions of r/σ, for $\theta = 0$; u_r is indicated by short dashes, w_r by long dashes, u_r - w_r by a solid line, and ϵ_{irr} by dash-dot. All quantities are normalized so that the intracellular displacement and strain are equal to one at the origin.

Figure 3 plots the intra- and extracellular displacements and their difference along the direction of the applied force. It also shows the intracellular strain, ϵ_{irr}. At large distances, the displacements fall as $1/r$, the strain as $1/r^2$, and the difference in the displacements as $1/r^3$. This result is a testable prediction. If mechanotransduction depends on the strain it decays relatively slowly, as $1/r^2$. If, however, mechanotransduction depends on $\mathbf{u}$ - $\mathbf{w}$ it decays relatively rapidly, as $1/r^3$.

IV. DISCUSSION

Most biomechanical models treat tissue as a single phase: a monodomain. These mathematical models are often valuable tools for predicting tissue displacements, stresses, and strains [4]. If, however, mechanotransduction is triggered by activation of integrins, and integrins are activated by differences between the displacements of the intra- and extracellular spaces, then a bidomain model is essential for predicting where mechanotransduc-

tion occurs. The activation of integrins could in principle be determined by measuring the intra- and extracellular displacements individually, and then taking their difference. In practice, however, this difference is very small compared to the displacements themselves, and a better strategy would be to measure a mechanotransduction effect caused by integrin activation, such as tissue growth, remodeling, or genetic changes associated with these processes.

The monodomain solution for a point source is $u_r = w_r = \frac{F}{8\pi(\nu+\mu)}\frac{2\cos\theta}{r}$ and $u_\theta = w_\theta = -\frac{F}{8\pi(\nu+\mu)}\frac{\sin\theta}{r}$. This solution is the same as the expression for the velocity caused by a point force in an incompressible fluid at low Reynolds number [10], sometimes referred to as a Stokeslet. When σ is small the Stokeslet approximates the displacements in the intra- and extracellular spaces, but it provides no information about where mechanotransduction occurs because it contributes nothing to $\mathbf{u}$ - $\mathbf{w}$. The monodomain term can be represented in Fig. 3 as a line that matches the $\mathbf{u}$ and $\mathbf{w}$ curves at large radii, and is extrapolated back linearly at smaller radii.

A key parameter in the model is the length constant σ, which depends on the bidomain constant K coupling the intra- and extracellular spaces. In monolayers of stem cells, σ is about 150 microns [1], which is larger than a cell and much larger than a nanoparticle, implying that a macroscopic model should be valid.

The mechanical bidomain model has many similarities to the electrical bidomain model [6] used to describe pacing and defibrillation of the heart. My analysis of the mechanical bidomain model's response to a point force is analogous to the calculation of the transmembrane potential produced by a point current using the electrical bidomain model [13]. In the electrical model, unequal anisotropy ratios for the intra- and extracellular conductivities plays a crucial role in determining the transmembrane potential distribution. Similar effects might arise in the mechanical model if it were made anisotropic.

What experiment can test the predictions of this

model? One suggestion is to grow a large cluster of epithelial cells, with a magnetic particle at its center. Alternatively, tissue engineering techniques could be used to grow cells in an extracellular substrate containing a magnetic particle. Then, a force could be applied to the particle, and the mechanotransduction response could be imaged by monitoring a second messenger activated by the integrins, or the turning on of a gene associated with cell growth.

The bidomain model has several limitations. It assumes a linear relationship between displacement and strain, which is only appropriate for small strains [10]. In my solution, the extracellular displacement and strain diverge at the origin, so the small strain assumption is violated there. However, the delta function is an approximation that breaks down on a distance scale similar to the radius of the magnetic nanoparticle used to exert the force. As long as the strains are small at this scale, the linear approximation should be valid. I assume the stress-strain relationships are linear, whereas in tissue these relationships can be nonlinear [4]. If the strains are small enough, however, a linear approximation should suffice. I assume that the tissue is isotropic, but tissues such as muscle are anisotropic and the model needs to be extended to account for anisotropy. I assume both the intra- and extracellular spaces are incompressible. Because both spaces contain mostly water, the incompressible assumption should be accurate [14]. My model is for steady-state. If the applied force varies with time, the solution might be invalid over short times because of the propagation of sound waves, or over long times because of viscoelasticity or tissue growth and remodeling. Finally, and fundamentally, I assume that mechanotransduction depends on the difference in the displacements, $\mathbf{u}$ - $\mathbf{w}$. If it depends on other factors, such as the intracellular stress or strain, or some microscopic behavior that is not included in this macroscopic model, the results might not describe mechanotransduction correctly.

The model could be extended to avoid some of my limiting assumptions, but in that case an analytical solution might not exist. Analytical solutions can provide insight into the model behavior and are valuable even when the model is only an approximation. Moreover, analytical solutions are useful for testing limiting cases of complex models and for evaluating the accuracy of numerical methods.

V. Conclusion

The mechanical bidomain model makes testable predictions about where mechanotransduction occurs. In particular, the model predicts that the distribution of mechanotransduction in response to a point source in the extracellular space falls off with distance more rapidly if mechanotransduction is driven by the difference in the intra- and extracellular displacements, and less rapidly if mechanotransduction is driven by intra- or extracellular strain. This prediction could be tested by measuring how the tissue responds to a force applied using a magnetic nanoparticle embedded in the extracellular space.

References

[1] Auddya D, Roth BJ (2017) A mathematical description of a growing cell colony based on the mechanical bidomain model. *J Phys D* 50:105401.

[2] Chiquet M (1999) Regulation of extracellular matrix gene expression by mechanical stress. *Matrix Biology* 18:417-426.

[3] Dabiri BE, Lee H, Parker KK (2012) A potential role for integrin signaling in mechanoelectrical feedback. *Prog Biophys Mol Biol* 110:196-203.

[4] Fung YC (1981) *Biomechanics: Mechanical Properties of Living Tissues*. Springer, New York.

[5] Gosse C, Croquette V (2002) Magnetic tweezers: Micromanipulation and force measurement at the molecular level. *Biophys J* 82:3314-3329.

[6] Henriquez CS (1993) Simulating the electrical behavior of cardiac tissue using the bidomain model. *Crit Rev Biomed Eng* 21:1-77.

[7] Hughes S, McBain S, Dobson J, El Haj AJ (2007) Selective activation of mechanosensitive ion channels using magnetic particles. *J R Soc Interface* 5:855-863.

[8] Ingber DE (2009) From cellular mechanotransduction to biologically inspired engineering. *Ann Biomed Eng* 38:1148 1161.

[9] Kresh JY, Chopra A (2011) Intercellular and extracellular mechanotransduction in cardiac myocytes. *Pflugers Arch Eur J Physiol* 462:75-87.

[10] Love AEH (1944) *A Treatise on the Mathematical Theory of Elasticity*. Dover, New York.

[11] Roth BJ (2013) The mechanical bidomain model: A review. *ISRN Tissue Engineering* 2013:863689.

[12] Schwarz US (2017) Mechanobiology by the numbers: A close relationship between biology and physics. *Nat Rev Mol Cell Biol* 18:711-712.

[13] Sepulveda NG, Roth BJ, Wikswo JP (1989) Current injection into a two-dimensional anisotropic bidomain. *Biophys J* 55:987-999.

[14] Sharma K, Roth BJ (2014) How compressibility influences the mechanical bidomain model. *BIOMATH* 3:141171.

[15] Sharma K, Al-asuoad N, Shillor M, Roth BJ (2015) Intracellular, extracellular, and membrane forces in remodeling and mechanotransduction: The mechanical bidomain model. *Journal of Coupled Systems and Multiscale Dynamics* 3:200-207.

[16] Sun Y, Chen CS, Fu J (2012) Forcing stem cells to behave: A biophysical perspective of the cellular microenvironment. *Annu Rev Biophys* 41:519-542.

12

Modelling cell-cell collision and adhesion with the filament based lamellipodium model

Nikolaos Sfakianakis[*], Diane Peurichard[†], Aaron Brunk[‡], Christian Schmeiser[§]

[*]Institute of Applied Mathematics, Heidelberg University
Im Neuenheimerfeld 205, 69120, Heidelberg, Germany,
sfakiana@math.uni-heidelberg.de

[†] Laboratoire Jacques-Louis Lions, INRIA, Sorbonne University
Place Jussieu 4, Paris, France
diane.a.peurichard@inria.fr

[‡] Institute of Mathematics, Johannes Gutenberg University
Staudingerweg 9, 55128, Mainz, Germany
abrunk@uni-mainz.de

[§] Faculty of Mathematics, University of Vienna
Oskar-Morgenstern-Platz 1, 1090, Vienna, Austria
christian.schmeiser@univie.ac.at

Abstract—**We extend the live-cell motility Filament Based Lamellipodium Model (FBLM) to incorporate the forces exerted on the lamellipodium of the cells due to cell-cell collision and *cadherin* induced cell-cell adhesion. We take into account the nature of these forces via physical and biological constraints and modelling assumptions. We investigate the effect these new components have in the migration and morphology of the cells through particular experiments. We exhibit moreover the similarities between our simulated cells and HeLa cancer cells.**

I. INTRODUCTION.

Cell adhesion is a key process in a wide range of biological phenomena. It usually acts along with *cell migration* and together they play a fundamental role in the development of the organism e.g. during the *gastrulation* and the *patterning* phases of a vertebrates' body. Cell adhesion and migration are important after the developmental phase in the maintenance and repair of the cell and tissue structure. On the other hand, the dysregulation of these processes has been associated to a number of *diseases* and *conditions* including *tumour metastasis*.

Cell adhesion is the result of interactions between specialized proteins found at the surface of the cells termed *cell-adhesion molecules* (CAM). The CAMs are divided into four main groups: *inte-*

grins, *immunoglobulins*, *cadherins*, and *selectins*. Of these, the *integrins* participate, primarily, in the *cell-extracellular matrix* (ECM) adhesion and play a pivotal role in the migration of the cells. The *cadherins* (calcium dependent adhesions) are fundamental in *cell-cell adhesion* and in the formation of *cell clusters* and *tissues*.

The *cadherin* proteins, in particular, are comprised of three domains, an *intracellular*, a *transmembrane*, and an *extracellular* domain. The *intracellular* domain is linked to the *actin filaments* (F-actin), whereas the *extracellular* domain binds to the *extracellular* domain of *cadherins* of neighbouring cells. The *extracellular* domain is highly binding specific and accordingly classifies the *cadherins* in several types (*E-*, *N-cadherins* etc.). Variable expression levels of these *cadherin* types lead to preferential adhesion organization of the cells and to the formation of different tissues.

In the current paper, our objective is to model *cadherin* induced cell-cell adhesion and combine it with a mathematical model of cell migration and cell-ECM adhesion. We focus on a particular type of cell migration in which the lamellipodium of the cell plays a pivotal role. It is termed *actin-based cell motility* and is employed by fast migrating cells such as *fibroblasts*, *keratocytes*, and *cancer cells*.

There have been several efforts to model and simulate this type of cell migration in the literature, e.g. [5], [12], [19], [1], [4], [22], [13], [9], [2], [20], [21]. Here, we use and build on the *Filament Based Lamellipodium Model* (FBLM). This is a two-dimensional, two-phase model that describes the lamellipodium at the level of actin-filaments. The FBLM was first derived in [18], [16] and later extended in [10]. When endowed with a particular and problem specific *Finite Element Method* (FEM), the resulting FBLM-FEM is able to reproduce biologically realistic, crawling-like lamellipodium driven cell motility [11], [3], [23].

Although the FBLM describes the dynamics of the actin-filaments and the lamellipodium, the deduced motility is understood as the motility of

the cell. This is primarily due to the predominant role of the lamellipodium in the motility of the model-biological cell (i.e. *fish keratocyte*) that we consider, [25]. So, for the rest of this work we will not distinguish between the two cases, and will use the term *cell motility* for both.

The extensions of the FBLM that we propose in this work, account for two phenomena: the exchange of *cadherin* mediated adhesion forces and physical collision forces between two neighbouring cells. The *cell-cell adhesion* forces are attractive/pulling whereas the *cell-cell collision* forces are repulsive/pushing. Both are introduced in the FBLM through an attractive-repulsive potential that depends, non-linearly, on the relative distance of the two cell membranes. When the cells come close enough, within a distance that justifies the deployment of *cadherin* adhesions, an attractive force is developed between the two membranes. As the distance between the cells decreases, the adhesion forces increase in magnitude and gradually collision repulsion forces between the cell membranes emerge. These increase in magnitude faster than the *cadherin* adhesion forces (which remain bounded) and an equilibrium between the two types of forces is quickly achieved. The collision forces are not bounded and, if they increase above a particular threshold (corresponding to an extremely small distance between the membranes), the polymerization of the filaments involved in the collision ceases. This ensures that the two cells will not overlap.

The rest of the paper is structured as follows: in Section II we briefly discuss the FBLM and some of its main components, including the polarization of the lamellipodium and the calibration of the polymerization rate. In Section III we present the new components of the FBLM. We derive in detail the (sub-)model for the collision and adhesion forces and justify it biologically. In Section IV we discuss the coupling of the FBLM with the extracellular environment and its response to chemical and haptotaxis stimuli. Finally, in Section V we present three numerical experiments. The first two exhibit and compare the effects of cell-cell

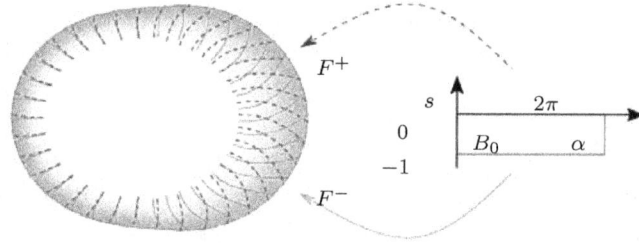

Fig. 1.   Graphical representation of the $\mathbf{F}^{\pm} : B_0 \to \mathbb{R}^2$ mappings that define the lamellipodium. The $s = 0$ boundary of B_0 is mapped to the membrane of the cell and the $s = -1$ to the minus-ends of the filaments inside the cell. The filaments and the rest of the functions of α are periodic with respect to α. The "filaments" plotted in the lamellipodium correspond to the discretization interfaces of B_0 along the α direction. The grey colour represents the density of F-actin inside the cell.

collision and cell-cell adhesion in the migration and morphology of the cells, and one that exhibits the first stages of cell-cluster formation and its response to a variable chemical and haptotaxis environment. In the last experiment we compare our deduced cell morphologies with the ones of HeLa cancer cells under *in vitro* cell-cell interaction and migration.

II. THE FBLM.

We present here only the main components of the FBLM and refer to [18], [16], [15], [10], [11], [3], [23] for more details.

The FBLM is a two-dimensional model that describes the lamellipodium of living cells by including key biomechanical processes of the actin-filaments, the interactions between them, and their interactions with the extracellular environment. The basic assumptions behind the FBLM are the following: the lamellipodium is a two-dimensional structure, comprised of actin filaments that are organized in two locally parallel families (which are denoted by the superscripts $\pm$). The two families of filaments cover a ring-shaped domain between the membrane of the cell and its interior. In the "inside" part of the cell, behind the lamellipodium, further cellular structures are to be found, e.g. *nucleus* and more. We will henceforth refer to the combined lamellipodium-intracellular space as "cell" or "FBLM-cell", see e.g. Figure 1.

The filaments of the two families are indexed by the continuum variable $\alpha \in [0, 2\pi)$, and are parametrised by their arclength

$$\left\{ \mathbf{F}^{\pm}(\alpha, s, t) : -L^{\pm}(\alpha, t) \leq s \leq 0 \right\} \subset \mathbb{R}^2, \quad (1)$$

where $L^{\pm}(\alpha, t)$ is the maximal length of the filament α at time t. The plus ends of the filaments (at $s = 0$) of every family define the outer boundary of the family and "coincide" with the membrane of the cell,

$$\begin{aligned} &\left\{ \mathbf{F}^{+}(\alpha, 0, t) : 0 \leq \alpha < 2\pi \right\} \\ &= \left\{ \mathbf{F}^{-}(\alpha, 0, t) : 0 \leq \alpha < 2\pi \right\}, \ \forall t \geq 0 \, . \ (2) \end{aligned}$$

For every (α, s, t) holds that

$$\left| \partial_s \mathbf{F}^{\pm}(\alpha, s, t) \right| = 1 \quad \forall (\alpha, s, t) \, . \qquad (3)$$

This arclength condition can be understood as an *inextensibility* constraint between the subsequent monomers that comprise the filaments. Moreover, we assume that filaments of the same family do not cross, i.e.

$$\det \left(\partial_{\alpha} \mathbf{F}^{\pm}, \partial_s \mathbf{F}^{\pm} \right) > 0 \qquad (4)$$

and that filaments of different families cross at most once

$$\begin{aligned} \Big\{ &\forall (\alpha^+, \alpha^-) \ \exists \text{ at most one } (s^+, s^-) : \\ &\mathbf{F}^{+}(\alpha^+, s^+, t) = \mathbf{F}^{-}(\alpha^-, s^-, t) \Big\}. \quad (5) \end{aligned}$$

The FBLM is comprised of the force balance

system

$$0 = \underbrace{\mu^B \partial_s^2 \left(\eta \partial_s^2 \mathbf{F} \right)}_{\text{bending}} - \underbrace{\partial_s \left(\eta \lambda_{\text{inext}} \partial_s \mathbf{F} \right)}_{\text{in-extensibility}} + \underbrace{\mu^A \eta D_t \mathbf{F}}_{\text{adhesion}}$$

$$+ \underbrace{\partial_s \left(p(\rho) \partial_\alpha \mathbf{F}^\perp \right) - \partial_\alpha \left(p(\rho) \partial_s \mathbf{F}^\perp \right)}_{\text{pressure}}$$

$$\pm \underbrace{\partial_s \left(\eta \eta^* \widehat{\mu^T} (\phi - \phi_0) \partial_s \mathbf{F}^\perp \right)}_{\text{twisting}}$$

$$+ \underbrace{\eta \eta^* \widehat{\mu^S} \left(D_t \mathbf{F} - D_t^* \mathbf{F}^* \right)}_{\text{stretching}}, \tag{6}$$

where $\mathbf{F}^\perp = (F_1, F_2)^\perp = (-F_2, F_1)$ and where the $\pm$ notation has been dropped here to focus on one of the two filament families. The other family, for which a similar equation holds, is indicated by the superscript *.

The function $\eta(\alpha, s, t)$ represents the local density of filaments of length at least $-s$ at time t with respect to α. Its evolution is dictated, along with $L(\alpha, t)$, by a particular submodel that includes the effects of *actin polymerization, filament nucleation, branching,* and *capping.* The derivation of this submodel is thoroughly discussed in [10].

The first term of the FBLM (6) describes the resistance of the filaments against bending, the second term describes the tangential tension force that enforces the inextensibility constraint (3) with the *Lagrange multiplier* $\lambda_{\text{inext}}(\alpha, s, t)$, and the third term describes the friction between the filament and the substrate. The *material derivative* operator

$$D_t := \partial_t - v \partial_s \tag{7}$$

describes the velocity of F-actin relative to the substrate, and $v(\alpha, t) \geq 0$ is the polymerization rate at the leading edge of the filaments. Similarly, $D_t^* := \partial_t - v^* \partial_s$ is the corresponding material derivative operator for the *-family. The pressure term in (6) encodes the Coulomb repulsion between neighbouring filaments of the same family, where the *pressure* $p(\rho)$ is given through the density of actin as

$$\rho = \frac{\eta}{|\det(\partial_\alpha \mathbf{F}, \partial_s \mathbf{F})|} . \tag{8}$$

The two last terms in (6) model the resistance of the cross-link proteins and branch junctions against changing the inter-filament angle

$$\phi = \arccos(\partial_s \mathbf{F} \cdot \partial_s \mathbf{F}^*)$$

away from the equilibrium angle ϕ_0, and against stretching.

The system (6) is also subject to the boundary conditions

$$- \mu^B \partial_s \left(\eta \partial_s^2 \mathbf{F} \right) - p(\rho) \partial_\alpha \mathbf{F}^\perp + \eta \lambda_{\text{inext}} \partial_s \mathbf{F}$$

$$\mp \eta \eta^* \widehat{\mu^T} (\phi - \phi_0) \partial_s \mathbf{F}^\perp \tag{9a}$$

$$= \begin{cases} \eta \left(f_{\text{tan}}(\alpha) \partial_s \mathbf{F} + f_{\text{inn}}(\alpha) \mathbf{V}(\alpha) \right), & \text{for } s = -L, \\ \pm \lambda_{\text{tether}} \nu, & \text{for } s = 0, \end{cases}$$

$$\eta \partial_s^2 \mathbf{F} = 0, \qquad \text{for } s = -L, 0 . \tag{9b}$$

The right-hand side of (9a) describes various forces applied to the filament ends. At $s = 0$ (cell membrane), the force in the direction ν orthogonal to the leading edge arises from the constraint (2) with the Lagrange parameter λ_{tether}. The forces at the inner end-point $s = -L$ model the contraction effect of actin-myosin interaction and are directed toward the interior of the cell, refer to [10] for details.

Lamellipodium polarization.

Fundamental to the motility of the cells is the polarization of the lamellipodium. The effective pulling force becomes stronger in the direction of the wider lamellipodium and the cell migrates accordingly.

This is also encoded in the FBLM where the maximal filament length $L(\alpha, t)$ (and hence the local width of the lamellipodium) depends directly on the local polymerization rate $v(\alpha, t)$. This was previously modelled in [10], where based on the *capping, severing,* and *filament nucleation* processes, it was deduced that

$$L(\alpha, t) = -\frac{\kappa_{\text{cap}}}{\kappa_{\text{sev}}} + \sqrt{\frac{\kappa_{\text{cap}}^2}{\kappa_{\text{sev}}^2} + \frac{2v(\alpha, t)}{\kappa_{\text{sev}}} \log \frac{\eta(0, t)}{\eta_{\text{min}}}} . \tag{10}$$

Note the monotonic relation between the polymerization rate $v(\alpha, t)$ and the lamellipodium width

$L(\alpha, t)$. This is employed in the FBLM to control the polarization of the lamellipodium and the migration of the cell.

Adjusting the polymerization rate.

We account for two different mechanisms that adjust the polymerization rate $v(\alpha, t)$. The first is the response of the polymerization machinery to *extracellular* chemical signals, as they are perceived by the cell through specialized transmembrane receptors. The second mechanism represents various (unspecified in this work) *intracellular* processes that might cut off, enhance, or otherwise destabilize the polymerization rate, independently of extracellular chemical or other stimuli.

In more detail, the first mechanism responds to the density of the chemoattractant c at the plus ends ($s = 0$) of the filaments

$$c^{\pm}(\alpha, t) = c\left(\mathbf{F}^{\pm}(\alpha, 0, t), t\right). \quad (11a)$$

We assume that the polymerization rate is adjusted between two biologically relevant minimum and maximum values $v_{\min}, v_{\max}$ in the following manner

$$v_{\text{ext}}^{\pm}(\alpha, t) = v_{\max} - (v_{\max} - v_{\min})e^{-\lambda_{\text{res}} c^{\pm}(\alpha, t)}, \quad (11b)$$

where the coefficient λ_{res} represents the response of the cell to changes of the extracellular chemical. The second mechanism describes the response of the polymerization machinery to internal destabilization processes that might lead to a plethora of phenomena such as persistent or abruptly changing very high or very low polymerization rates, etc. We understand the biological significance and distinctive functionality of these mechanisms and employ them both. Overall, the polymerization rate $v^{\pm}$ is given by

$$v^{\pm}(\alpha, t) = \mathcal{D}_{\text{stb}}\left(v_{\text{ext}}^{\pm}(\alpha, t)\right), \quad (12)$$

where $\mathcal{D}_{\text{stb}}$ describes the internal controlling mechanism that can potentially depend on a large number of cellular processes.

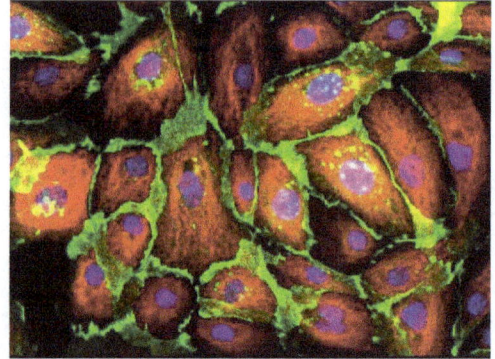

Fig. 2.   Cryopreserved human mammary epithelial cells stained visualize the calcium-dependent cell-cell adhesion glycoprotein E-cadherin in green. Image by N. Prigozhina (2015) CIL:48102q doi:10.7295/W9CIL48102.

III. CELL-CELL ADHESION AND COLLISION.

The FBLM is developed in a modular way in which every contribution accounts for the potential energy stored in the lamellipodium by the action of the corresponding biological component, see e.g. [18], [17], [10]. In a similar fashion, cell-cell adhesion and collision are incorporated in the FBLM as additional potential energies acting at the plus-ends of the filaments. To that end we make the following simplifying modelling assumptions:

Assumption 1: When two cells come in adhesion proximity (a given parameter of the model), the extracellular domains of their *cadherins* attach and bind to each other. This introduces attractive/pulling forces exerted on the plus ends of the actin-filaments on which the intracellular domain of the *cadherins* are linked to. These adhesion forces increase to a maximum value (a given parameter of the model) with the decrease of the *cadherin* binding length,

Assumption 2: Upon collision, repulsion/pushing forces are developed between the two cells and increase rapidly. By nature, these forces can be unbounded, and they soon counteract the effect of the *cadherin* adhesion forces. We model the collision forces pro-actively, i.e. they appear shortly before the two cells collide (a given distance parameter of the model). Furthermore, the polymerization of actin ceases

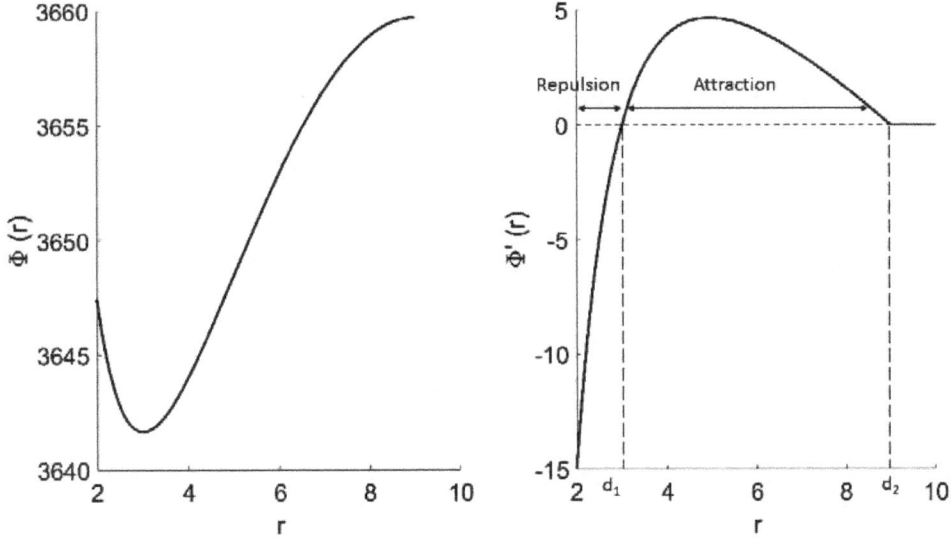

Fig. 3. Left: the potential $\Phi(r)$ (13b) for $d_1 = 3$, $d_2 = 9$. Right: the corresponding potential force $\Phi'(r)$ with a cut-off for $r > d_2$.

when the collision forces become too strong (a given parameter of the model), c.f. [7]..

We combine the assumptions on adhesion and collision forces, and introduce an attraction-repulsion potential of the form:

$$U_{ar}[F] = \int_{-\pi}^{\pi} \eta(\alpha,0,t)\Phi\left(\left|F(\alpha,0,t) - \tilde{F}(\alpha,0,t)\right|\right) d\alpha,$$ (13a)

where $\tilde{F}(\alpha,0,t)$ is the projection of point $F(\alpha,0,t)$ on the other cells' membrane, and

$$\Phi(r) = \frac{\mu^R}{2}\begin{cases} -(r-r_1)^2 + (\frac{1}{r}-r_2)^2, & \text{if } r \leq d_2, \\ 0, & \text{otherwise,} \end{cases}$$ (13b)

where μ^R represents the intensity of the attraction-repulsion force, d_2 is the maximal distance for adhesion attraction and r_1 and r_2 read:

$$\begin{cases} r_2 = \frac{1}{1/d_1^2 - 1/d_2^2}(d_1 - d_2 + \frac{1}{d_1^3} - \frac{1}{d_2^3}), \\ r_1 = d_1 + \frac{1}{d_1^3} - \frac{r_2}{d_1^2} \end{cases}$$ (13c)

Thus defined, the function $\Phi(r)$ is as depicted in Figure 3, d_1 being the size of the repulsion zone, d_2 the maximal attraction distance. Note that by (13a) the combined adhesion-collision force is applied on the membranes of the cells and is

compactly supported, in the sense that the two cells will only interact as long as their membranes are at a distance smaller than d_2.

To incorporate this new mechanical feature in the FBLM, we compute the variation of U_{ar} from (13a):

$$\delta U_{ar}\delta F = \int_{-\pi}^{\pi} \eta(\alpha,0,t)\Phi'\left(\left|F - \tilde{F}\right|_{(\alpha,0,t)}\right)$$

$$\frac{(F - \tilde{F})(\alpha,0,t)}{\left|F - \tilde{F}\right|_{(\alpha,0,t)}} \cdot \delta F(\alpha,0,t)d\alpha,$$ (14)

and include its contribution in the (membrane) boundary conditions at $s = 0$. In effect that Eqs. (9a)-(9b) recast into

$$-\mu^B\partial_s\left(\eta\partial_s^2 F\right) - p(\rho)\partial_\alpha F^\perp + \eta\lambda_{\text{inext}}\partial_s F$$
$$\mp \eta\eta^*\widehat{\mu^T}(\phi - \phi_0)\partial_s F^\perp$$ (15a)
$$= \begin{cases} \eta\left(f_{\tan}(\alpha)\partial_s F + f_{\text{inn}}(\alpha)V(\alpha)\right), & \text{for } s = -L, \\ \pm\lambda_{\text{tether}}\nu - \eta\Phi'\left(\left|F - \tilde{F}\right|\right)\frac{F-\tilde{F}}{|F-\tilde{F}|}, & \text{for } s = 0, \end{cases}$$

$$\eta\partial_s^2 F = 0, \qquad \text{for } s = -L, 0.$$ (15b)

Furthermore, we assume that the polymerization machinery is destabilized by cell-cell interactions. In particular, when the collision repulsion forces become too large (above a given threshold $\Phi^* > 0$), we set the local polymerization rate to 0. On the contrary, when the combined adhesion-collision is attractive, we increase the polymerization rate locally. These considerations are supported by biological studies showing the effects of pulling forces on actin polymerization such as in [7]. More specifically, we adjust the polymerization rate locally by setting:

$$v_*^{\pm}(\alpha) = \begin{cases} 0, & \text{if } \Phi'\left(\left|F - \tilde{F}\right|\right) \leq -\Phi^*, \\ 3.5v^{\pm}(\alpha), & \text{if } \Phi'\left(\left|F - \tilde{F}\right|\right) \geq 0. \end{cases} \quad (16)$$

IV. Cell-environment interactions.

To account for more biologically realistic situations, we embed the FBLM in a complex and adaptive extracellular environment. The particular coupling of the FBLM with the extracellular environment that we consider here was previously proposed in [23]. We give here a brief description.

We consider an extracellular environment that is comprised of the ECM —represented by the density of the *glycoprotein vitronectin* v onto which the FBLM cells adhere through the binding of the *integrins*— an extracellular chemical component c that serves as *chemoattractant* for the FBLM cell(-s), and the *matrix degrading metalloproteinases* (MMPs) m that are secreted by the cell and participate in the degradation of the matrix. In our formulation, these environmental components are represented by the *density* of the corresponding (macro-)molecules. Overall the model of the environment reads:

$$\begin{cases} \dfrac{\partial c}{\partial t}(\mathbf{x},t) = D_c \Delta c(\mathbf{x},t) + \alpha \, \mathcal{X}_{\mathcal{P}(t)}(\mathbf{x}) - \gamma_1 c(\mathbf{x},t) \\ \qquad\qquad - \delta_1 \mathcal{X}_{\mathcal{C}(t)}(\mathbf{x}) \\ \dfrac{\partial m}{\partial t}(\mathbf{x},t) = D_m \Delta m(\mathbf{x},t) + \beta \mathcal{X}_{\mathcal{C}(t)}(\mathbf{x}) - \gamma_2 m(\mathbf{x},t) \\ \dfrac{\partial v}{\partial t}(\mathbf{x},t) = -\delta_2 m(\mathbf{x},t) v(\mathbf{x},t) \end{cases}$$
$$\qquad\qquad\qquad\qquad\qquad\qquad\qquad (17)$$

where $\mathbf{x} \in \Omega \subset \mathbb{R}^2$, $t \geq 0$, $D_c, D_m, \alpha, \beta, \gamma_i, \delta_i \geq 0$, and where $\mathcal{X}_-$ is the characteristic function of the corresponding set. $\mathcal{P}$ denotes the support of the pipette(-s) that inject the chemical c in the

environment, and the FBLM cell(-s) influence the environment through $\mathcal{X}_{\mathcal{C}(t)}(\mathbf{x})$, where $\mathcal{C}(t) \subset \mathbb{R}^2$ represents the *full cell* (lamellipodium and internal structures).

The model of the environment (17) and the FBLM (6) are coupled at three different places: at the characteristic function $\mathcal{X}_{\mathcal{C}}$ in (17), where the cell $\mathcal{C}$ produces MMPs and degrades the chemical, at the adhesion coefficient μ^A in (6) which reflects the density of the ECM influences the migration of the cell, and in the polymerization rates $v_{\text{ext}}^{\pm}$ of the filaments in (11b) which are primarily adjusted according to the density of the extracellular chemical c.

Despite the simple structure of the model (17), and the numerous biological simplifications we have made, we are able to reconstruct with the FBLM-environment combination, realistic and complex biological phenomena, see e.g. Experiment 3.

V. Experiments and simulations.

We present three indicative experiments to study the effect of the collision and adhesion components of the FBLM on the migration and morphology of the cells. The first experiment highlights the mechanical effect of cell-cell collisions. In the second experiment, we include the adhesion effect of the *cadherin* protein. In the third experiment, we embed several FBLM cells in the same environment and study the first stages of a cell cluster development. In this experiment, we also compare our results with a particular biological setting involving the migration of HeLa cells.

Experiment 1 (Cell-cell collision). We embed two FBLM cells in an environment that it is adhesion and chemically uniform and fixed. Initially, both cells are rotationally symmetric, with diameter 50, and lamellipodia of thickness 8. They are centred at (50,4) and (-50,-4) respectively and the length of their filaments is 10. The environment is such that the adhesion coefficient μ^A of the FBLM (common for both cells) is uniform and fixed

$$\mu^A = 0.4101,$$

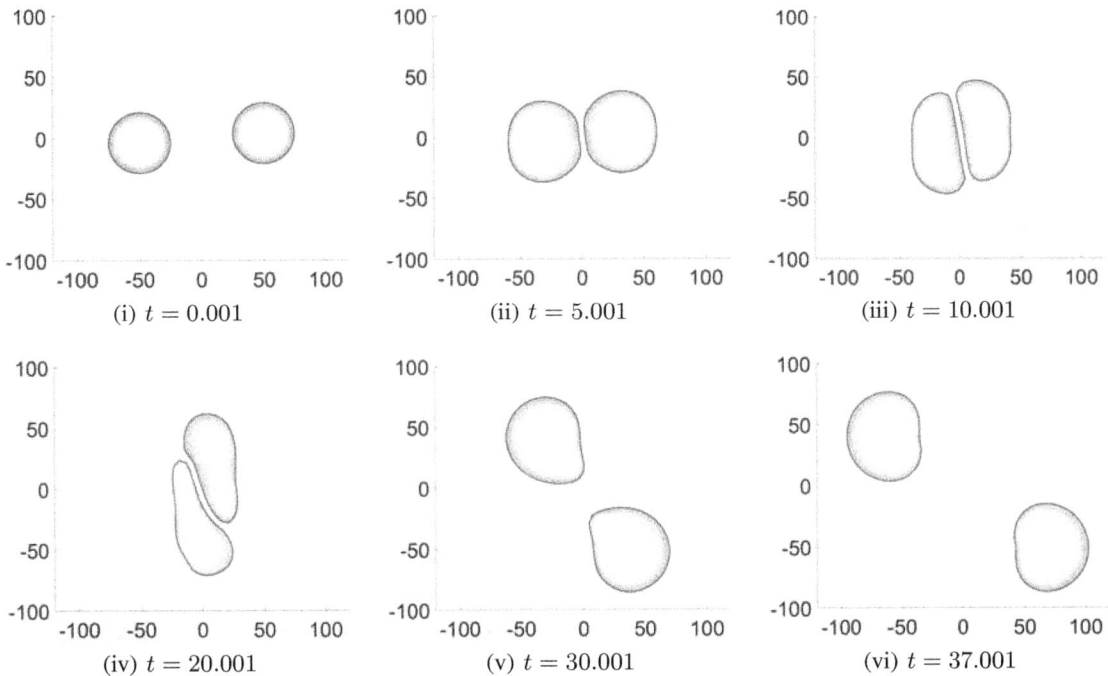

Fig. 4. Experiment 1 (Cell-cell collision). (i): Two FBLM cells migrate in opposing east-west directions. (ii)-(iv): The cells collide and deform due to the exchange of repulsive collision forces. The cells slip by each other. (v)-(vi): The deformation of the cells is elastic and the cells recover their pre-collision morphology.

and the polymerization rates of the filaments are given by (11b) and vary in a smooth sinusoidal manner between a minimum $v_{\min} = 1.5$ and a maximum $v_{\max} = 8$ value from the posterior to the anterior side of the cell. The direction of the cell centred at $(50, 4)$ is directed eastwards, and of the cell centred at $(-50, -4)$ is directed westwards. This brings the two cells in a collision path.

To avoid physical overlapping of the cells, the collision forces act proactively, i.e. when the cells come closer than a pre-defined threshold distance. In this experiment, this distance is set to 5. When this occurs, the collision forces increase rapidly in magnitude, and when they become very strong (stronger than a predefined threshold), the polymerization of the corresponding filaments ceases. This threshold force is set to be 0.01 in this experiment; the rest of the parameters are given in Table I.

In Figure 4 we present the corresponding simulation results. After a short time, during which the size of the cells is adjusted to the environmental

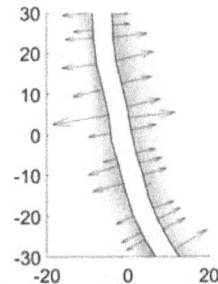

Fig. 5. Experiment 1 (Cell-cell collision) In a close-up we visualize the repulsive collision forces in action. The magnitude of the forces increases rapidly when the cells come in proximity (closer than a user-defined threshold). When the forces become too large, the polymerizaiton of the corresponding filaments ceases.

conditions, the cells collide. The forces that the cells exchange are repulsive and applied symmetrically on the plus ends of the filaments of the two cells; their effect is seen in the deformation of the cells. When the collision forces become very strong (stronger than a predefined threshold),

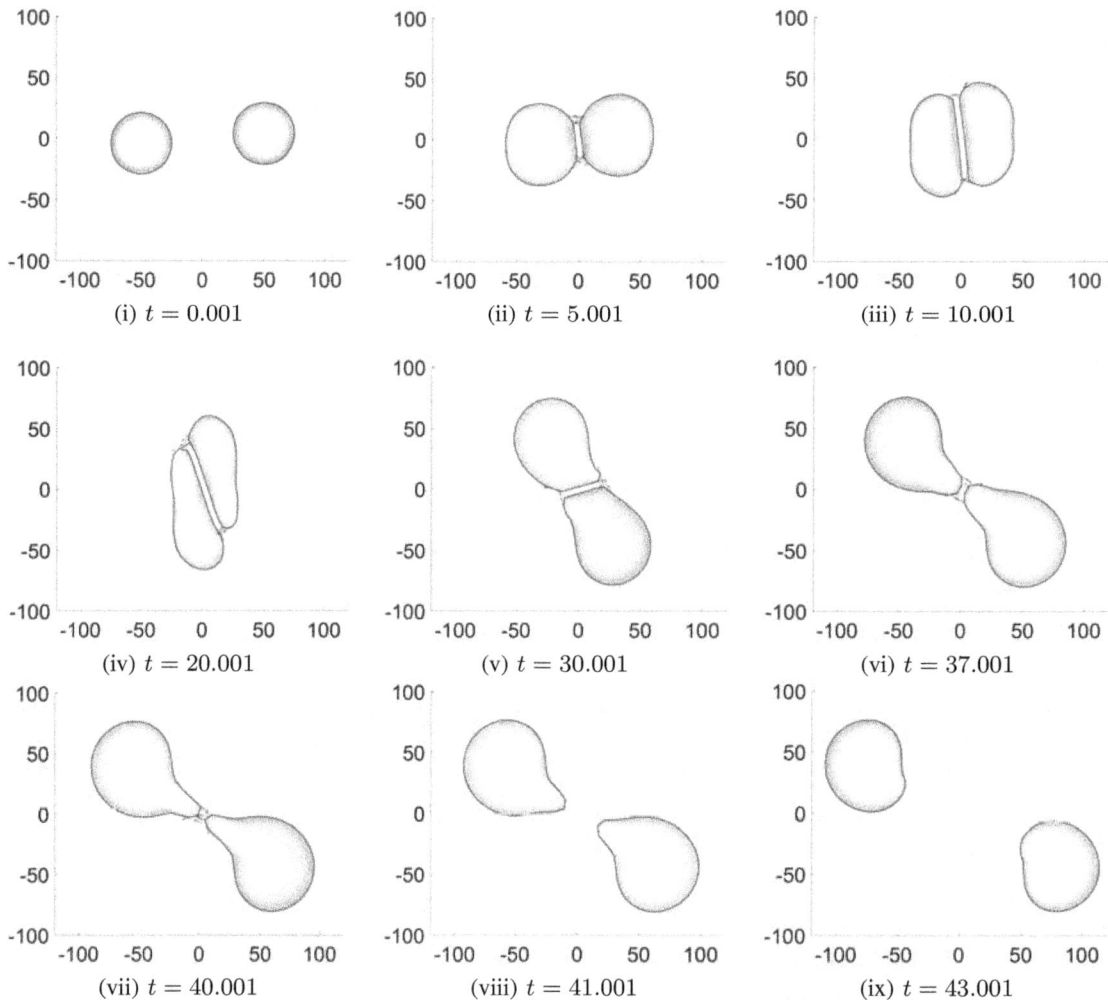

Fig. 6. Experiment 2 (Cell-cell adhesion) As in Figure 4, two cells are found in opposing colliding paths. This time though, they are able to develop cadherin induced cell-cell-adhesions. This has an impact in the deformation of the cells, their migrations, and their tendency to stick with each other and to resist their separation. (ii)—(v): The adhesive forces are stronger at the ends of the colliding parts of their membranes than the middle parts of it. (vi)—(ix): Note the elastic retraction of the "tail"/rear part of the cell. (ix): Note also the larger time that is needed for the cells to reach the boundary of the domain, as opposed to the cell-cell collision experiment in Figure 4.

the polymerization of the corresponding filaments ceases. At the non-colliding regions, the polymerization continuous and as a result the cells slip by each other. After moving away from each other, the cells recover the morphology they had before the collision. This implies that the deformation due to collision is *elastic*. This remark can serve as a starting point to measure the *elastic modulus* of the lamellipodium when cell-type specific experimental evidence is considered.

In Figure 5 we visualize the force exchange between the two cells. When the distance of the two cells becomes shorter than the (predefined) threshold, the repulsive forces are applied at the plus ends of the corresponding filaments. The magnitude of the forces increases as the distance between the filaments decreases. When the forces reach a maximum value, the corresponding polymerization rates cease. The overall effect is that the cells have the tendency to maintain the threshold distance between each other.

Experiment 2 (Cell-cell adhesion). In this experiment, the setting, the initial conditions, and the parameters considered are the same as in the Experiment 1. We augment this time the FBLM with the effect of cadherin forces. These forces are complementary to the cell-cell collision forces and are incorporated in the FBLM in a similar way, see Section III.

When the distance between the two cells reaches the cell-cell collision threshold, the repulsive collision forces are introduced and counterbalance the attractive adhesion forces. Unless the relative position of the cells changes (possibly due to other reasons), the equilibrium between the adhesion and collision forces is maintained. The adhesion threshold distance in this experiment is set to 15, whereas the collision threshold distance is set to 5. When the collision forces become larger than 0.01 the polymerization of the filaments ceases.

In Figure 6 we visualize the simulation results of the combined effect of collision and adhesion in the deformation of the cells and their tendency to "stick together". It can be seen that at the end of the contact zones the adhesion forces are more eminent whereas, in the middle of these zones, no forces are visible. There, the adhesion and collision forces are in equilibrium. As the cells continue their migration, they slip by each other and their contact zones get stretched due to the adhesion between them. As a result, each cell develops a tail that quickly retracts when the adhesions break.

We can quantify the effect of cadherin forces, by comparing the average speed of the cells in the two experiment. In the cell-cell collision Experiment 1, the cells collide at time $t = 5$ at $x = 0$ and reach $x = 100$ at time $t = 37$ i.e. with an average speed $100/(37 - 5) = 3.125$. Similarly, the approximate speed in the cell-cell adhesion case is estimated by $100/(43 - 5) = 2.625$. The difference between the two speeds (although not precisely measured) is another effect of the adhesion in the migration of the cells.

In Figure 7 we visualize a close-up in the tails that the cells develop; there the adhesion

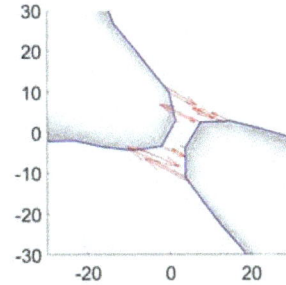

Fig. 7.   Experiment 2 (Cell-cell adhesion). With a close-up in the adhesion zone, we visualize the cadherin adhesion forces. They are exerted at the plus ends of the filaments and are opposite to each other. In the middle region, the adhesion forces have been balanced by the repulsive collision force.

forces are clearly visualized. As noted previously, these forces come in pairs, are contractile, and mostly visible at the ends of the contact zone. The adhesion forces exerted on the filaments in the middle of the zone have been counterbalanced by the repulsive collision forces.

Experiment 3 (Cluster formation). In this experiment, we embed several FBLM cells in the same extracellular environment. They collide and adhere with each other, they form a *cell cluster* and we study the first steps of its migration under the influence of an adaptive adhesion and chemical environment.

We consider 14 cells that are initially the same and rotationally symmetric and reside in the same extracellular environment. The initial extracellular adhesion landscape and the chemical environment are variable and given respectively by

$$v_0(\mathbf{x}) = \sin^2\left(2\,\frac{x + 200}{400} - \left(\frac{y + 150}{350}\right)^3\right)\pi + 1,$$

(18a)

$$c_0(\mathbf{x}) = e^{-5 \cdot 10^{-4}(10^{-2}(x-30)^2 + (y-40)^2)},$$

(18b)

where $\mathbf{x} = (x, y) \in [-200, 200] \times [-150, 200]$.

We assume that the cells respond to the chemical and haptotaxis gradients of the environment while at the same time colliding and adhering to each other. The overall model is comprised of 14 FBLM equations of the form (6), one for each cell,

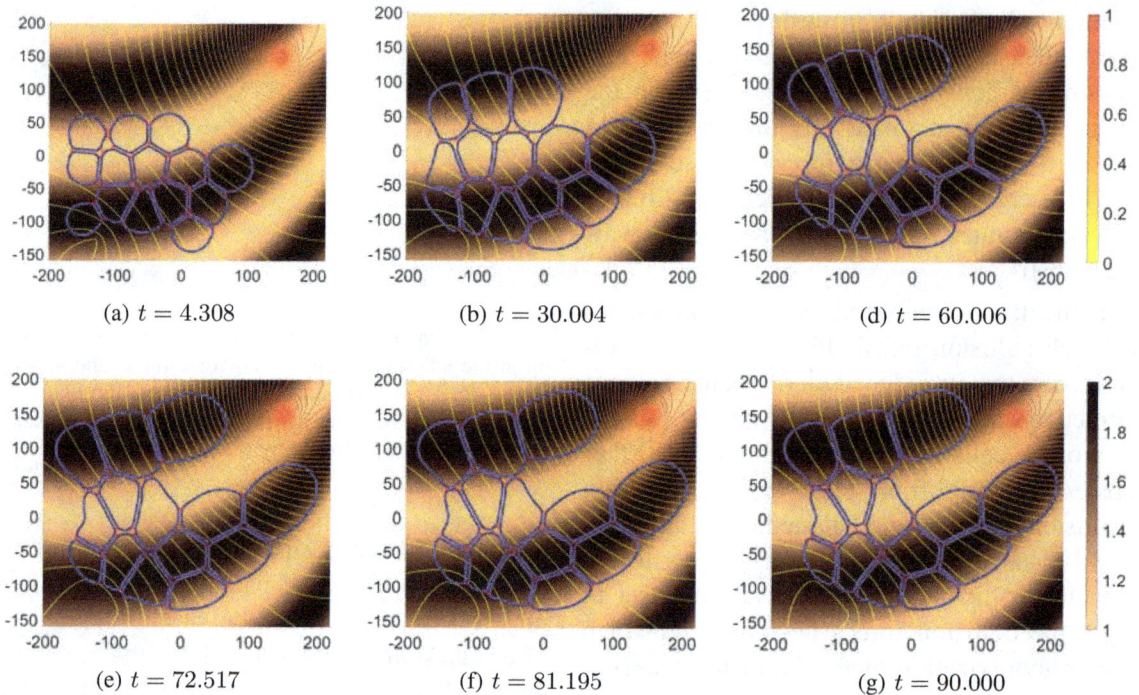

Fig. 8. Experiment 3 (Cluster formation). A number of 14 FBLM cells are placed in a non-uniform and adaptive environment. The cells collide and adhere with each other, and respond haptotactically to the gradient of the ECM (shown here as the background landscape with the corresponding colorbar in the second row) and chemotactically to the chemical gradient (shown as isolines with the colorbar in the first row).

and one system for the environment (17) in which the characteristic function $\mathcal{X}_{\mathcal{C}(t)}$, in the degradation of the chemical and the production of the MMPs, is replaced by

$$\mathcal{X}_{\cup_i \mathcal{C}_i(t)},$$

where $C_i(t)$, $i = 1, \ldots, 14$ represent the support of the cells, i.e. the area occupied by the lamellipodium and the inner part of the cells. We assume that all the cells are of the same type and satisfy the FBLM (6) with the same parameters; these are given in Table I. Their adhesion and collision threshold distances have been set to 15 and 5, respectively, and the collision force threshold to 0.01. The parameters for the environment (17) are given in Table II.

In Figure 8 we present several snapshots of the time evolution of the cluster. The cells respond to the gradient of the ECM v, they elongate and align themselves with the higher density of the ECM. The effect of cell-cell adhesion is evident

primarily in the cells that are found in the ridges of the ECM. As they are pulled by the neighbouring cells that have already climbed on the higher ECM density regions, they get stretched and elongate in a way "perpendicular" to the direction of the ECM. At the same time the cells, and primarily the leading ones, are directed towards the source of the chemical; due to the cell-cell adhesion, the whole cluster moves slowly in the same direction.

We do not reproduce in this experiment a particular biological experimental setting. Still, the resulting cell morphologies are very close to the biological reality. We exhibit this remark in Figure 9 where we compare our simulation results, taken from Figure 8 (g), with a specific biological experiment of HeLa cells. In particular, from one frame of the video [26] —where the time evolution of a (relatively large) cluster of HeLa cells is observed in-vitro— we "cut out" some of the HeLa cells and superimpose them on our simulations.

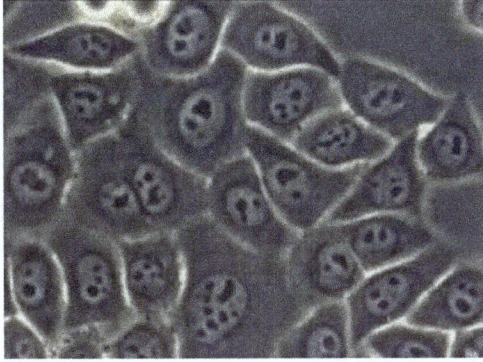

(a) A video frame from [26] shows a number of *in vitro* migrating HeLa cells. We "extract" the cells by cutting along their common interfaces.

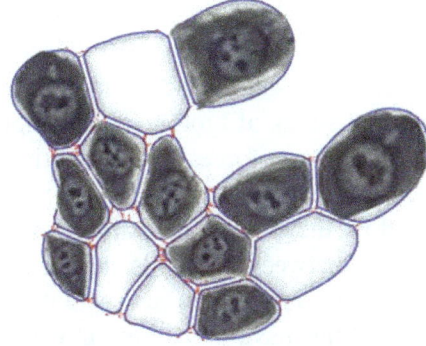

(b) We superimpose the cut HeLa cells extracted from (a) on the simulation results from Figure 8 (g).

Fig. 9. Experiment 3 (Cluster formation). We compare the simulation results of Experiment 3, and in particular the morphology of the resulting cells, with *in vitro* culture of the HeLa cancer cells studied in [26]. (a): The single frame from the video in [26] from which HeLa cells were "extracted". (b): The fit between the HeLa cells from (a) and our numerical simulations from Figure 8 (g). The comparison follows after properly rotating and scaling the HeLa cells and superimposing them on the simulation results.

TABLE I

BASIC SET OF PARAMETER VALUES USED IN THE NUMERICAL SIMULATIONS OF THE FBLM IN ALL THE EXPERIMENT OF THIS WORK. THESE PARAMETERS HAVE BEEN ADOPTED FROM [11], [23].

symb.	description	value	comment
μ^B	bending elasticity	$0.07\,\text{pN}\,\mu\text{m}^2$	[6]
μ^A	adhesion	$0.4101\,\text{pN}\,\text{min}\,\mu\text{m}^{-2}$	[8], [14] & [18], [16], [15]
μ^T	cross-link twisting	$7.1 \times 10^{-3}\,\mu\text{m}$	
μ^S	cross-link stretching	$7.1\times10^{-3}\,\text{pN}\,\text{min}\,\mu\text{m}^{-1}$	
ϕ_0	crosslinker equil. angle	70°	[15]
μ^{IP}	actin-myosin strength	$0.1\,\text{pN}\,\mu\text{m}^{-2}$	
$v_{\min}$	minimal polymerization	$1.5\,\mu\text{m}\,\text{min}^{-1}$	in biological range
$v_{\max}$	maximal polymerization	$8\,\mu\text{m}\,\text{min}^{-1}$	in biological range
μ^P	pressure constant	$0.1\,\text{pN}\,\mu\text{m}$	
A_0	equilibrium inner area	$650\,\mu\text{m}^2$	[27], [24]
λ_{inext}	inextensibility	20	
λ_{tether}	membrane tethering	1×10^{-3}	

TABLE II

PARAMETER SETS USED FOR THE SIMULATION OF THE ENVIRONMENT (17) IN THE EXPERIMENT 3 (CLUSTER FORMATION).

symb.	description	value
D_c	diffusion of the chemical	$3 \times 10^3\,\text{cm}^2\,\text{min}^{-1}$
D_m	diffusion of the MMPs	$3 \times 10^3\,\text{cm}^2\,\text{min}^{-1}$
α_1	production rate of chemical	$10^2\,\text{mol}\,\text{min}^{-1}$
β	production of MMPs	$0.1\,\text{mol}\,\text{min}^{-1}$
γ_1	decay of the chemical	$10\,\text{mol}\,\text{min}^{-1}$
γ_2	decay of the MMPs	$10\,\text{mol}\,\text{min}^{-1}$
δ_1	degr. chemical by the cell	$10^4\,\text{mol}\,\text{min}^{-1}$
δ_2	degr. of the ECM by the MMPs	$0\,\text{cm}^2\,\text{mol}^{-1}\,\text{min}^{-1}$

VI. DISCUSSION.

We propose in this work an extension of the actin-based cell motility model (6), termed FBLM, to account also for the collisions and the adhesions between cells. This is achieved by modelling the effect of these two phenomena on the lamellipodium through a single attractive-repulsive potential, (13a), which is then incorporated in the FBLM.

We deduce the adhesion-collision potential (13a) based on a series of biological assumptions, namely: the adhesion forces are attractive and appear when the cells are in proximity, in a distance justified by the size of the *cadherin* protein. As the distance between the cells decreases, the magnitude of the adhesion forces increases. The adhesion forces can have a maximum value that represents the maximum "pulling" strength of the *cadherin* protein. When the cells come closer, repulsive collision forces appear. The collision forces increase rapidly as the distance between the cells decreases. They are unbounded in magnitude and soon counteract the adhesive effect of the *cadherins*. Both forces are exerted on the plus-end of the filaments and through them are transferred to the cytoskeleton and the rest of the cell. Accordingly, they participate in the $s = 0$ boundary conditions of the FBLM, (15a).

We study the cell-cell collision and adhesion through three particular experiments: we first simulate the elastic deformation of two cells when only collision is considered. We notice there, the restoration of the cells to their previous morphology after the collision forces cease. We then incorporate and simulate the effect *cadherins* in the FBLM. We notice the differences in the deformation of the cells as opposed to the collision-only case, the tendency of the cell to "stick together" and the elastic retraction fo their "tails" when eventually the adhesion forces break. We then embed a number of cells in a non-uniform (haptotaxis and chemotaxis wise) environment while allowing them to collide and adhere with each other. We then compare the results with a *in vitro* experiment of migrating HeLa-cell cluster. We notice the striking similarity of between the simulated and the experimental.

Overall, the cell-cell collision and adhesion extensions of the FBLM that we propose in this paper is of utmost importance for a large number of biologically relevant studies, ranging from *cell-cluster* and *monolayer* formation to *cancer* invasion.

ACKNOWLEDGEMENTS.

N.S was partially supported by the German Science Foundation (DFG) under the grant SFB 873 "Maintenance and Differentiation of Stem Cells in Development and Disease".
The work of C.S. has been supported by the Vienna Science and Technology Fund, Grant no. LS13-029, and by the Austrian Science Fund, Grants no. W1245, SFB 65, and W1261.

REFERENCES

[1] W. Alt and E. Kuusela. Continuum model of cell adhesion and migration. *J. Math. Biol.*, 58(1-2)::135, 2009.

[2] D. Ambrosi and A. Zanzottera. Mechanics and polarity in cell motility. *Physica D*, 330:58–66, 2016.

[3] A. Brunk, N. Kolbe, and N. Sfakianakis. Chemotaxis and haptotaxis on a cellular level. *Proc. XVI Int. Conf. Hyper. Prob.*, 2016.

[4] L. Cardamone, A. Laio, V. Torre, R. Shahapure, and A. DeSimone. Cytoskeletal actin networks in motile cells are critically self-organized systems synchronized by mechanical interactions. *PNAS*, 108:1397813983, 2011.

[5] J. Fuhrmann and A. Stevens. A free boundary problem for cell motion. *Diff. Integr. Eq.*, 28:695–732, 2015.

[6] F. Gittes, B. Mickey, J. Nettleton, and J. Howard. Flexural rigidity of microtubules and actin filaments measured from thermal fluctuations in shape. *J. Cell Biol.*, 120(4):923–34, 1993.

[7] M.M. Kozlov and A.D. Bershadsky. Processive capping by formin suggests a force-driven mechanism of actin polymerization. *J. Cell Biol.*, 167(6):1011–1017, 2004.

[8] F. Li, S.D. Redick, H.P. Erickson, and V.T. Moy. Force measurements of the $\alpha5\beta1$ integrin-fibronectin interaction. *Biophys. J.*, 84(2):1252–1262, 2003.

[9] A. Madzvamuse and U.Z. George. The moving grid finite element method applied to cell movement and deformation. *Finite Elem. Anal. Des.*, 74:76 – 92, 2013.

[10] A. Manhart, D. Oclz, C. Schmeiser, and N. Sfakianakis. An extended Filament Based Lamellipodium: Model produces various moving cell shapes in the presence of chemotactic signals. *J. Theor. Biol.*, 382:244–258, 2015.

[11] A. Manhart, D. Oelz, C. Schmeiser, and N. Sfakianakis. Numerical treatment of the Filament Based Lamellipodium Model (FBLM). *Book chapter in Modelling Cellular Systems*, 2016.

[12] W. Marth, S. Praetorius, and A. Voigt. A mechanism for cell motility by active polar gels. *J. R. Soc. Interface*, 12:20150161, 2015.

[13] C. Möhl, N. Kirchgessner, C. Schäfer, B. Hoffmann, and R. Merkel. Quantitative mapping of averaged focal adhesion dynamics in migrating cells by shape normalization. *J. Cell Sci.*, 125:155–165, 2012.

[14] A.F. Oberhauser, C. Badilla-Fernandez, M. Carrion-Vazquez, and J.M. Fernandez. The mechanical hierarchies of fibronectin observed with single-molecule AFM. *J. Mol. Biol.*, 319(2):433–47, 2002.

[15] D. Oelz and C. Schmeiser. *Cell mechanics: from single scale-based models to multiscale modeling*, chapter How do cells move? Mathematical modeling of cytoskeleton dynamics and cell migration. Chapman and Hall, 2010.

[16] D. Oelz and C. Schmeiser. Derivation of a model for symmetric lamellipodia with instantaneous cross-link turnover. *Arch. Ration. Mech. An.*, 198:963–980, 2010.

[17] D. Oelz and C. Schmeiser. Simulation of lamellipodial fragments. *Journal of Mathematical Biology*, 64:513–528, 2012.

[18] D. Oelz, C. Schmeiser, and J.V. Small. Modeling of the actin-cytoskeleton in symmetric lamellipodial fragments. *Cell Adhes. Migr.*, 2:117–126, 2008.

[19] B. Rubinstein, M.F. Fournier, K. Jacobson, A.B. Verkhovsky, and A. Mogilner. Actin-myosin viscoelastic flow in the keratocyte lamellipod. *Biophys. J.*, 97(7):1853–1863, 2009.

[20] B. Sabass and U.S. Schwarz. Modeling cytoskeletal flow over adhesion sites: competition between stochastic bond dynamics and intracellular relaxation. *J. Phys.:*

Condens. Matter, 22:194112 (10pp), 2010.

[21] U.S. Schwarz and M.L. Gardel. United we stand - integrating the actin cytoskeleton and cell-matrix adhesions in cellular mechanotransduction. *J. Cell Sci.*, 125:3051–3060, 2012.

[22] M. Scianna, L. Preziosi, and K. Wolf. A cellular potts model simulating cell migration on and in matrix environments. *Math. Biosci. Engng.*, 10:235–261, 2013.

[23] N. Sfakianakis and A. Brunk. Stability, convergence, and sensitivity analysis of the FBLM and the corresponding FEM. *Bull. Math. Biol.*, 2018.

[24] J.V. Small, G. Isenberg, and J.E. Celis. Polarity of actin at the leading edge of cultured cells. *Nature*, bf 272:638–639, 1978.

[25] J.V. Small, T. Stradal, E. Vignal, and K. Rottner. The lamellipodium: where motility begins. *Trends Cell Biol.*, 12(3):112–20, 2002.

[26] Th. Tlsty and exploratorium.edu (producer). Cancer: Cellular misfits run amok – a scientist's view (motion picture), May 18, 2008.

[27] A.B. Verkhovsky, T.M. Svitkina, and G.G. Borisy. Self-polarisation and directional motility of cytoplasm. *Curr. Biol.*, 9(1):11–20, 1999.

13

Inverse problem of the Holling-Tanner model and its solution

Adejimi Adesola Adeniji, Igor Fedotov, Michael Y. Shatalov
Department of Mathematics and Statistics, Tshwane University of Technology
adejimi.adeniji@gmail.com, fedoptovi@tut.ac.za, shatalovm@tut.ac.za

Abstract—In this paper we undertake to consider the inverse problem of parameter identification of nonlinear system of ordinary differential equations for a specific case of complete information about solution of the Holling-Tanner model for finite number of points for the finite time interval. In this model the equations are nonlinearly dependent on the unknown parameters. By means of the proposed transformation the obtained equations become linearly dependent on new parameters functionally dependent on the original ones. This simplification is achieved by the fact that the new set of parameters becomes dependent and the corresponding constraint between the parameters is nonlinear. If the conventional approach based on introduction of the Lagrange multiplier is used this circumstance will result in a nonlinear system of equations. A novel algorithm of the problem solution is proposed in which only one nonlinear equation instead of the system of six nonlinear equations has to be solved. Differentiation and integration methods of the problem solution are implemented and it is shown that the integration method produces more accurate results and uses less number of points on the given time interval.

Keywords-Parameter estimation, Goal function, Absolute error curves, Inverse method, Holling-Tanner model, Least square method, Differentiation method, Integration method

I. INTRODUCTION

The numerical evaluation of known coefficient of a dynamical system i.e. the problem of dynamical system identification, is one of the most important problem of the mathematical biology [1], ecology [2], [3], [4], etc. Usually, to identify a dynamics of a system, it is necessary to have certain statistical information for time values about the unknown functions of this system. In the present paper we consider the inverse problem of parameter identification of the Holling-Tanner predator-prey model [5], [6]. This model is widely used in mathematical biology, for example, in the study of transmissible disease [7]. Several investigations have been done by various researchers on the mite-spider-mite, lynx-hare and sparrow-hawk-sparrow competition [8], [9], [10]. In [11], the authors proposed a method consisting in the direct integration of a given dynamical system with the subsequent application of quadrature rules and the least square method [12], [13] provided that there is complete statistical information about the unknown function. In this paper, we assume that the complete information about the competing species is available and the two methods of solution, differentiation and integration methods, are proposed. The problem of the Holling-Tanner model identification has its specifics, because it nonlinearly depends on the unknown parameters. It is possible to transform this model to a new form where the equations of the system linearly depends on the set of new parameters. These new parameters are not independent and we need to consider the constraint between the parameters, which are nonlinear. The Holling-Tanner model has only one constraint and hence, can be simply treated by a novel method developed by the authors. The theoretical considerations are accompanied by numerical examples where the developed algorithm is tested for both differentiation and integration methods of solution. It is shown that the integration methods is more accurate than the differentiation one and needs less amount of experimental information.

II. MAIN RESULTS

In our paper we consider the Holling-Tanner model [9], [14] described by the following system of equations:

$$\begin{cases} \dot{x} = b_1 x - b_2 x^2 - b_3 \frac{x \cdot y}{b_4 + x}, \\ \dot{y} = b_5 y - b_6 \frac{y^2}{x}, \\ t = 0, \quad x(0) = x_0, \quad y(0) = y_0 \end{cases} \quad (1)$$

where $x = x(t)$, $y = y(t)$, $\dot{x} = \frac{dx(t)}{dt}$, $\dot{y} = \frac{dy(t)}{dt}$, t is time and $b_1, \cdots b_6$ are positive constant parameters [15]. Initial conditions for this system are formulated so that at $t = 0 : x(t = 0) = x_0 > 0$ and $y(t = 0) = y_0 > 0$. The main results relating to solution of this initial value problem were obtained in [10], [16], [17], [18] as

- Solution of the initial value problem (1) $\{x(t), y(t)\}$ with positive initial conditions is positive, i.e. $x(t) > 0$ and $y(t) > 0$ for $t \geq 0$.

- Initial value problem (1) has the positive steady-state solution [15] $(\tilde{x}, \tilde{y})$ which cor-

responds to either stable focus or stable node critical point depending on $b_1, \cdots b_6$ so that:

$$\tilde{x} = \frac{b_1 b_6 - b_3 b_5 - b_2 b_4 b_6 + \sqrt{\Delta}}{2 b_2 b_6} > 0,$$

$$\tilde{y} = \frac{b_5 (b_1 b_6 - b_3 b_5 - b_2 b_4 b_6 + \sqrt{\Delta})}{2 b_2 b_6^2} > 0. \quad (2)$$

where

$$\Delta = (b_1 b_6 - b_3 b_5 - b_2 b_4 b_6)^2 + 4 b_1 b_2 b_4 b_6^2.$$

- Initial value problem (1) has unstable steady-state solution

$$(\tilde{\tilde{x}}, \tilde{\tilde{y}}) = (\frac{b_1}{b_2}, 0),$$

which corresponds to the saddle critical point.

III. ON SOLVABILITY OF IDENTIFICATION PROBLEM

Assume that solution of initial problem (1), $x(t)$ and $y(t)$ is given on the finite time interval $t \in [0, T]$ with initial $t = 0$ and terminal $t = T$ time instants in $N + 1$ equispaced time instants $t_i = \frac{T}{N} i \in [0, T]$:

$$x_i = x(t_i), \quad y_i = y(t_i) \quad (i = 0, \cdots, N) \quad (3)$$

Lets us formulate the identification problem for parameters $b_1, \cdots b_6$ from the known solution (3) This problem can be solved if the conditions of the following theorem are satisfied:

Theorem 1. *Parameters b_1, b_2, b_3, b_4 of model (1) can be identified by the least squares method if $(N + 1) \times 1$-vector columns $[x_i], [x_i^2], [x_i^3], [\dot{x}_i], [x_i, y_i]$ are linearly independent. Parameters b_5 and b_6 of the above mentioned model can be identified by the mentioned method if $(N + 1) \times 1$-vector columns $[y_i]$ and $[\frac{y_i^2}{x_1}]$ are linearly independent.*

Proof: By multiplying the first equation of system (1) by $(b_4 + x)$ and grouping the resulting terms we obtain

$$C_1(-x^3(t)) + C_2(-x(t)y(t)) + C_3(-\dot{x}(t))$$
$$+ C_4(x(t)) + C_5(x^2(t)) + (-x(t)\dot{x}(t)) = 0, \quad (4)$$

where $C_1 = b_2$, $C_2 = b_3$, $C_3 = b_4$, $C_4 = b_1 b_4$, $C_5 = b_1 - b_2 b_4$ are new unknown parameters. It is easy to check that the parameters C_1, C_3, C_4, C_5 satisfy the following constrains: $C_1 C_3^2 + C_3 C_5 - C_4$. Considering $x(t)$ and $y(t)$ in time instants $t = t_i$ we obtain the following overdetermined system of $N + 1$ linear algebraic equations:

$$C_1 \vec{f_1} + C_2 \vec{f_2} + C_3 \vec{f_3} + C_4 \vec{f_4} + C_5 \vec{f_5} - \vec{f_6} = 0, \quad (5)$$

where

$$\vec{f_1} = [f_{1i}] = [-x_i^3], \vec{f_2} = [f_{2i}] = [-x_i y_i],$$
$$\vec{f_3} = [f_{3i}] = [-\dot{x}_i], \vec{f_4} = [f_{4i}] = [x_i],$$
$$\vec{f_5} = [f_{5i}] = [x_i^2], \text{ and } \vec{f_6} = [f_{6i}] = [x_i \dot{x}_i]$$

are $(N + 1) \times 1$-vector columns. Hence, the unknown parameters C_1, C_2, C_3, C_4 and C_5 can be found by, for example, the least squares method [19] by means of the constrained minimization of function G_1:

$$G_1 = G_1(C_1, C_2, C_3, C_4, C_5, \lambda) =$$
$$= \frac{1}{2}(C_1 \vec{f_1} + C_2 \vec{f_2} + C_3 \vec{f_3} + C_4 \vec{f_4} + C_5 \vec{f_5} - \vec{f_6})^T$$
$$(C_1 \vec{f_1} + C_2 \vec{f_2} + C_3 \vec{f_3} + C_4 \vec{f_4} + C_5 \vec{f_5} - \vec{f_6})$$
$$+ \lambda(C_1 C_3^2 + C_3 C_5 - C_4) \longrightarrow \min \quad (6)$$

This problem can be solved providing that vectors $\vec{f_1}, \cdots, \vec{f_5}$ are linearly independent in (6). The last term contains the Lagrange multiplier λ and the constraint between coefficients $C_1, \cdots, C_5$. Moreover, the second equation of system (1) can be rewritten in time instants $t = t_i$ as the following overdetermined system of $N + 1$ linear algebraic equations:

$$C_6 \vec{f_7} + C_7 \vec{f_8} - \vec{f_9} = 0, \quad (7)$$

where

$$\vec{f_7} = [f_{7i}] = [y_i], \vec{f_8} = [f_{8i}] = \left[\frac{-y_i^2}{x_i}\right],$$
$$\vec{f_9} = [f_{9i}] = [\dot{y}_i], C_6 = b_5, C_7 = b_6.$$

That is why coefficients C_6, C_7 can be found by application of the least square method by means

of minimization of function G_2

$$G_2 = G_2(C_6, C_7) = \frac{1}{2}\left(C_6 \vec{f_7} + C_7 \vec{f_8} + \vec{f_9}\right)^T$$
$$(C_6 \vec{f_7} + C_7 \vec{f_8} + \vec{f_9}) \longrightarrow \min \quad (8)$$

This problem can be solved providing that vectors $\vec{f_7}$ and $\vec{f_8}$ are linearly independent of (8).

Remark 2. *In vectors $\vec{f_3}, \vec{f_6}$ the component $\dot{x}_i$, and in vector $\vec{f_9}$ the components $\dot{y}_i$ are calculated by means of numerical differentiation of x_i, y_i with respect to time t and that is why the proposed method is called the differential method of identification.*

Corollary 3. *Parameters b_1, b_2, b_3, b_4 of the model (1) can be identified by the least square method [19] if $(N + 1) \times 1$-vector columns*

$$\left[\int_0^{t_i} x(\tau) d\tau\right], \left[\int_0^{t_i} x^2(\tau) d\tau\right], \left[\int_0^{t_i} x^3(\tau) d\tau\right],$$
$$[x_i - x_0], \left[\int_0^{t_i} x(\tau) y(\tau) d\tau\right]$$

are linearly dependent. Parameters b_5 and b_6 of the abovementioned model can be identified by the abovementioned method if $(N + 1) \times 1$-vector columns $\left[\int_0^{t_i} y(\tau) d\tau\right]$ and $\left[\int_0^{t_i} \frac{y^2(\tau)}{x(\tau)} d\tau\right]$ are linearly dependent.

Proof: Integrating expression (4) with respect to time $t \in [0, T]$ we obtain

$$C_1 \left(-\int_0^t x^3(\tau) d\tau\right) + C_2 \left(-\int_0^t x(\tau) y(\tau) d\tau\right)$$
$$+ C_3 (x_0 - x(t)) + C_4 \left(\int_0^t x(\tau) d\tau\right)$$
$$+ C_5 \left(\int_0^t x^2(\tau) d\tau\right) - \left(\frac{1}{2}(x^2(t) - x_0^2)\right) = 0. \quad (9)$$

Integrating second equation of system 5 with respect to time $t \in [0, T]$ we have

$$C_6 \left(\int_0^t y(\tau) d\tau\right) + C_7 \left(-\int_0^t \frac{y^2(\tau)}{x(\tau)} d\tau\right)$$
$$- C_3 (y(t) - y_0) = 0. \quad (10)$$

Performing all integrations in (9) and (10) from 0 to $t_j \in [0, T]$ we obtain the following overdetermined systems of $N+1$ linear algebraic equations

$$C_1\vec{g_1} + C_2\vec{g_2} + C_3\vec{g_3} + C_4\vec{g_4} + C_5\vec{g_5} - \vec{g_6} = 0,$$
$$C_6\vec{g_7} + C_7\vec{g_8} - \vec{g_9} = 0, \qquad (11)$$

where

$$\vec{g_1} = \left[-\int_0^{t_i} x^3(\tau)d\tau\right], \quad \vec{g_2} = \left[-\int_0^{t_i} x(\tau)y(\tau)d\tau\right],$$

$$\vec{g_3} = [x_0 - x_i], \quad \vec{g_4} = \left[\int_0^{t_i} x(\tau)d\tau\right],$$

$$\vec{g_5} = \left[\int_0^{t_i} x^2(\tau)d\tau\right], \quad \vec{g_6} = \left[\frac{1}{2}(x_i^2 - x_0^2)\right],$$

$$\vec{g_7} = \left[\int_0^{t_i} y(\tau)d\tau\right], \quad \vec{g_8} = \left[-\int_0^{t_i} \frac{y^2(\tau)}{x(\tau)}d\tau\right],$$

$$\vec{g_9} = [y_i - y_0]$$

are the $(N+1) \times 1$-vector columns. Now applying the method used in Theorem 1 we prove the Corollary.

Remark 4. *In vector $\vec{g_1}, \vec{g_2}, \vec{g_4}, \vec{g_5}, \vec{g_7}, \vec{g_8}$ the integrals are calculated by means of numerical integration of x_i, y_i and their combinations with respect to time t and that is why the proposed method is called the integration method of identification.*

Remark 5. *Note that expressions (5), (7) and (11) are linear with respect to unknown constants $C_1, \cdots, C_7$. Direct use of the constraint minimization using the Lagrange multiplier with constraint:*

$$C_1C_3^2 + C_3C_5 - C_4 = 0 \qquad (12)$$

produces nonlinear system of equations for determination of six unknowns $C_1, C_2, C_3, C_4, C_5, \lambda$. Thus the search is performed in six-dimensional space of parameters and hence this method substantially complexifies the solution procedure. Determination of parameters and C_6 and C_7 needs solution of linear system of two algebraic equations. In the next section we describe an original problem solution algorithm reducing the search space dimension to one and using only linear matrix manipulations in the process of solution, *which substantially simplifies and accelerates the problem solution.*

IV. SOLUTION OF THE PARAMETER IDENTIFICATION PROBLEM

There are four original independent parameters (b_1, b_2, b_3, b_4) in the first equation of (1). First four C- parameters (C_1, C_2, C_3, C_4) depend on b-parameters so that there is one-to-one correspondence between them. The parameter C_5 depends on the first four C-parameter as follows:

$$C_5 = \frac{C_4}{C_3} - C_1C_3^2. \qquad (13)$$

Hence, it is possible to consider (C_1, C_2, C_3, C_4) as independent parameters and introduce new name for the dependent parameter $C_5 = -\lambda$. The novel algorithm will be considered in detail for the differentiation method of solution, i.e. with $\vec{f}_{1,\cdots,9}$ - vector columns(see expression (5) and (7). The integration method of solution uses the same algorithm in which $\vec{f}_{1,\cdots,9}$ - vector columns are changed to $\vec{g}_{1,\cdots,9}$ -ones (see (11)). Parameter λ will be selected from the given interval $\lambda \in [\lambda_{min}, \lambda_{max}]$ and substituted in goal function G_3 which is composed as follows

$$G_3 = G_3(C_1, C_2, C_3, C_4, \lambda)$$
$$= \frac{1}{2}\left(C_1\vec{f_1} + C_2\vec{f_2} + C_3\vec{f_3} + C_4\vec{f_4} - (\lambda\vec{f_5} + \vec{f_6})\right)^T$$
$$\left(C_1\vec{f_1} + C_2\vec{f_2} + C_3\vec{f_3} + C_4\vec{f_4} - (\lambda\vec{f_5} + \vec{f_6})\right) \quad (14)$$

and subjected to minimization. In expression (14), parameter λ is considered as constant at every minimization and minimization itself is performed with respect to parameters C_1, C_2, C_3, C_4. Solution of this problem is given by the following formula

$$C(\lambda) = [C_1(\lambda), \ C_2(\lambda), \ C_3(\lambda), \ C_4(\lambda)]^T$$
$$= \left((L_1^T L_1)^{-1}L_1^T\right) R(\lambda), \quad (15)$$

where

$$L_1 = \left[\vec{f_1} \ \ \vec{f_2} \ \ \vec{f_3} \ \ \vec{f_4}\right]^T, \quad R(\lambda) = \lambda\vec{f_5} + \vec{f_6}. \quad (16)$$

In expression (15) it is possible to calculate $1 \times (N+1)$- vector row $\left((L_1^T L_1)^{-1}L_1^T\right)$ only once and after that perform its multiplication by $(N+1) \times 1$- vector row $R(\lambda)$, which is very fast operation. Components of vector $C(\lambda)$ and $C_5 = -\lambda$ are substituted in the constraint (12) to obtain the following nonlinear scalar equation

$$C_1(\lambda)C_3^2(\lambda) - \lambda C_3(\lambda) - C_4(\lambda) = 0, \qquad (17)$$

which is solved with respect to λ. All roots of Equation (17) are found (sometimes to find all the roots it is necessary to expand the interval $\lambda \in [\lambda_{min}, \lambda_{max}]$ to the left or to the right or to both sides). After finding a particular root λ the corresponding b-parameters are calculated as follows:

$$b_1(\lambda) = \frac{C_4(\lambda)}{C_3(\lambda)}, \quad b_2(\lambda) = C_1(\lambda),$$
$$b_3(\lambda) = C_2(\lambda), \quad b_4(\lambda) = C_3(\lambda). \qquad (18)$$

(See (4)). The estimations of b-parameters are obtained from the proper selection of root $\lambda = \bar{\lambda}$:

$$\bar{b}_1 - b_1(\bar{\lambda}), \quad \bar{b}_2 = b_2(\bar{\lambda}),$$
$$\bar{b}_3 = b_3(\bar{\lambda}), \quad \bar{b}_4 = b_4(\bar{\lambda}) \qquad (19)$$

(one of the criteria of the correct choice of $\bar{\lambda}$ must be positiveness of all estimated $\bar{b}$ parameters, see **Numerical Examples**). Parameters b_5 and b_6 are estimated by means of minimization of the goal function of Equation 8. Solution of this problem is given by the formulas:

$$\begin{bmatrix} \bar{b}_5 \\ \bar{b}_6 \end{bmatrix} = (L_2^T L_2)^{-1} L_2^T \vec{f}_9 \qquad (20)$$

where $L_2 = \begin{bmatrix} \vec{f}_7 & \vec{f}_8 \end{bmatrix}$ is $(N+1) \times 2$- matrix, (See (7)).

Expression (15)-(20) give solution to the identification problem by means of the differentiation method. To find solution of the problem by the integral method it is necessary to consider vectors $\vec{g}_{1,\dots,9}$ (See expression 11) instead of $\vec{f}_{1,\dots,9}$. In the next section you will find more information about application of the differentiation and integration methods.

V. NUMERICAL EXAMPLES

Let us solve the initial problem of Equation 1 with the following parameters:

$$b_1 = 0.2 \quad b_2 = 0.01 \quad b_3 = 0.05$$
$$b_4 = 1 \quad b_5 = 0.062 \quad b_6 = 0.0223 \qquad (21)$$

and initial conditions: $[x_0 \ y_0]^T = [10 \ 5]^T$. The stable critical point has coordinates $(\tilde{x}, \tilde{y}) \approx (7.77064, 21.4066)$ (see Equation 2) and it is the stable focus (eigenvalues of the linearized system in the vicinity of the critical point are $\nu_{1,2} \approx -0.0138 \pm 0.0735i$, where $i^2 = -1$). The unstable saddle has coordinates $(\tilde{\tilde{x}}, \tilde{\tilde{y}}) = (20,0)$. Numerical solution $x = x(t)$ on the time interval $t \in [0, T = 150]$ in $N + 1 = 25$ points is shown in Figure 1 and solution $y = y(t)$ is shown in Figure 2. Performing solution by means of the differential

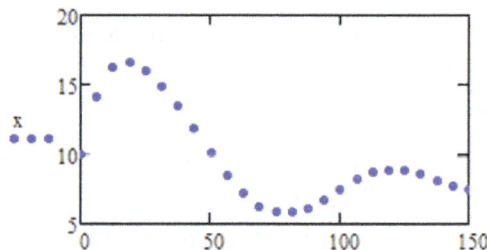

Fig. 1.   Graph of solution $x = x(t)$

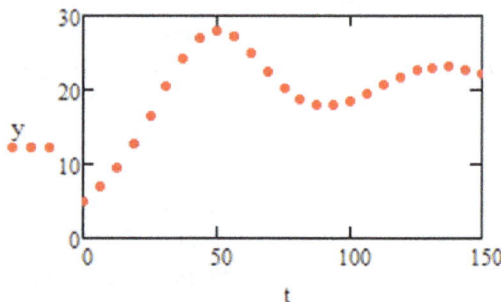

Fig. 2.   Graph of solution $y = y(t)$

method in accordance with the described algorithm we obtain nonlinear Equation (12) from which the parameters are calculated: $\lambda_1 \approx -0.3282$, $\lambda_2 \approx -0.1091$ and $\lambda_3 \approx 0.2087$. As we see, only λ_3 parameter can be selected from three

TABLE I
VALUES OF b-PARAMETERS, CORRESPONDING TO DIFFERENT ROOTS OF EQUATION (12) FOR $N + 1 = 25$
(DIFFERENTIATION METHOD)

Original Values	$\lambda_1 \approx -0.3282$	$\lambda_2 \approx -0.1091$	$\lambda_3 \approx 0.2087$
$b_1 = 0.2000$	-0.0580	-0.0621	0.2129
$b_2 = 0.0100$	-0.0150	-0.0045	0.0108
$b_3 = 0.0500$	-0.0295	-0.0026	0.0491
$b_4 = 1.0000$	-18.0380	-10.5185	0.3906

roots, because λ_1 and λ_2 generate the negative values of b-parameters. The relative error of the b-parameters corresponding to λ_3-parameter are as follows:

$$ERROR\%(b_1) \approx 6.437\%$$
$$ERROR\%(b_2) \approx 7.725\%$$
$$ERROR\%(b_3) \approx 1.832\%$$
$$ERROR\%(b_4) \approx 60.943\% \quad (22)$$

Estimation of parameters b_5 and b_6 gives coincidence with the original values of the parameters in four decimals with the following relative errors:

$$ERROR\%(b_5) \approx 0.029\%$$
$$ERROR\%(b_6) \approx 0.028\% \quad (23)$$

Comparison of original graphs with graphs obtained by numberical solution of initial problem (1) with the same initial conditions but with estimated parameters is shown in Figure 3 and Figure 4.

As we see the estimated parameters gives quite good estimation of the process dynamics. The estimated values of the steady states are as follows $(\tilde{\tilde{x}}, \tilde{\tilde{y}}) \approx (7.7143, 21.4282)$ with relative errors:

$$ERROR\%(\tilde{x}) \approx 0.102\%$$
$$ERROR\%(\tilde{y}) \approx 0.101\% \quad (24)$$

Estimation of the parameters with $N + 1 = 49$ points gives $\lambda_1 \approx -0.2585$, $\lambda_2 \approx -0.0878$, $\lambda_3 \approx 0.1914$ and the following values of parameters (see Table 2)

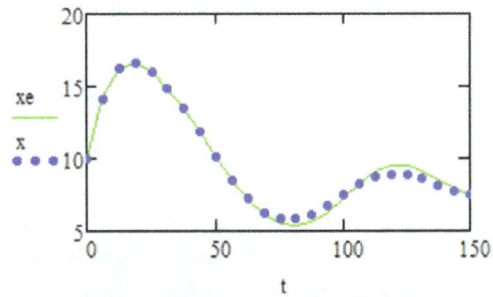

Fig. 3. Graph of original solution $x = x(t)$ (dots) and solution with estimated parameters (solid line)

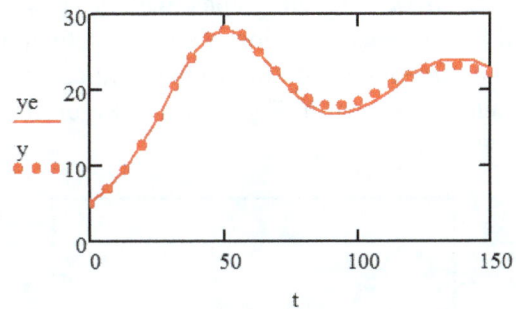

Fig. 4. Graph of original solution $y = y(t)$ (dots) and solution with estimated parameters (solid line)

As we see, only λ_3 parameter can be selected from the three roots, because λ_1 and λ_2 generate the negative values of b-parameters. One can see the substantial improvement of the parameters estimations. The relative errors of the b-parameters

TABLE II
VALUES OF b-PARAMETERS, CORRESPONDING TO DIFFERENT ROOTS OF EQUATION (12) FOR $N + 1 = 49$
(DIFFERENTIATION METHOD)

Original Values	$\lambda_1 \approx -0.2585$	$\lambda_2 \approx -0.0878$	$\lambda_3 \approx 0.1914$
$b_1 = 0.2000$	-0.0405	-0.0485	0.2010
$b_2 = 0.0100$	-0.0119	-0.0036	0.0101
$b_3 = 0.0500$	-0.0271	-0.0022	0.0500
$b_4 = 1.0000$	-18.3099	-10.9975	0.9553

corresponding to λ_3- parameter are as follows:

$$ERROR\%(b_1) \approx 0.496\%$$
$$ERROR\%(b_2) \approx 0.583\%$$
$$ERROR\%(b_3) \approx 0.087\%$$
$$ERROR\%(b_4) \approx 4.469\% \qquad (25)$$

Estimation of parameters b_5 and b_6 gives coincidence with the original ones in four decimals with the following relative errors:

$$ERROR\%(b_5) \approx 0.002\%$$
$$ERROR\%(b_6) \approx 0.002\% \qquad (26)$$

Comparison of original graphs with graphs obtained by numerical solution of initial problem (1) with the same initial conditions but with estimated parameters is shown in Figure 5 and Figure 6.

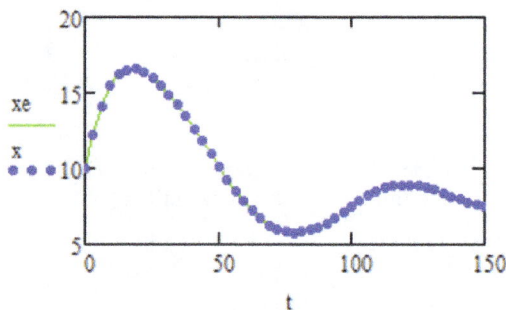

Fig. 5.   Graph of original solution $x = x(t)$ (dots) and solution with estimated parameters (solid line)

As we see the estimated parameters give very good estimation of the process dynamics. The

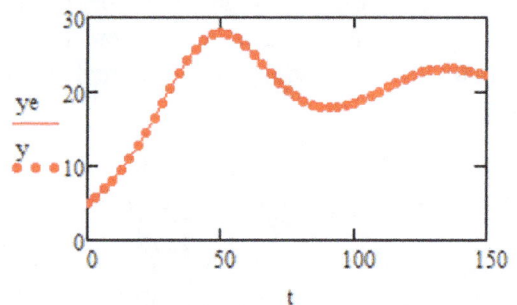

Fig. 6.   Graph of original solution $y = y(t)$ (dots) and solution with estimated parameters (solid line)

estimated values of the steady states are as follows $(\tilde{\tilde{x}}, \tilde{y}) \approx (7.7077, 21.4102)$ with relative errors:

$$ERROR\%(\tilde{x}) \approx 0.017\%$$
$$ERROR\%(\tilde{y}) \approx 0.017\% \qquad (27)$$

Absolute errors in calculation of $x = x(t)$ and $y = y(t)$ in the differentiation method for $N + 1 = 25$ and $N + 1 = 49$ points are shown in Figure 7 and Figure 8. Performing solution by means of the integration method in accordance with the described algorithm we obtain three roots of nonlinear equation (12): $\lambda_1 \approx -0.2391$, $\lambda_2 \approx -0.0725$, $\lambda_3 \approx 0.1899$.

As we see, only λ_3 parameter can be selected from the three roots, because λ_1 and λ_2 generate the negative values of b-parameters. The relative errors of the b-parameters corresponding to λ_3-

TABLE III

VALUES OF b-PARAMETERS, CORRESPONDING DIFFERENT ROOTS OF EQUATION (12) FOR $N + 1 = 25$ (INTEGRATION METHOD)

Original Values	$\lambda_1 \approx -0.2391$	$\lambda_2 \approx -0.0725$	$\lambda_3 \approx 0.1899$
$b_1 = 0.2000$	-0.0302	-0.0386	0.1999
$b_2 = 0.0100$	-0.0115	-0.0032	0.0100
$b_3 = 0.0500$	-0.0282	-0.0022	0.0500
$b_4 = 1.0000$	-18.1074	-10.6869	0.9997

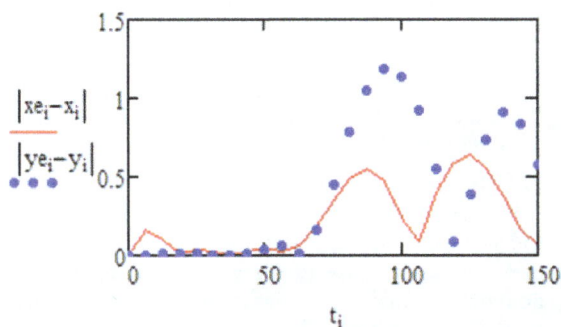

Fig. 7. Absolute Errors of Calculation for $N+1 = 25$ points (Differentiation method)

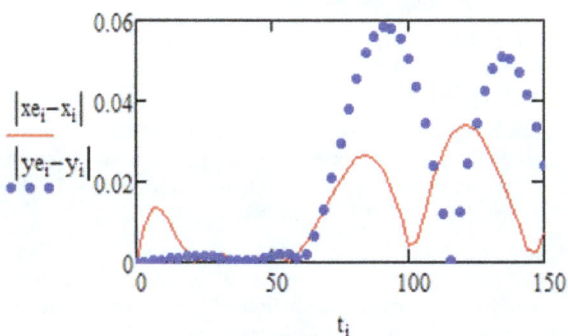

Fig. 9. Absolute errors of calculation for $N+1 = 25$ points (Integration method)

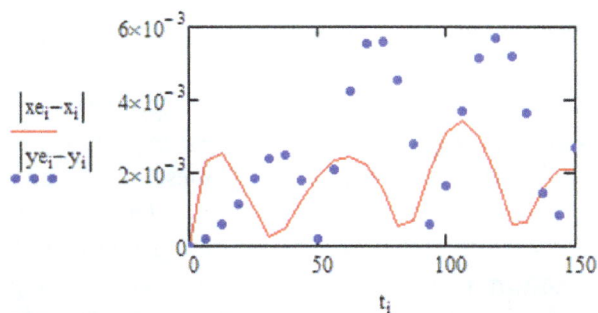

Fig. 8. Absolute Errors of Calculation for $N+1 = 49$ points (Differentiation method)

parameter are as follows:

$$ERROR\%(b_1) \approx 0.052\%$$
$$ERROR\%(b_2) \approx 0.044\%$$
$$ERROR\%(b_3) \approx 0.059\%$$
$$ERROR\%(b_4) \approx 0.033\% \tag{28}$$

Estimation of parameters b_5 and b_6 gives coincidence with the original values of b-parameter in four decimals with the following relative errors:

$$ERROR\%(b_5) \approx 0.008\%$$
$$ERROR\%(b_6) \approx 0.007\% \tag{29}$$

Comparison of original graphs with graphs obtained by numerical solution of initial problem (1) with the same initial conditions but with estimated parameters are visually indistinguishable from Figure 5 and Figure 6. Absolute errors in calculation of $x = x(t)$ and $y = y(t)$ in the integration method for $N + 1 = 25$ points are shown in Figure 9.

The parameters are estimated with very high accuracy at $N + 1 = 25$ points. The estimated values of the steady states are as follows ($\tilde{\tilde{x}}, \tilde{\tilde{y}}$) $\approx$ (7.7062, 21.4061) with relative errors:

$$ERROR\%(\tilde{x}) \approx 0.002\%$$
$$ERROR\%(\tilde{y}) \approx 0.002\% \tag{30}$$

VI. Conclusion

Two methods of solution of the inverse problem on parameter identification of the Holling-Tanner model with complete information are discussed. These are the differentiation and integration methods of solution. The conditions are indicated at which all parameters of the model can be identified. The main disadvantage of the conventional method of constraint minimization by means of the Lagrange multipliers is that the method generates a system of six nonlinear equations with unknown initial guess values. Proposed is the novel method of the problem solution in which the six dimensional space of search is reduced to one dimensional space and the procedure of the initial guess value is performed by fast vector multiplication. Numerical examples of the proposed algorithm implementation are demonstrated for the differentiation and integration methods. It is shown that the integration method generates more accurate results than the differentiation one. The integration method also needs less number of points on the fixed time interval to produce accurate results than the differentiation method.

VII. Acknowledgment

The authors acknowledge the department of Mathematics and Statistics of the Tshwane University of Technology towards the research.

References

[1] M Yu Shatalov, AS Demidov, and IA Fedotov. Estimating the parameters of chemical kinetics equations from the partial information about their solution. *Theoretical Foundations of Chemical Engineering*, 50(2):148–157, 2016.

[2] Dmitrii Logofet. *Matrices and Graphs Stability Problems in Mathematical Ecology: 0*. CRC press, 2018.

[3] John Pastor. *Mathematical ecology of populations and ecosystems*. John Wiley & Sons, 2011.

[4] Richard McGehee and Robert A Armstrong. Some mathematical problems concerning the ecological principle of competitive exclusion. *Journal of Differential Equations*, 23(1):30–52, 1977.

[5] Gilles Clermont and Sven Zenker. The inverse problem in mathematical biology. *Mathematical biosciences*, 260:11–15, 2015.

[6] Andreas Kirsch. *An introduction to the mathematical theory of inverse problems*, volume 120. Springer Science & Business Media, 2011.

[7] Mainul Haque and Ezio Venturino. The role of transmissible diseases in the Holling–Tanner predator–prey model. *Theoretical Population Biology*, 70(3):273–288, 2006.

[8] James T Tanner. The stability and the intrinsic growth rates of prey and predator populations. *Ecology*, 56(4):855–867, 1975.

[9] David J Wollkind, John B Collings, and Jesse A Logan. Metastability in a temperature-dependent model system for predator-prey mite outbreak interactions on fruit trees. *Bulletin of Mathematical Biology*, 50(4):379–409, 1988.

[10] Peter A Braza. The bifurcation structure of the Holling–Tanner model for predator-prey interactions using two-timing. *SIAM Journal on Applied Mathematics*, 63(3):889–904, 2003.

[11] M Shatalov and I Fedotov. On identification of dynamic system parameters from experimental data. 2007.

[12] Michael Shatalov, Igor Fedotov, and Stephan V Joubert. Novel method of interpolation and extrapolation of functions by a linear initial value problem. In *Buffelspoort TIME 2008 Peer-reviewed Conference Proceedings*. Buffelspoort TIME 2008 Peer-reviewed Conference Proceedings, 2008.

[13] Joubert S.V Shatalov M, Greeff J.C and Fedotov I. Parametric identification of the model with one predator and two prey species. *Buffelspoort TIME2008 Peer-reviewed Conference Proceedings,*, 2008.

[14] Rashi Gupta. Dynamics of a Holling–Tanner model. *American Journal of Engineering Research (AJER)*, page 2, 2017.

[15] Rui Peng and Mingxin Wang. Positive steady states of the Holling–Tanner prey–predator model with diffusion. *Proceedings of the Royal Society of Edinburgh Section A: Mathematics*, 135(1):149–164, 2005.

[16] Guirong Liu, Sanhu Wang, and Jurang Yan. Positive periodic solutions for neutral delay ratio-dependent predator-prey model with holling-tanner functional response. *International Journal of Mathematics and Mathematical Sciences*, 2011, 2011.

[17] Sze-Bi Hsu and Tzy-Wei Huang. Global stability for a class of predator-prey systems. *SIAM Journal on Applied Mathematics*, 55(3):763–783, 1995.

[18] Shanshan Chen and Junping Shi. Global stability in a diffusive Holling–Tanner predator–prey model. *Applied Mathematics Letters*, 25(3):614–618, 2012.

[19] Charles L Lawson and Richard J Hanson. *Solving least squares problems*, volume 15. Siam, 1995.

14

Bayesian inference of a dynamical model evaluating Deltamethrin effect on Daphnia survival

Abdoulaye Diouf*, Baba Issa Camara†, Diène Ngom*, Héla Toumi†‡, Vincent Felten†,
Jean-François Masfaraud†, Jean-François Férard†

*Département de Mathmatiques, UMI 2019-IRD & UMMISCO-UGB,
Université Assane Seck de Ziguinchor, B.P. 523 Ziguinchor, Sénégal
a.diouf1269@zig.univ.sn, dngom@univ-zig.sn

†Université de Lorraine - CNRS UMR 7360, Laboratoire Interdisciplinaire des Environnements
Continentaux, Campus Bridoux - 8 Rue du Général Delestraint, 57070 Metz, France
baba-issa.camara@univ-lorraine.fr, vincent.felten@univ-lorraine.fr,
jean-francois.masfaraud@univ-lorraine.fr, jean-francois.ferard@univ-lorraine.fr

‡Laboratoire de Bio-surveillance de l'Environnement (LBE), Université de Carthage,
Faculté des Sciences de Bizerte, 7021 Zarzouna, Bizerte, Tunisie
toumihela@yahoo.fr

Abstract—The toxicokinetic and toxicodynamic models (TK-TD) are very well-known for their ability, at both the individual and the population level, to accurately describe life cycles such as the growth, reproduction and survival of sentinel organisms under the influence of an ecological biomarker. Being dynamics, the consistent inference of life history and environmental traits parameters that engender them is sometimes very complex numerically, especially as these parameters vary from one individual to another. In this paper, we estimate the parameters of a survival model TK-TD already applied and validated by the implementation of the R package GUTS (the General Unified Threshold Model of Survival) by another coding applied to another very recent implementation of Bayesian inference with the R package deBInfer in order to evaluate the survival effects of our ecotoxicological biomarker called Deltamethrin on our Daphnia sample. The study allowed us to evaluate from a population point of view especially the threshold concentration not to be exceeded to observe a survival effect commonly known NEC (No effect Concentration) and possibly determine the correlations between different variables of life history and the environment traits.

Keywords-Bayesian inference; parameter correlations; Daphnia survival; deBInfer, Deltamethrin; dynamic; NEC

I. Introduction

Statistical methods for the analysis of survival data have continued to flourish over the last two decades [7], [31]. Since then, there have been many publications that deal with this hot topic in various fields such as medicine [13], [19], [34], epidemiology [8], [21], criminology [7], [23], business reliability research [11], [29], [35], and the social and behavioral sciences [25], [27], [31], [36]. They are intensively used in biology particulary ecotoxicology as in [4], [6], [12], [15], [16], [24], [32], pharmacology and medical research globally for example in [5], [26], [33].

The simulation of the temporal evolution of processes leading to toxic effects on organisms is the major role of the use of toxicokinetic-toxicodynamic models (TK-TD models) [17]. There is a diversity of TK-TD models for modeling seemingly simple survival according to the underlying assumptions (individual tolerance or stochastic death, speed of toxicodynamic damage recovery, threshold distribution). The General Unified Threshold Model for Survival (GUTS) is the more general survival TK-TD model from which a wide range of existing models can be inferred as special cases [17]. It has special cases of very appropriate model that can be adjusted to the survival data. As a result, it is actively contributing to increasing its application in ecotoxicology research as well as in the assessment of environmental risks related to chemicals.

However, it is known that in toxicokinetics and pharmacokinetics the evolution of xenobiotics (toxic or therapeutic) in a living organism is qualitative and quantitative. By means of a realistic description (ie anatomical, physiological and biochemical) of the absorption processes (inhalation, skin contact, ingestion or intravenous injection), distribution, metabolism and excretion (ADME process), the mechanistic models , which will result, allow the understanding and the simulation of this evolution of the dose of a substance in the various organs and fluids of the body [9]. The action of the organism on the substance defines the toxicokinetics (TK) whereas the opposite effect

translates the toxicodynamics (TD). The equations that govern them are differential equations.

To answer why some individuals survive after exposure of chemicals while others die, Ashauer and al., 2015 [2] established the General Unified Threshold Model of Survival (GUTS), a mathematical relationship. In GUTS, there is two assumptions: the threshold of tolerance is individually distributed and that its overcoming causes sudden death among the individuals of a population and the existence of a certain threshold, above which death occurs stochastically, which all people share. As a result, GUTS appeared to be a promising development in the analysis of traditional survival curves and dose-response models.

Recently, Roman Ashauer and al., 2017 [3] treated the paradigm "dose is poison". They illustrated that it is not only the dose that makes the poison but also the sequence of exposure taking into account the toxicokinetic recovery assumptions (the lack of effect that once a chemical is removed from organism) and toxicodynamic recovery (the neglect of the other homeostasis recovery process may be rapid or slow depending on the chemical). To do this, they tested four toxic substances acting on different targets (diazinon, propiconazole, 4,6-dinitro-o-cresol, 4-nitrobenzyl chloride) on the freshwater crustacean Gammarus pulex.

In this study, special consideration is given to the application of Bayesian inference to the evaluation of the effects of Deltamethrin (a pesticide) on a toxicokinetic and toxicodynamic (TK-TD) survival model. Bayesian inference can be a very sophisticated tool for survival data analysis. It is well known for its ability to process data of any sample size, especially small samples as opposed to conventional methods.

Many statistical methods are currently too complex to be fitted using classical statistical methods, but they can be fitted using Bayesian computational methods [14], [23], [28]. However, it may be reassuring that, in many cases, Bayesian inference gives answers that numerically closely match those obtained by classical methods.

In this article, it is mainly to use, from another angle, a new approach to very recent Bayesian implementation [30] allowing the inference of parameters of the model TK-TD GUTS applicable to the adjustment of our survival data collected at the Interdisciplinary Laboratory of Continental Environments (LIEC). It is a very rigorous methodology insofar as it makes it possible to detect the different relations that can exist between the observable quantities of the unobservable quantities, the states and the parameters of the model. Simple to implement, it requires a differential equation TK-TD or DEBtox model, experimental data for the calculation of the likelihood on these data and a prior distribution assumption. A Markov Chain Monte Carlo procedure (MCMC) describes these inputs to estimate the posterior distributions of the parameters and any derived error variables, including model trajectories. This approach is designed with a MCMC diagnosis of inference, the visualization of posterior distributions of the parameters and trajectories of the model used. This manuscript assesses the long-term survival effects of a toxic substance (a pesticide) called Deltamethrin via the use of the highly reputable GUTS model for assessing the survival of living organisms under stressors such as toxic or pharmaceuticals. The plan adopted for the organization of this article is as follows: in the second section (II), we explain the experimental protocol established in the laboratory and present the model TK-TD GUTS used to translate our experimental protocol. In the third section (III), we discuss the results of the Bayesian analysis. We end in section (IV) with a conclusion and discussion.

II. MATERIALS AND METHODS

A. Organism test

One of the three most widely used biological models for the ecotoxicological risk assessment of toxic substances, Daphnia is a major invertebrate of freshwater aquatic ecosystems. The experiments were conducted with clone A of Daphnia magna Straus 1820 (identified by Professor Calow, Uni-

versity of Sheffield, United Kingdom). They are more than 40 years old at LIEC (University of Lorraine, France) [38]. Parthenogenetic cultures were carried out in 1L aquaria with LCV medium: a mixture (20/80) of LefevreCzarda (LC) medium and French mineral water called Volvic (V). This medium is supplemented with i) Ca and Mg in order to obtain a total hardness of 250 mg.L^{-1} and a Ca/Mg molar ratio of 4/1, and ii) a mixture of vitamins (0.1 mL.L^{-1}) containing thiamine HCl (750 mg.L^{-1}), vitamin B12 (10 mg.L^{-1}) and biotin (7.5 mg.L^{-1}). Parthenogenetic cultures of daphnids were maintained under a temperature of $20°C$, a photoperiod of $16 - 8$ h lightdark and at a density of 40 organism per liter of culture medium. The Daphnia medium was renewed at least three times weekly and daphnids were fed with a mixture of three algal species (5×10^6 Pseudokirchneriella subcapitata, 2.5×10^6 Desmodesmus subspicatus, and 2.5×10^6 Chlorella vulgaris/Daphnia/day). These algae were also continuously cultivated in the laboratory using a nutrient LC medium.

B. Test chemical

Intensely used in agriculture, Deltamethrin is a class II pyrethroid insecticide that is harmful to freshwater ecosystems, especially the cladoceran Daphnia magna (Straus 1820) [37], [38]. The Deltamethrin ($C_{22}H_{19}Br_2NO_3$) used in the experiments is the technical active substance of the formulation DECIS EC25 (25 g.L^1) commercialized by Bayer (Germany). Stock solutions were prepared by dissolving the toxicant in acetone immediately prior to each experiment.

C. Data sample

The experimental protocol was carried out during 21 days of observation. Without the control, five different doses of Deltamethrin (9, 20, 40, 80 and 160 ng.L-1, respectively) were administered to Daphnia magna, with a replicate of 10 for each dose submitted. The survivor count has allowed us to summarize our data sample in the table I.

TABLE I
CHRONIC TEST SUMMARY TABLE (21 DAYS) OF
DELTAMETHRIN EFFECTS SURVIVAL.

Time (*day*)	mean $\pm$ standard deviation (SD) of the survivors number during 21 days
Control	10 ± 0
9 $ng.L^{-1}$	9.667 ± 0.913
20 $ng.L^{-1}$	9.619 ± 0.921
40 $ng.L^{-1}$	9.429 ± 1.121
80 $ng.L^{-1}$	9.333 ± 1.238
160 $ng.L^{-1}$	8 ± 1.761

D. Model Used

GUTS is part of mathematical modeling to quantify the temporal evolution of the survival of an organism population, statistically speaking. It is highly reputed for its ability to assess a population survival effects due to a chemical stressor presence (toxicity in other words) responsible for the individuals mortality in this population. Indeed, the toxicokinetic model criterion is explained by the fact that the ingested chemicals will affect a target site within the body before exerting a toxic effect thus causing damage over time. All TK-TD models including a damage state use either the assumption of individual tolerance or SD hypothesis (ie the existence of a single threshold not to be exceeded for all individuals). The modeling assumptions are not the same, it is obviously clear that the results and interpretations that will follow will differ thereafter. Let us not forget that the term "hazard" and specific terms of parametrization of the different models (such as killing rate, recovery rate constant or elimination rate constant) will be misinterpreted in both cases [17]. But GUTS was designed to overcome these confusions because playing a unifying role that merges different concepts of existing models. GUTS is a synthesis of all these models by mixing the aforementioned hypotheses. More complete documentation of GUTS formulation hypotheses can be found in [17]. For all these reasons, we take the GUTS model to adopt it to our survival data study. As in [1], the GUTS model considered is as follows: (1).

$$\dot{D}(t) = k_e\Big(C(t) - D(t)\Big), \qquad (1)$$

Where $C(t)$ represent the toxic dose subjected linearly causing the time course of damage $D(t)$. The dominant rate constant denoted k_e (in days^{-1} units) models the slowest process inducing the recovery of the exposed organism. In fact, the more slow the recovery in the individual, the more vulnerable he is to the damage. Note that in the body, there are systematically compensation mechanisms and damage repair. The assumption made in this GUTS model is that damage noted $D(t)$ ("*damage*") is considered to be the same for all individuals while knowing that once we exceed a certain threshold. The death considered at individual level as a stochastic event will occur and whose probability increases linearly with the damage. At the population level, this threshold is assumed to vary stochastically over the whole population. The hazard rate $h_z(t)$ (*days*$^{-1}$) for individual with threshold z or *NEC* (No-Effect Concentration) in equation (2) below represents the "instantaneous probability to die" at individual level. The NEC define the concentration threshold not to be exceeded in the body, an amount that we would like to estimate on average. Once it is reached, it affects the health of the living organism.

$$h_z(t) = k_k \max\Big(0, D(t) - NEC\Big) + h_b, \qquad (2)$$

where the proportionality constant k_k (in $ng.L^{-1}.days^{-1}$ units) is well known called killing rate and h_b (in *days*$^{-1}$ units) is the background mortality rate, that is, the control mortality rate, which is assumed to be constant over time [$days^{-1}$]. The equation (3) expresses the probability $S(t)$ that an individual of the population considered will survive until time t conditionally at the threshold z or NEC.

$$\dot{S}_z(t) = -h_z(t)S_z(t), \qquad (3)$$

Additional information on GUTS model modeling assumptions can be found on [1], [2], [3], [17].

E. Statistical method

In contrast to visual estimation methods, which are often considered biased and not robust,

TABLE II
SURVIVAL MODEL PARAMETERS INFERENCE.

Parameter	Symbol	Units	Prior distribution	Initial value
Elimination rate	k_e	$days^{-1}$	$\mathscr{G}(1;1)$ [1]	0.001
Killing rate	k_k	$ng.L^{-1}.days^{-1}$	$\mathscr{G}(1;\frac{1}{4})$ [1]	0.015
Threshold for effects	NEC	$ng.L^{-1}$	$\mathscr{G}(6;1)$ [1]	1.5
Background hazard rate	h_b	$days^{-1}$	$\mathscr{B}(0.1;0.15)$ [2]	0.001
Control correction	$ec1$	$[-]$	$\mathscr{LN}(0;1)$ [3]	0.005
9 $ng.L^{-1}$ correction	$ec2$	$[-]$	$\mathscr{LN}(0;1)$ [3]	0.005
20 $ng.L^{-1}$ correction	$ec3$	$[-]$	$\mathscr{LN}(0;1)$ [3]	0.005
40 $ng.L^{-1}$ correction	$ec4$	$[-]$	$\mathscr{LN}(0;1)$ [3]	0.005
80 $ng.L^{-1}$ correction	$ec5$	$[-]$	$\mathscr{LN}(0;1)$ [3]	0.005
160 $ng.L^{-1}$ correction	$ec6$	$[-]$	$\mathscr{LN}(0;1)$ [3]	0.005

Bayesian statistics using kinetic data have been very successful over the last two decades [9]. For all these reasons, we use in this paper the Bayesian approach often considered from a practical point of view as a descriptive statistical analysis technique among the others [22]. In Bayesian statistics, any unknown entity is considered as a random variable, in particular parameters of the model used. An assumption of a prior distribution, assigned to each parameter to be estimated, is necessary before the experimental data analysis. Via the famous Bayes theorem, these prior information will be updated with the experimental data in order to retrieve posterior information. Only the Bayesian approach allows to integrate the knowledge that one has of a system by taking advantage of the experimental information [22]. It is a conjunction of the information provided by the probabilistic model by a prior distribution and experimental data. The R package used for our model parameters inferring is deBInfer [30]. We use the R package deSolve [20], [39] as underlined in [30] for the resolution of the implemented TK-TD model. To extrapolate likelihood on our experimental data, we use the Poisson log-likelihood function as defined in the equation (4). The log-likelihood of the data given the parameters, underlying model, and initial conditions is then a sum over the n observations at each time point in t':

$$\mathscr{L}(\mathscr{Y}|\theta) = \sum_t^n N_t \log \lambda - n\lambda \qquad (4)$$

Here we use small corrections $(ec_i)_{i=1,\cdots,6}$ that are needed because of the differential equations solutions can equal zero, whereas the parameter lambda of the poison likelihood must be strictly positive. We infer them later as suggested in [18], [20]. We set 20,000 iterations for the MCMC procedure, $cnt = 500$ worth only 1231.06 seconds of execution with an Intel (R) Core (TM) i3-2350M CPU processor running at 2.30 GHz. The prior distributions assumptions as well as the parameters measures units are presented in the table II.

III. RESULTS AND DISCUSSION

The inference results are presented in tables III and IV. They were obtained using the major functions ode() of the R package deSolve [20] and de_mcmc() of the R package deBInfer [30]. Tables III and IV respectively give the empirical mean and standard deviation for each variable, plus standard error of the mean and the quantiles for each variable.

The threshold concentration above which there are effects on the survival of our test species (Daphnia magna) commonly called NEC is estimated cap 6.042 ± 2.418 $ng.L^{-1}$. It is similar to that estimated in one of our studies on the risk assessment of Deltamethrin on growth and reproduction treated separately [10]. This result is

[1]The Gamma distribution
[2]The Beta distribution
[3]The Log-normal distribution

TABLE III
EMPIRICAL MEAN AND STANDARD DEVIATION FOR EACH VARIABLE, PLUS STANDARD ERROR OF THE MEAN.

	Mean	SD	Naive SE	Time-series SE
ke	0.428169	0.738795	5.224e-03	0.0827944
kk	0.003064	0.010340	7.311e-05	0.0013127
NEC	6.041585	2.418373	1.710e-02	0.1636959
hb	0.002097	0.003396	2.401e-05	0.0001835
ec1	0.723883	0.459832	3.252e-03	0.0303711
ec2	0.593472	0.379989	2.687e-03	0.0211085
ec3	0.662622	0.431101	3.048e-03	0.0245585
ec4	0.699403	0.430309	3.043e-03	0.0244237
ec5	0.938278	0.556387	3.934e-03	0.0313650
ec6	0.808771	0.581670	4.113e-03	0.0385683

TABLE IV
QUANTILES FOR EACH VARIABLE.

	2.5%	25%	50%	75%	97.5%
ke	7.848e-04	1.588e-02	9.558e-02	0.542053	2.66691
kk	1.352e-04	2.854e-04	5.183e-04	0.001508	0.02582
NEC	2.150e+00	4.184e+00	5.729e+00	7.690363	10.81920
hb	6.774e-18	1.279e-07	9.653e-05	0.003075	0.01194
ec1	1.023e-01	3.698e-01	6.445e-01	0.997039	1.83111
ec2	9.134e-02	3.152e-01	5.145e-01	0.791141	1.52596
ec3	1.150e-01	3.398e-01	5.604e-01	0.883916	1.73556
ec4	1.089e-01	3.752e-01	6.185e-01	0.936333	1.74968
ec5	1.666e-01	5.078e-01	8.321e-01	1.252371	2.27439
ec6	1.106e-01	3.811e-01	6.464e-01	1.111107	2.26996

very consistent in that death stops any evolution process. While the recovery process under the toxic effect is estimated to be around 0.43 ± 0.74. The dominant rate constant is not so negligible as that in contrast to the killing rate and the control mortality rate constants whose respective values are close to 0.003 ± 0.01 and 0.002 ± 0.03. These different estimated values would translate faithfully our experimental realities as shown in the data table I. With 10 replicates for each Deltamethrin dose, few deaths were observed in this experimental protocol. The GUTS model again reflects the reality of the facts in this study. The density plots for the various inferred parameters can be read in figure 1. In this image, some chain trajectories are reasonable and consistent over time in that their posterior distributions are unimodal, sometimes resembling that of a normal distribution. We can cite for example the parameters NEC and the small corrections $ec_{i=1,\cdots,6}$. Their prior distributions were those of a log-normal

distribution. Unlike the distributions of the k_e, k_k and h_b unimodal parameters, but suspect because they include a large number of outliers. This aberration would confirm the inference complexity of these types of studies. Let's not forget that these are constant rate. The study results are very consistent overall. The figure 2 perfectly shows a lack of detected correlation between parameters. The highest correlation value is 0.36 between k_k and NEC, the two most important parameters in GUTS [17].

For proof purposes, we remove a burnin period of 1500 samples and examine parameter correlations in the figure 2 and overlap between prior and posterior densities. The figure 2 reflects the correlation lack between the different parameters of our dynamics evaluating Daphnia survival in the presence of our Deltamethrin stressor.

From the posterior, we simulate 500 trajectories of our TK-TD model while calculating at 95%HDI (Highest Posterior Density Intervals) for the de-

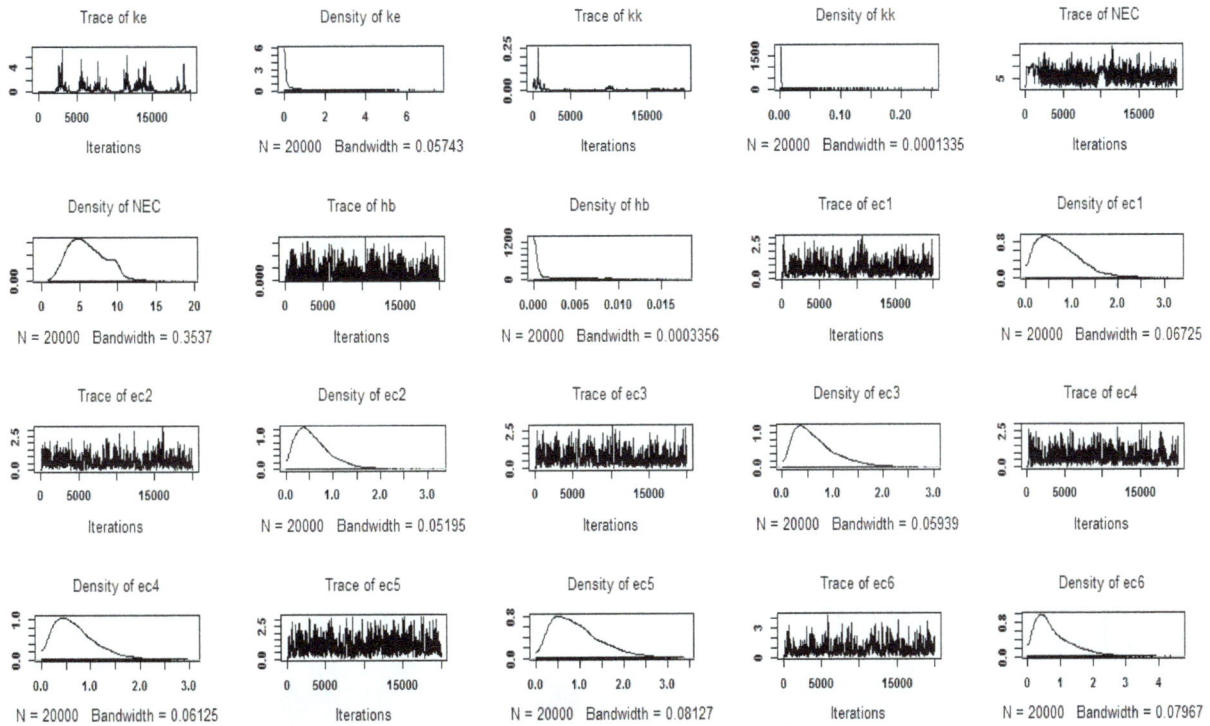

Fig. 1.   MCMC chains plotting & summarizing

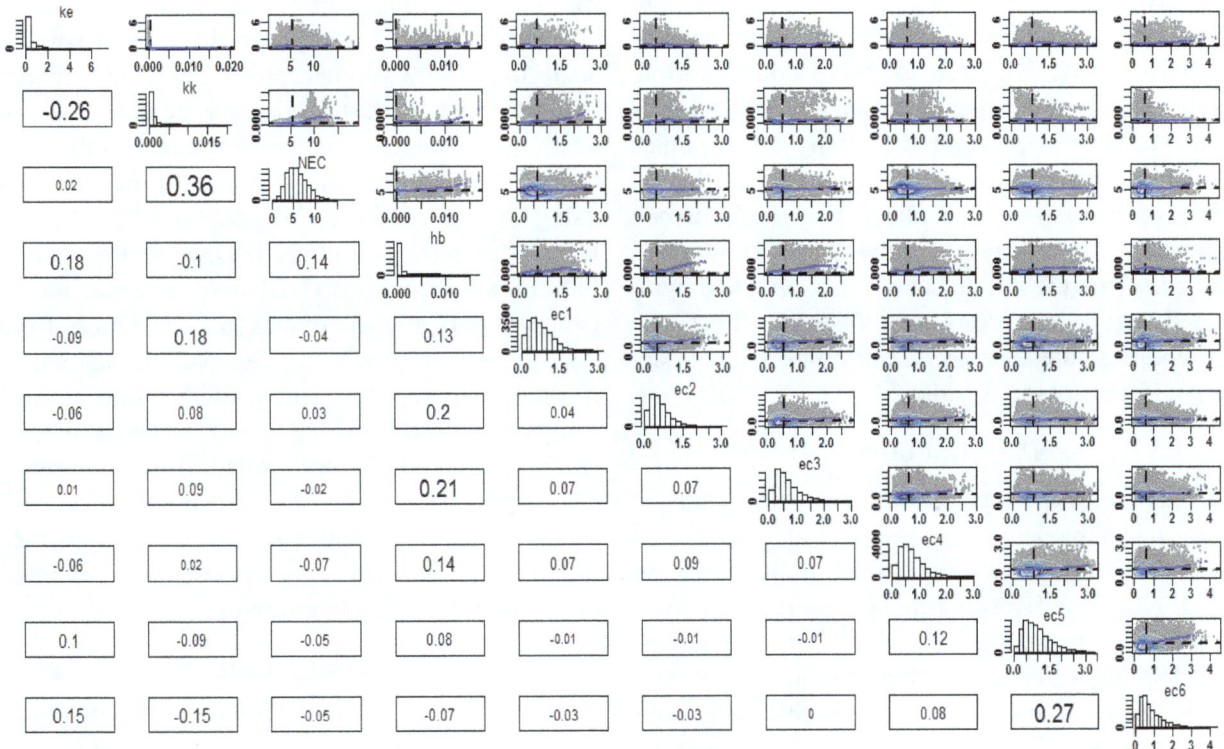

Fig. 2.   Parameter correlations plotting

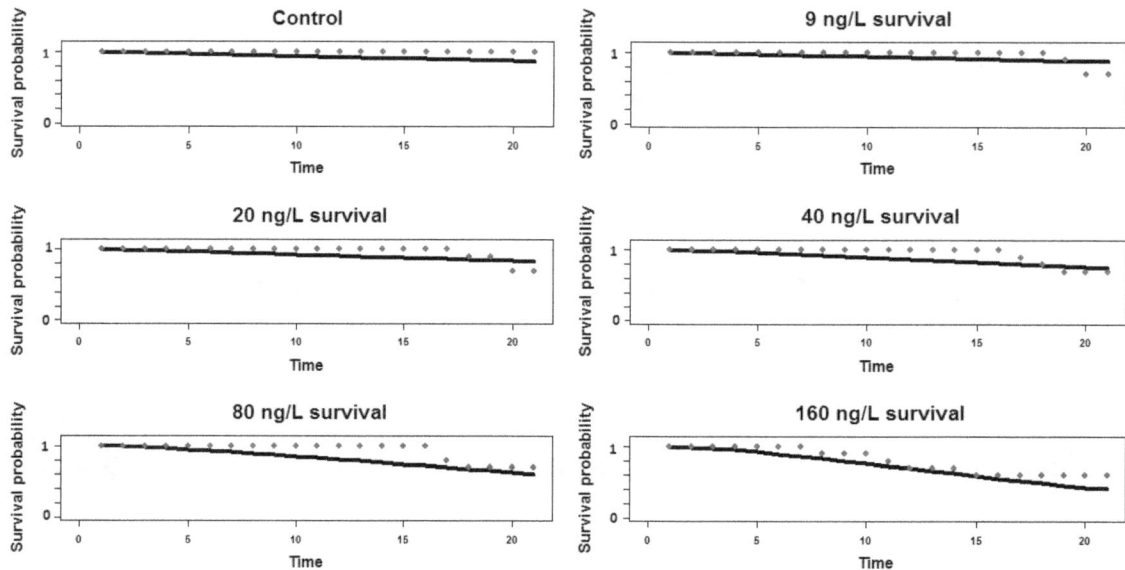

Fig. 3. Median survival fitting with 95% HDI where the black lines represent the solution trajectories of the survival dynamics and the red points reflect the experimental data.

terministic part of the model. HDI sets (intervals) contain all values of the parameter θ such that the posterior density $f_{\theta|y}$ is larger than some constant c_α, where c_α ensures that the coverage probability will be $1 - \alpha$. For each exposure concentration, figure 3 shows in the same graph, the experimental data and the model output describing the dynamics of alive Daphnia magna number during the 21 days of experience. It confirms that the inference procedure actually retrieves the model to our data. In addition, the fitted curves are obtained with small estimation errors, see figure 3. Post hoc trajectories adjust very well our observational data for different pesticide doses.

IV. CONCLUSION

This paper is very instructive in that it adapts the GUTS model to our survival data collected at LIEC through a more recent implementation of Bayesian inference (the R library deBInfer). Thus we ignored the use of GUTS (R package GUTS) implementation. With this new R library, it is easy to encode any toxicico-kinetic and toxicodynamic dynamics (TK-TD) then infer the parameters that compose it. Once differential en-

coding is complete, the R package deBInfer has a function named de_mcmc() where is integrated that of ode() function of the R package deSolve specially designed for system differential solving such that ordinary, partial or delay differential equations. These last facilitate access to a lot of users types whether they are specialists in the field or not. Most of the life phenomena are modeled using Ordinary Differential Equations (EDO), Partial Differential Equations (PDE), or the Delay-Differential Equations (DDE). As a result, this R package deBInfer facilitates the transition from determinism to stochastic. As part of our study, it allowed us to consistently address our survival analysis with the GUTS TK-TD model use. It really facilitated the manipulation and inference of the parameters of a mechanistic model to describe the bioaccumulation kinetics and dynamics of survival effects in a contaminated environment of the pesticide Deltamethrin.

ACKNOWLEDGMENTS

This study was funded by the French cooperation and the African Center of Excellence in Mathematics, Computer Sciences and TIC (CEA-

MITIC). We are very grateful for their financial supports.

REFERENCES

[1] Albert C, Vogel S, Ashauer R (2016) Computationally Efficient Implementation of a Novel Algorithm for the General Unified Threshold Model of Survival (GUTS). PLoS Comput Biol 12(6): e1004978. doi:10.1371/journal.pcbi.1004978

[2] Ashauer, R., O'Connor, I., Hintermeister, A., & Escher, B. I. (2015). Death dilemma and organism recovery in ecotoxicology. Environmental science & technology, 49(16), 10136-10146.

[3] Ashauer, R., O'Connor, I., & Escher, B. I. (2017). Toxic Mixtures in Time The Sequence Makes the Poison. Environmental Science & Technology, 51(5), 3084-3092.

[4] Ankley, G. T., Daston, G. P., Degitz, S. J., Denslow, N. D., Hoke, R. A., Kennedy, S. W., ... & Tyler, C. R. (2006). Toxicogenomics in regulatory ecotoxicology.

[5] Backhaus, T. (2014). Medicines, shaken and stirred: a critical review on the ecotoxicology of pharmaceutical mixtures. Phil. Trans. R. Soc. B, 369(1656), 20130585.

[6] Baguer, A. J., Jensen, J., & Krogh, P. H. (2000). Effects of the antibiotics oxytetracycline and tylosin on soil fauna. Chemosphere, 40(7), 751-757.

[7] Benda, B. B. (2005). Gender differences in life-course theory of recidivism: A survival analysis. International Journal of Offender Therapy and Comparative Criminology, 49(3), 325-342.

[8] Beyersmann, J., Latouche, A., Buchholz, A., & Schumacher, M. (2009). Simulating competing risks data in survival analysis. Statistics in medicine, 28(6), 956-971.

[9] Brochot, C., Willemin, M. E., Zeman, F., & Halatte, F. (2014). La modélisation toxico/pharmacocinétique à fondement physiologique: son rôle en évaluation du risque et en pharmacologie. Modéliser & simuler. Epistémologies et pratiques de la modélisation et de la simulation, 2, 455-493.

[10] Camara B. I., Diouf A., Toumi H., Ngom D., Felten V., & Férard J. F., Estimation of deltametrin bio-accumulation kinetics and dynamics of effects on daphnia population, submitted in Nonlinear Dynamics NODY-D-18-01984.

[11] Chandler, G. N., & Hanks, S. H. (1993). Measuring the performance of emerging businesses: A validation study. Journal of Business venturing, 8(5), 391-408.

[12] Diao, X., Jensen, J., & Hansen, A. D. (2007). Toxicity of the anthelmintic abamectin to four species of soil invertebrates. Environmental pollution, 148(2), 514-519.

[13] Delen, D., Walker, G., & Kadam, A. (2005). Predicting breast cancer survivability: a comparison of three data mining methods. Artificial intelligence in medicine, 34(2), 113-127.

[14] Delignette-Muller, M. L., Ruiz, P., & Veber, P. (2017). Robust Fit of ToxicokineticToxicodynamic Models Using Prior Knowledge Contained in the Design of Sur-

[15] Fent, K., Weston, A. A., & Caminada, D. (2006). Ecotoxicology of human pharmaceuticals. Aquatic toxicology, 76(2), 122-159.

[16] Forfait-Dubuc, C., Charles, S., Billoir, E., & Delignette-Muller, M. L. (2012). Survival data analyses in ecotoxicology: critical effect concentrations, methods and models. What should we use?. Ecotoxicology, 21(4), 1072-1083.

[17] Jager, T., Albert, C., Preuss, T. G., & Ashauer, R. (2011). General unified threshold model of survival-a toxicokinetic-toxicodynamic framework for ecotoxicology. Environmental Science & Technology, 45(7), 2529-2540.

[18] Johnson, L. R., Pecquerie, L., & Nisbet, R. M. (2013). Bayesian inference for bioenergetic models. Ecology, 94(4), 882-894.

[19] Kantoff, P. W., Schuetz, T. J., Blumenstein, B. A., Glode, L. M., Bilhartz, D. L., Wyand, M., ... & Dahut, W. L. (2010). Overall survival analysis of a phase II randomized controlled trial of a Poxviral-based PSA-targeted immunotherapy in metastatic castration-resistant prostate cancer. Journal of Clinical Oncology, 28(7), 1099.

[20] Karline Soetaert, Thomas Petzoldt, R. Woodrow Setzer (2010). Solving Differential Equations in R: Package deSolve. Journal of Statistical Software, 33(9), 1–25. http://www.jstatsoft.org/v33/i09/DOI10.18637/jss.v033.i09

[21] Kelly, P. J., & Lim, L. L. Y. (2000). Survival analysis for recurrent event data: an application to childhood infectious diseases. Statistics in medicine, 19(1), 13-33.

[22] Kéry, M. (2010). Introduction to WinBUGS for ecologists: Bayesian approach to regression, ANOVA, mixed models and related analyses. Academic Press.

[23] Lee, E. T., & Go, O. T. (1997). Survival analysis in public health research. Annual review of public health, 18(1), 105-134.

[24] Lee, E. Y., Hwang, K. Y., Yang, J. O., & Hong, S. Y. (2002). Predictors of survival after acute paraquat poisoning. Toxicology and industrial health, 18(4), 201-206.

[25] Masse, L. C., & Tremblay, R. E. (1997). Behavior of boys in kindergarten and the onset of substance use during adolescence. Archives of general psychiatry, 54(1), 62-68.

[26] Mazet, J. A., Newman, S. H., Gilardi, K. V., Tseng, F. S., Holcomb, J. B., Jessup, D. A., & Ziccardi, M. H. (2002). Advances in oiled bird emergency medicine and management. Journal of Avian Medicine and Surgery, 16(2), 146-149.

[27] Nelson, C. M., Ihle, K. E., Fondrk, M. K., Page Jr, R. E., & Amdam, G. V. (2007). The gene vitellogenin has multiple coordinating effects on social organization. PLoS biology, 5(3), e62.

[28] Ott, R. L., & Longnecker, M. T. (2015). An introduction

to statistical methods and data analysis. Nelson Education.

[29] Peña, E. A., & Hollander, M. (2004). Models for recurrent events in reliability and survival analysis. In Mathematical reliability: An expository perspective (pp. 105-123). Springer, Boston, MA.

[30] Philipp H Boersch-Supan, Sadie J Ryan, and Leah R Johnson (2017). deBInfer: Bayesian inference for dynamical models of biological systems in R. Methods in Ecology and Evolution 8:511-518. https://doi.org/10.1111/2041-210X.12679

[31] Rosenbaum, P. R. (2002). Observational studies. In Observational studies (pp. 1-17). Springer, New York, NY.

[32] S.A.L.M. Kooijman and J.J.M. Bedaux, Analysis of toxicity test on Daphnia survival and reproduction, Water Res. 30 (1996) 1711-1723.

[33] Santos, L. H., Arajo, A. N., Fachini, A., Pena, A., Delerue-Matos, C., & Montenegro, M. C. B. S. M. (2010). Ecotoxicological aspects related to the presence of pharmaceuticals in the aquatic environment. Journal of hazardous materials, 175(1-3), 45-95.

[34] Shojania, K. G., Sampson, M., Ansari, M. T., Ji, J., Doucette, S., & Moher, D. (2007). How quickly do systematic reviews go out of date? A survival analysis.

Annals of internal medicine, 147(4), 224-233.

[35] Stepanova, M., & Thomas, L. (2002). Survival analysis methods for personal loan data. Operations Research, 50(2), 277-289.

[36] Strawbridge, W. J., Shema, S. J., Cohen, R. D., & Kaplan, G. A. (2001). Religious attendance increases survival by improving and maintaining good health behaviors, mental health, and social relationships. Annals of Behavioral Medicine, 23(1), 68-74.

[37] Toumi H., Boumaiza M., Millet M., Radetski C. M., Camara B. I., Felten V., & Ferard J. F. (2015). Investigation of differences in sensitivity between 3 strains of Daphnia magna (crustacean Cladocera) exposed to malathion (organophosphorous pesticide). Journal of Environmental Science and Health, Part B, 50(1), 34-44.

[38] Toumi, H., Boumaiza, M., Millet, M., Radetski, C. M., Felten, V., & Frard, J. F. (2015). Is acetylcholinesterase a biomarker of susceptibility in Daphnia magna (Crustacea, Cladocera) after Deltamethrin exposure?. Chemosphere, 120, 351-356.

[39] Wickham, H. & Chang, W. (2016). Devtools: Tools to make developing r packages easier.

15

Which Matrices Show Perfect Nestedness or the Absence of Nestedness? An Analytical Study on the Performance of NODF and WNODF

N. F. Britton*, M. Almeida Neto [†], Gilberto Corso [‡]
*Department of Mathematical Sciences and Centre for Mathematical Biology
University of Bath, Bath BA2 7AY, UK. Email: n.f.britton@bath.ac.uk
[†]Departmento de Ecologia, Universidade Federal de Goiás,
74001-970 Goiânia-GO, Brazil, Email: marioeco@gmail.com
[‡]Departamento de Biofísica e Farmacologia, Centro de Biociências,
Universidade Federal do Rio Grande do Norte,
59072-970 Natal-RN, Brazil, Email: corso@cb.ufrn.br

Abstract—Nestedness is a concept employed to describe a particular pattern of organization in species interaction networks and in site-by-species incidence matrices. Currently the most widely used nestedness index is the NODF (Nestedness metric based on Overlap and Decreasing Fill), initially presented for binary data and later extended to quantitative data, WNODF. In this manuscript we present a rigorous formulation of this index for both cases, NODF and WNODF. In addition, we characterize the matrices corresponding to the two extreme cases, (W)NODF=1 and (W)NODF=0, representing a perfectly nested pattern and the absence of nestedness respectively. After permutations of rows and columns if necessary, the perfectly nested pattern is a full triangular matrix, which must of course be square, with additional inequalities between the elements for WNODF. On the other hand there are many patterns characterized by the total absence of nestedness. Indeed, any binary matrix (whether square or rectangular) with uniform row and column sums (or marginals) satisfies this condition: the chessboard and a pattern reflecting an underlying annular ecological gradient, which we shall call gradient-like, are symmetrical or nearly symmetrical examples from this class.

Keywords-biogeography, interaction networks, nestedness, bipartite networks

I. INTRODUCTION

Observing nature is one of the most fascinating experiences in life. A honeybee visits a daisy, a rosemary, and other ten different species. Another bee of the same family is specialized in just one flower that by its turn is visited by twenty diverse pollinators. Once we put together the community of pollinators and flowers an intricate mutualist network arises [5]. In the opposite side of life a caterpillar feed on two asteraceae species which are eaten by another couple of insects, the full set of herbivorous and plants forms a complex antagonist network. An central quest in ecology of communities today is the search for patterns in networks that can distinguish between mutualist and antagonist webs [13, 21]. One network pattern that is part of this answer is nestedness, the subject of this manuscript.

Nestedness is a concept used in ecology to study a specific formation pattern in species interaction networks and in site-by-species incidence matrices. In general terms, nestedness is a specific kind of topological organization in adjacency matrices of bipartite networks where any vertex S, with m links, tend to be connected to a subset of the vertices connected to any other vertex with n links, where $n > m$. The nestedness concept was first introduced by [8] to characterize species distribution pattern in a spatial set of isolated habitats such as islands. In a perfectly nested pattern site-by-site incidence matrix there is a hierarchy of sites such that the set of species inhabiting any site is a subset of the set inhabiting any site further up the hierarchy. When applied to describe the topological organization in ecological interaction networks this new nestedness concept was first used to networks formed by pollinators and flowering plants and by seed dispersers and flesh-fruited plants [4, 12]. In cases a network is perfectly nested if (i) there is a hierarchy of plant species such that the set of animal (pollinator or seed disperser) species interacting with any plant is a subset of the set of animals interacting with any plant further up the hierarchy, and (ii) there is a similar hierarchy of animals. It is clear that in such a network generalist species interact with specialists and generalists, but specialists do not interact with each other.

The proper mathematical framework for introducing nestedness is in the context of bipartite networks. From a general perspective we consider a bipartite network formed by two sets S_1 and S_2. Nestedness is characterized by several indices [22, 18] and it is not the objective of this work to compare them. Here we focus on the $NODF$ index, which has a clear mathematical definition that allows further analytic developments. The $NODF$ index, an acronym for Nestedness metric based on Overlap and Decreasing Fill, is an index that was introduced in [2] and that has been widely used in the literature. An extension of this index to quantitative networks, $WNODF$, was recently proposed [3], and we include it in our analysis because of the importance of quantitative networks, specially for networks of interacting species [9, 13].

Null models are an important methodological tool widely used in ecology to test model fitting, perform statistical tests or test the validity of indices and measures [10]. In order to assess an index a large set of empirical or artificial data is used as a data bank to explore its limitations and fragility. This process has already been used to test a set of nestedness indices [22]. Null models are necessary because statistical tests are otherwise always questionable by limitation in the range of tested parameters, interpretation bias of the results, or equivocal choice of random models. These studies emphasis the necessity of analytic results to strength confidence about nestedness indices and their applications.

The original definition of the $NODF$ index depends on how the rows and columns are ordered, and a frequently used software for calculating NODF explicitly asks the user if they would like to order the matrix according to row and column sums (or marginals) [11]. In this paper we employ a definition of (W)NODF in which the matrix is previously sorted before the computation of the index.

In this paper we give rigorous definitions of $NODF$ and $WNODF$ and prove two mathematical theorems in each case. For the sake of clarity, and for historical reasons, we explore separately qualitative (binary) and quantitative (weighted) networks. The treatment of the qualitative case is more intuitive and helps the reader to follow the analytic developments. In section 2 we start with a formal definition of $NODF$ and $WNODF$ and present two theorems that characterize the extreme cases, $NODF = 0$ and $WNODF = 0$ corresponding to absence of nestedness, and $NODF = 1$ and $WNODF = 1$ corresponding to the perfectly nested arrangement. In section 3 we summarize the main ideas of the work and put the results in a broader context.

II. Analytic treatment

We shall consider a bipartite network of set S_1, containing m elements, and set S_2, containing n elements, with quantitative data for the frequency w_{ij} of the interactions between element i of set S_1 and element j of set S_2. In the simplest case $w_{i,j}$ is equal to 1 or 0, a situation corresponding to the binary network, qualitative network or presence/absence matrix. The adjacency matrix for the network is the $m \times n$ matrix $A = (a_{ij})$, where a_{ij} is defined by:

$$a_{ij} = \begin{cases} 1 & \text{if } w_{ij} \neq 0, \text{ so that element } i \text{ of } S_1 \\ & \text{and element } j \text{ of } S_2 \text{ are linked} \\ 0 & \text{if } w_{ij} = 0, \text{ so that they are not linked.} \end{cases} \quad (1)$$

We define the row and column marginal totals MT_i^r and MT_l^c by

$$MT_i^r = \sum_{j=1}^{n} a_{ij} \quad \text{and} \quad MT_l^c = \sum_{k=1}^{m} a_{kl}, \quad (2)$$

so that MT_i^r is the number of elements of S_2 interacting with element i of S_1, and MT_l^c is the number of elements of S_1 interacting with element l of S_2. Define the row and column decreasing-fill indicators DF_{ij}^r and DF_{kl}^c by

$$DF_{ij}^r = \begin{cases} 1 & \text{if } MT_i^r > MT_j^r, \\ 0 & \text{if } MT_i^r \leq MT_j^r, \end{cases} \quad (3)$$

$$DF_{kl}^c = \begin{cases} 1 & \text{if } MT_k^c > MT_l^c, \\ 0 & \text{if } MT_k^c \leq MT_l^c. \end{cases} \quad (4)$$

Note that, if $i < j$, so that row i is above row j, then $DF_{ij}^r = 1$ if and only if element i of set S_1 is linked with more elements of set S_2 than element j of S_1; similarly, if $k < l$, so that column k is to the left of column l, then $DF_{kl}^c = 1$ if and only if element k of S_2 is linked with more elements of set S_1 than element l of S_2. It is always possible to permute the rows and columns of the matrix so that $MT_i^r \geq MT_j^r$ whenever $i < j$, and $MT_k^c \geq MT_l^r$ whenever $k < l$, but the definition does not require this to be done.

A. Qualitative matrices, the case $NODF$

In order to define $NODF$ we start with the row paired-overlap quantifier PO_{ij}^r as the fraction of unit elements in row j that are matched by unit elements in row i, and the column paired-overlap quantifier PO_{kl}^c as the fraction of unit elements in column l that are matched by unit elements in row k, so that

$$PO_{ij}^r = \frac{\sum_{p=1}^{n} a_{ip} a_{jp}}{\sum_{p=1}^{n} a_{jp}}, \quad PO_{kl}^c = \frac{\sum_{q=1}^{n} a_{kq} a_{lq}}{\sum_{q=1}^{n} a_{lq}}. \quad (5)$$

Note that PO_{ij}^r is the fraction of elements of S_2 linked to element j of S_1 that are also linked to element i of S_1, and similarly for PO_{kl}^c. Define the row paired nestedness NP_{ij}^r between rows i and j, and the column paired nestedness NP_{kl}^c between columns k and l, by

$$NP_{ij}^r = DF_{ij}^r PO_{ij}^r + DF_{ji}^r PO_{ji}^r, \quad (6)$$

$$NP_{kl}^c = DF_{kl}^c PO_{kl}^c + DF_{lk}^c PO_{lk}^c. \quad (7)$$

Note that these definitions are valid whatever the signs of $MT_i^r - MT_j^r$ and $MT_k^c - MT_l^c$. Finally, define the row and column nestedness metrics $NODF^r$ and $NODF^c$ by

$$NODF^r = \frac{\sum_{i=1}^{m} \sum_{j=i+1}^{m} NP_{ij}^r}{\frac{1}{2}m(m-1)}, \quad (8)$$

$$NODF^c = \frac{\sum_{k=1}^{n} \sum_{l=k+1}^{n} NP_{kl}^c}{\frac{1}{2}n(n-1)}, \quad (9)$$

and the overall nestedness metric $NODF$ as a weighted average of these, by

$$NODF = \frac{\sum_{i=1}^{m} \sum_{j=i+1}^{m} NP_{ij}^r + \sum_{k=1}^{n} \sum_{l=k+1}^{n} NP_{kl}^c}{\frac{1}{2}m(m-1) + \frac{1}{2}n(n-1)}. \quad (10)$$

1) Conditions for $NODF = 0$: Our objective is to characterize all matrices for which $NODF = 0$. It is clear that $NODF = 0$ if and only if both $NODF^r = 0$ and $NODF^c = 0$, so let us first consider the conditions for which $NODF^r = 0$. This is true if and only if $NP^r_{ij} = 0$ for all pairs (i, j) of rows. From equation (6), $NP^r_{ij} = 0$ if and only if either $MT^r_i = MT^r_j$, so that $DF^r_{ij} = DF^r_{ji} = 0$, or $\sum_{p=1}^n a_{ip}a_{jp} = 0$, so that $PO^r_{ij} = PO^r_{ji} = 0$. In other words, either rows i and j have the same number of unit elements, so that elements i and j of S_1 interact with the same number of elements of S_2, or there is no p in S_2 that interacts with both i and j. If our bipartite network is connected, then it is possible to move from any i in S_1 to any other j in S_1 by following a path composed of edges of the network from S_1 to S_2 to S_1 and so on. Hence, in this connected case, $NODF^r = 0$ if and only if *all* elements of S_1 are linked to the same number of elements of S_2. Similarly, for a connected network, $NODF^c = 0$ if and only if all elements of S_2 are linked to the same number of elements of S_1, and $NODF = 0$ if and only if both these conditions hold. If our network is disconnected, then $NODF = 0$ if and only if all elements of S_1 are linked to the same number of elements of S_2, and all elements of S_2 are linked to the same number of elements of S_1 within each connected component, or compartment. This is a necessary and sufficient condition for $NODF = 0$. There are many networks that satisfy this condition. For example in figure 1 we show a 9×6 network where each of the nine elements of S_1 interact with a different pair of elements of S_2, so that each element of S_2 interacts with three elements of S_1. Figure 1(c) does not resemble any of the $NODF = 0$ configurations exhibited in the literature [4, 15], which are all (including the chessboard after row and column permutation) compartmented with full connectivity within the compartments. Case 1(d) seems to reflect an underlying cyclic ecological gradient [15], and we call it gradient-like. The requirement that the gradient be cyclic is manifest in the occupied cell at the bottom left of the matrix,

and it is occupied to fulfil the rule that there should be two nonzero elements in each row and three in each column. It is interesting that the dimensions (m, n) of the adjacency matrix obey a constraint in the $NODF = 0$ case. The total number of matrix elements that is distributed along columns and rows should follow the relation:

$$\sum_{i=1}^n MT^c_i = \sum_{j=1}^m MT^r_j. \tag{11}$$

As MT^c_i and MT^r_j are constants we can rewrite 11 in the form $nMT^c = mMT^r$.

2) Conditions for $NODF = 1$: We now wish to characterize all matrices for which $NODF = 1$, see figure 2. It is clear that $NODF = 1$ if and only if both $NODF^r = 1$ and $NODF^c = 1$, so let us first consider the conditions under which $NODF^r = 1$. This is true if and only if $NP^r_{ij} = 1$ for all pairs (i, j) of rows. From equation (6), $NP^r_{ij} = 1$ implies that $MT^r_i \neq MT^r_j$, so that either $DF^r_{ij} = 1$ or $DF^r_{ji} = 1$. If there are more elements of S_2 interacting with element i in S_1 than with j in S_1, then $MT^r_i > MT^r_j$, so that $DF^r_{ij} = 1$, $DF^r_{ji} = 0$. Then we also require that $\sum_{p=1}^n a_{ip}a_{jp} = \sum_{p=1}^n a_{jp}$, so that $PO^r_{ij} = 1$, in other words that $a_{ip} = 1$ whenever $a_{jp} = 1$. Thus all elements of S_2 interacting with element j in S_1 also interact with element i in S_1, or the set of elements of S_2 interacting with j in S_1 is nested within (or a proper subset of) the set of elements of S_2 interacting with i in S_1. Similarly, if there are more elements of S_2 interacting with j in S_1 than with i in S_1, then the set of elements of S_2 interacting with i in S_1 must be nested within the set of elements of S_2 interacting with j in S_1. Similar results hold for $NODF^c = 1$, so that the set of elements of S_1 interacting with any k in S_2 must be a proper subset or superset of the set of S_1 elements interacting with any other l in S_2. For $NODF = 1$, all (S_1 and S_2) interaction sets must be proper sub- or supersets, so that by the pigeonhole principle we must have $m = n$, and it must be possible to permute the rows and columns of the matrix A so that $a_{ij} = 1$ if $i \geq j$, $a_{ij} = 0$ otherwise. The matrix with $NODF = 1$ is the

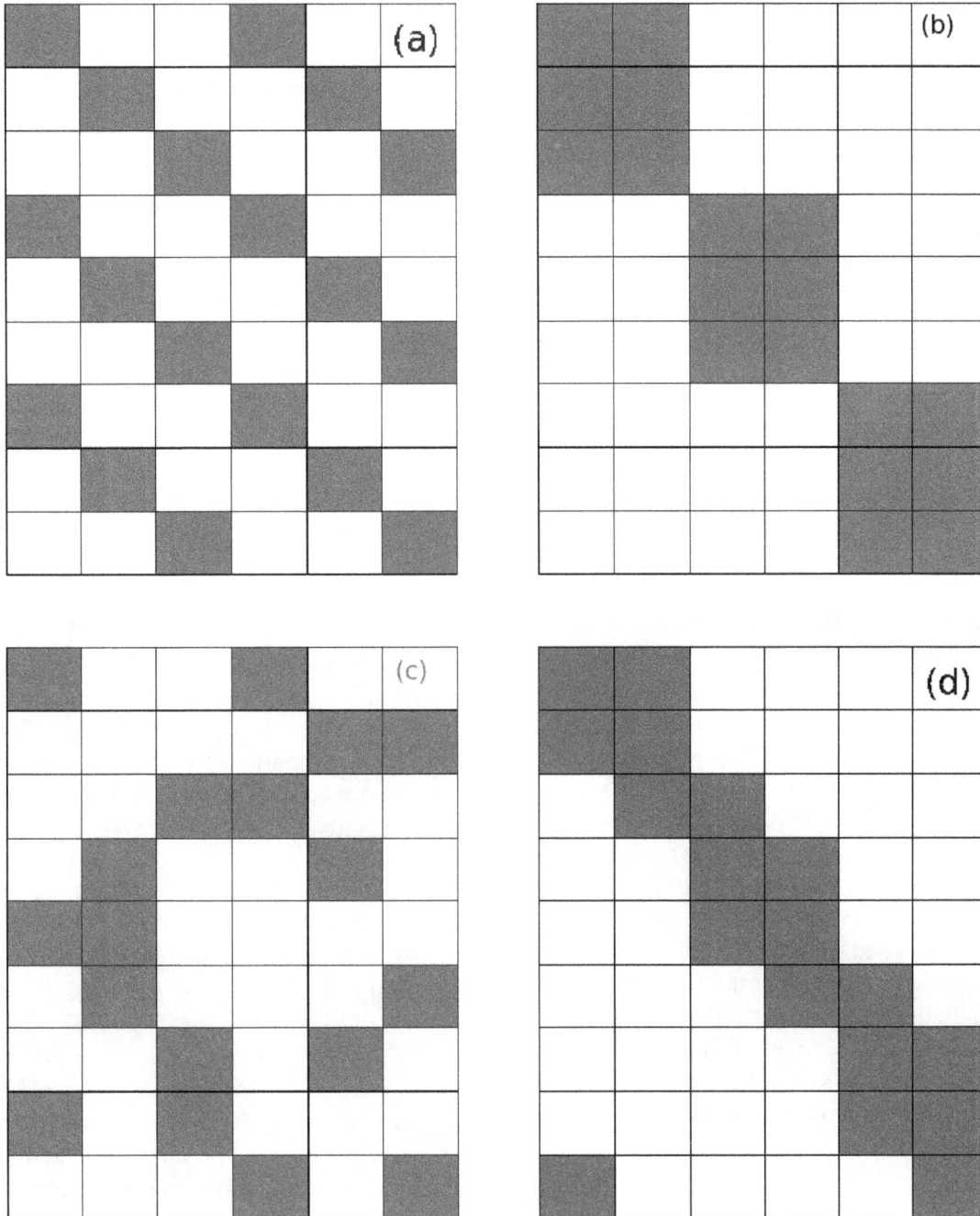

Fig. 1: Some $NODF = 0$ patterns. Panels (a) and (b) represent the same matrix after permutation of lines and columns; this non-chessboard tiling is a composition of three disconnected networks. Panels (c) and (d) show two connected networks that have $NODF = 0$, since $MT_i^c = 3$ and $MT_j^r = 2$ for all i and j respectively. Case (d) represents a gradient-like structure.

full triangular matrix, unique up to permutation of rows and columns.

B. Quantitative matrix, the case $WNODF$

To construct the $WNODF$ index we define the row-pair dominance quantifier D_{ij}^r as the fraction of non-zero weights in row j that are dominated by (less than) the corresponding weight in row i, and the column-pair dominance quantifier D_{kl}^c as the fraction of non-zero weights in column l that are dominated by the corresponding weight in column k, so that

$$D_{ij}^r = \frac{\sum_{p=1}^{n} H(w_{ip} - w_{jp}) H(w_{jp})}{MT_j^r}, \quad (12)$$

$$D_{kl}^c = \frac{\sum_{q=1}^{m} H(w_{qk} - w_{ql}) H(w_{ql})}{MT_l^c}, \quad (13)$$

where H is the Heaviside step function with $H(0) = 0$. Note that D_{ij}^r is the fraction of elements of S_2 interacting with j in S_1 that interact more strongly with i in S_1, and similarly for D_{kl}^c. Note that, when calculating $NODF$ for qualitative networks, the quantity corresponding to D_{ij}^r is the row-pair overlap quantifier PO_{ij}^r which is the fraction of elements of S_2 interacting with j in S_1 that also interact with i in S_1, and similarly for D_{kl}^c; the requirement that the interaction be stronger is not (and cannot be) applied. This is the essential difference between the index $WNODF$ for quantitative networks and the index $NODF$ for qualitative ones. Now define the row-pair dominance nestedness between rows i and j, and the column-pair dominance nestedness between columns k and l, by

$$DN_{ij}^r = DF_{ij}^r D_{ij}^r + DF_{ji}^r D_{ji}^r, \quad (14)$$

$$DN_{kl}^c = DF_{kl}^c D_{kl}^c + DF_{lk}^c D_{lk}^c. \quad (15)$$

Note that these definitions are valid whatever the signs of $MT_i^r - MT_j^r$ and $MT_k^c - MT_l^c$. For example, (i) if $MT_i^r > MT_j^r$ then $DF_{ij}^r = 1$ and $DF_{ji}^r = 0$, so $DN_{ij}^r = D_{ij}^r$, (ii) if $MT_i^r < MT_j^r$

then $DF_{ij}^r = 0$ and $DF_{ji}^r = 1$, so $DN_{ij}^r = D_{ji}^r$, and (iii) if $MT_i^r = MT_j^r$ then $DF_{ij}^r = DF_{ji}^r = 0$, and $DN_{ij}^r = 0$. Finally, define the row and column weighted nestedness metrics $WNODF^r$ and $WNODF^c$ by

$$WNODF^r = \frac{\sum_{i=1}^{m} \sum_{j=1}^{m} DN_{ij}^r}{m(m-1)}, \quad (16)$$

$$WNODF^c = \frac{\sum_{k=1}^{n} \sum_{l=1}^{n} DN_{kl}^c}{n(n-1)}, \quad (17)$$

and the overall weighted nestedness metric $WNODF$ as a weighted average of these, by

$$WNODF = \frac{\sum_{i=1}^{m} \sum_{j=1}^{m} DN_{ij}^r + \sum_{k=1}^{n} \sum_{l=1}^{n} DN_{kl}^c}{m(m-1) + n(n-1)}.$$

1) Conditions for $WNODF = 0$: The treatment of $WNODF = 0$ shares some similarities with the previous analysis of $NODF = 0$. To characterize all matrices for which $WNODF = 0$ we proceed as follows. It is clear that $WNODF = 0$ if and only if both $WNODF^r = 0$ and $WNODF^c = 0$, so let us first consider the conditions for which $WNODF^r = 0$. This is true if and only if $DN_{ij}^r = 0$ for all pairs (i, j) of rows. From equation (14), $DN_{ij}^r = 0$ if and only if either (i) $MT_i^r = MT_j^r$, so that $DF_{ij}^r = DF_{ji}^r = 0$, or (ii) $MT_i^r > MT_j^r$ and $\sum_{p=1}^{n} H(w_{ip} - w_{jp}) H(w_{jp}) = 0$, so that $D_{ij}^r = DF_{ji}^r = 0$, or (iii) $MT_i^r < MT_j^r$ and $\sum_{p=1}^{n} H(w_{jp} - w_{ip}) H(w_{ip}) = 0$, so that $D_{ji}^r = DF_{ij}^r = 0$. In case (i), the elements i and j of S_1 interact with the same number of S_2 elements. In case (ii), i in S_1 interacts with more elements of S_2 than does j in S_1, but any interaction between j and any element p of S_2 is at least as strong as the corresponding interaction between i and p. Although i in S_1 strictly dominates j in S_1 in terms of the number of its interactions, j in S_1 (not necessarily strictly) dominates i in S_1 in terms of the strength of the interactions it does have. Case

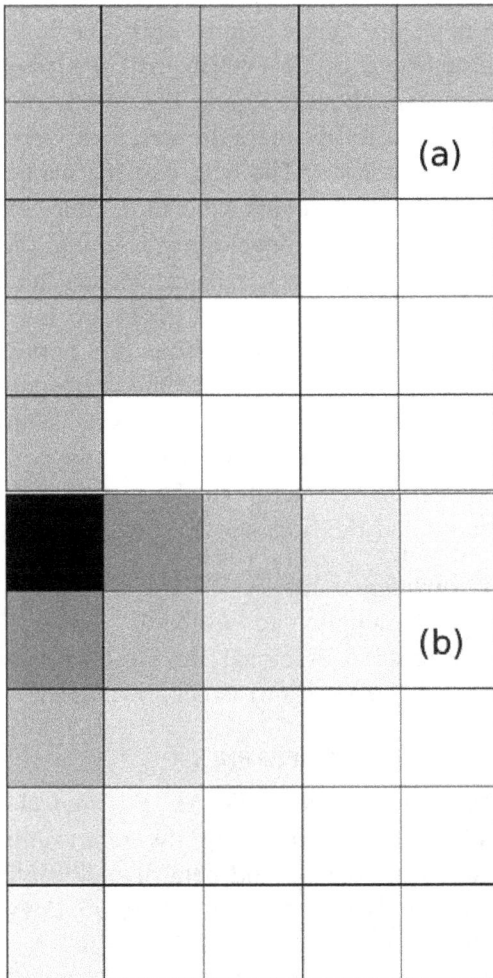

Fig. 2: The maximal nestedness pattern exemplified for qualitative (a) and quantitative (b) cases. In the second situation the weight of the link between species is indicated by grey tones.

(iii) is analogous, with i and j interchanged. There are many possible ways to obtain $WNODF^r = 0$, and similarly $WNODF^c = 0$ and $WNODF = 0$. In particular any connected bipartite network in which all elements of S_1 interact with the same number of elements of S_2, and all elements of S_2 interact with the same number of elements of S_1, has $WNODF = 0$, as does any network in which each element of W is either 0 or 1. Note that $WNODF$ is not a continuous function of the

elements of W; for example, if W is a 2×2 matrix with $w_{11} = 1 + \varepsilon$, $w_{12} = w_{21} = 1$, $w_{22} = 0$, then $WNODF(W) = 0$ if $\varepsilon = 0$ but $WNODF(W) = 1$ if ε is positive, however small it is.

C. Conditions for $WNODF = 1$

We now wish to characterize all matrices for which $WNODF = 1$, see figure 2. This demonstration has some points in common with the case $NODF = 1$. It is clear that $WNODF = 1$ if and only if both $WNODF^r = 1$ and $WNODF^c = 1$, so let us first consider the conditions under which $WNODF^r = 1$. This is true if and only if $DN_{ij}^r = 1$ for all pairs (i, j) of rows. From equation (15), $DN_{ij}^r = 1$ implies that $MT_i^r \neq MT_j^r$, so that either $DF_{ij}^r = 1$ or $DF_{ji}^r = 1$. If there are more elements of S_2 interacting with i in S_1 than with j in S_1, then $MT_i^r > MT_j^r$, and $DF_{ij}^r = 1$, $DF_{ji}^r = 0$. Then we also require that $\sum_{p=1}^n H(w_{ip} - w_{jp})H(w_{jp}) = MT_j^r$, so that $D_{ij}^r = 1$, in other words that $w_{ip} \geq w_{jp}$ whenever $w_{jp} \neq 0$. Thus all elements of S_2 interacting with j in S_1 not only interact with i in S_1, but interact more strongly with i than with j. The set of elements of S_2 interacting with j in S_1 not only has to be nested within (or a proper subset of) the set of S_2 elements interacting with i in S_1, but all the interactions with i in S_1 must be stronger than the corresponding interaction with j in S_1. Similarly, if there are more S_2 elements interacting with j in S_1 than with i in S_1, then the set of S_2 elements interacting with i in S_1 must be nested within the set of S_2 elements interacting with j in S_1, and each interaction with j in S_1 must be stronger than the corresponding interaction with i in S_1. Similar results hold for $WNODF^c = 1$, so that the set of elements of S_1 interacting with any k in S_2 must be a proper subset or superset of the set of S_1 elements interacting with any other l in S_2, corresponding interactions in subsets must be weaker, and corresponding interactions in supersets stronger. For $WNODF = 1$, all (S_1 and S_2) interaction sets must be proper sub- or supersets, so that by the pigeon-hole principle we must have $m = n$, and it must be possible to permute the rows and columns of the matrix W so that $w_{ij} > 0$ if $i + j \leq n + 1$, $w_{ij} = 0$ otherwise. Any matrix with $WNODF = 1$ has the same adjacency matrix, up to permutation of rows and columns, and also satisfies the row and column strict dominance properties $w_{ik} > w_{jk}$ for all $i < j$ whenever $w_{jk} > 0$, $w_{ki} > w_{kj}$ for all $i < j$ whenever $w_{kj} > 0$.

III. Final Remarks

This work focuses on probably the most commonly used nestedness index: the Nestedness metric based on Overlap and Decreasing Fill. Initially we introduce a rigorous formulation for $NODF$ and $WNODF$. We then elucidate the patterns of maximal and minimal nestedness, $(W)NODF = 1$ and $(W)NODF = 0$. The maximal nestedness pattern is already known in the literature [15, 2], but an understanding of the minimum nestedness pattern is substantially extended in this work. The literature usually presents the chessboard pattern as the prototype of the zero nestedness arrangement; but this work shows that there is in fact a large class of matrices that fulfil this condition. We cite the completely compartmented networks with equal modules (of which the chessboard is a special case) and gradient-like matrices. But there is another class of non-symmetrical matrices that also have zero nestedness as long as the row and column sums of the adjacency matrix are uniform.

The theoretical discussion about nestedness today resembles the debate around diversity and its measurements [14, 16, 17]. In both cases the community of ecologists is aware of the importance of the concept in understanding and quantifying patterns in ecological processes. In both contexts, also, there is a dynamic debate about the true meaning of the concepts, and the most adequate way to transform them into an index [1, 18, 20]. Intriguingly, the comparison between diversity and nestedness is not just a curiosity in the story of theoretical ecology, but also a challenging aspect of theory itself, because beta diversity and nestedness show common similarities and dissimilarities [6, 19].

We hope that this rigorous work that highlight the nestedness of (W)NODF will contribute to the discussion about the general meaning of nestedness by clarifying the extreme cases: zero and maximal nestedness. The basics of the mathematical framework presented here is flexible enough to encourage further developments using alternative pairwise nestedness indices. Despite the large number of nestedness indices, there are few analytic results relating the properties of a nestedness index and the characteristics of the corresponding adjacent matrix; an exception is [7]. With the exact results shown in this manuscript we add new elements to the debate about the real meaning of nestedness and the best way to measure it.

Acknowledgements

Financial support to Gilberto Corso from CNPq (Conselho Nacional de Desenvolvimento Científico e Tecnológico) is acknowledged.

References

[1] M. Almeida-Neto, D. M. B. Frensel, and W. Ulrich. Rethinking the relatioship between nestedness and beta diversity: a comment on Baselga(2010). *Global Ecology and Biogeography*, 21:772–777, 2012.

[2] M. Almeida-Neto, P.R. Guimarães, P. R. Guimarães Jr, R. D. Loyola, and W. Ulrich. A consistent metric for nestedness analysis in ecological systems: reconciling concept and measurement. *Oikos*, 117:1227, 2008.

[3] M. Almeida-Neto and W. Ulrich. A straightforward computational approach for measuring nestedness using quantitative matrices. *Enviromental Modeling & Software*, 26:1713, 2011.

[4] J. Bascompte, P. Jordano, C. J. Melián, and J. M. Olesen. The nested assembly of plant-animal mutualistic networks. *Proc. Natl Acad. Sci USA*, 100:9383, 2003.

[5] Jordi BAscompte and Pedro Jordano. *Mutualistic Networks*. Princeton University Press, 2013.

[6] A. Baselga. Partitioning the turnover and nestedness components of beta diversity.

Global Ecology and Biogeography, 19:134–143, 2010.

[7] G. Corso, A. L. de Araujo, and A. M. de Almeida. Connectivity and nestedness in bipartite networks from community ecology. *Journal of Physics: Conference Series*, 285:012009, 2011.

[8] J. P. Darlington. *Zoogeography: the geographical distribution of animals*. Wiley, 1957.

[9] Luis J. Gilarranz, Juan M. Pastor, and Javier Galeano. The architecture of weighted mutualistic networks. *Oikos*, 121:1154, 2011.

[10] N. J. Gotelli and G.R. Graves. *Null models in ecology*. Smithsonian Institution Press, Washington, D.C., 1996.

[11] P. R. Guimarães Jr. and P. Guimarães. Improving the analyses of nestedness for large sets of matrices. *Environmental Modeling & Software*, 21:1512, 2007.

[12] P. R. Guimarães Jr., V. Rico-Gray, S. F. dos Reis, and J. N. Thompson. Asymmetries in specialisation in ant–plant mutualistic networks. *Proc. R. Soc B*, 273:2041, 2006.

[13] T. C. Ings, J. M. Montoya, J. Bascompte, N. Bluthgen, L. Brown, C. F. Dormann, F. Edwards, D. Figueroa, U. Jacob, J. I. Jones, R. B. Lauridsen, M. E. Ledger, H. M. Lewis, J. M. Olesen, F.J. Frank van Veen, P. H. Warren, and G. Woodward. Ecological networks - beyond food webs. *Journal of Animal Ecology*, 78:253, 2009.

[14] L. Jost. Entropy and diversity. *Oikos*, 113:363–375, 2006.

[15] T. M. Lewinsohn, P. I. Prado, P. Jordano, J. Bascompte, and J. M. Olesen. Structure in plant-animal interaction assemblages. *Oikos*, 113:174, 2006.

[16] A. E. Magurran. *Measuring Biological Diversity*. Blackwell Publishing Company, 2004.

[17] O. Parkash and A. K. Thukral. Statistical measures as measures of diversity. *International Journal of Biomathematics*, 3:173, 2010.

[18] J. Podani and D. Schmera. A comparative evaluation of pairwise nestedness measures. *Ecography*, 35:1, 2012.

[19] D. Schmera and J. Podani. Comments on separating components of beta diversity. *Community Ecology*, 12:153–160, 2011.

[20] P. P. A. Staniczenko, J. C. Kopp, and S. Allesina. The ghost of nestedness in ecological networks. *Nature communications*, 4:1391, 2013.

[21] E. Thébault and C. Fontaine. Stability of ecological communities and the architecture of mutualistic and trophic networks. *Science*, 329:853, 2010.

[22] W. Ulrich, M. Almeida-Neto, and N. J. Gotelli. A consumer's guide to nestedness analysis. *Oikos*, 118:3, 2009.

16

Coupled cell networks: Boolean perspective

Katarzyna (Kasia) Świrydowicz
Department of Mathematics
Virginia Polytechnic Institute
Blacksburg, VA, USA
kswirydo@vt.edu

Abstract—In this paper we use Boolean framework to redefine coupled cell networkss, originally described in [2]. We also analyze some of the important properties of Boolean coupled cell networkss.

In the second part of this paper we focus on properties of a quotient networks. We redefine the concept of a quotient to suit Boolean network framework. Next, we investigate in details the networks in which two-cell bidirectional ring and three-cell bidirectional ring arise as quotients.

Keywords-Boolean networks; coupled cell networks; discrete models

I. Introduction

During the 1980s and early 1990s, Martin Golubitsky and Ian Stewart formulated and developed a theory of *coupled cell networkss* (CCNs) [2]. Their research was primarily focused on quadrupeds' gaits. Since they were particularly interested in the change of synchrony between four legs of an animal, they needed a special framework to describe this phenomenon. For example, they were interested in how does the synchrony of four legs change when the animal speeds up from walk to gallop.

The most important concept in the CCN theory is a *cell*. The cell captures the dynamics of one unit of the system (for example, one leg of an animal) and the dynamical system consists of many identical cells connected to each other. Each cell has its own state space and evolution equation(s). Even though models based on identical cooperating units are common in many areas – especially in biology, ecology and sociology, [4], [5], [6], [7], [8] – the CCN setup helps to formulate questions in terms of symmetry and synchrony rather than system evolution as a whole.

In this paper, we redefine coupled cell networkss using the framework of Boolean networks [9], [10]. This moves the theory to a new setting. As expected, some phenomena turns out to be very similar as for continuous networks and some others do not. In addition we study the phenomena specific to Boolean networks and not arising in continuos dynamical systems.

We note that the Boolean coupled cell networkss are a subclass of Boolean networks, which differs both from the original Kaufman's Boolean switching nets [9], and from cellular automata [23], [24]. In his work [9], Kaufman focused on networks with topology based on $k-$regular graph, which makes the topology similar to that of CCNs, however, he chose update rules for each of the nodes randomly. In contrast, in this paper we assume that the update rules for each cell-node in the network are identical. In Cellular Automata the update rules are the same for every cell-node in the network, and this makes Cellular Automata similar to Boolean CCNs. The difference is that a state of a cell in a cellular automaton depends only on the states of its immediate neighbors, and this formulation does not allow for any irregularity in the network structure. A cell in a cellular automaton might not depend on the cells that are far away from its physical location on the grid. In contrast, such a dependence can occur in Boolean CCNs. Hence, Boolean CCNs share some characteristics with Boolean switching nets and cellular automata, and in fact, can be treated as a cross between these two species of Boolean networks.

The main result in this paper is the analysis of the Boolean CCNs in which two cell bidirectional ring and three-cell bidirectional ring arise as a quotient.

This paper is organized as follows. Section II contains the definitions of the concepts needed to describe a Boolean network, and it also contaits the definition of coupled cell networks. Section III describes the main problem we want to address in this paper. Section IV points out the differences between continuous and Boolean dynamics. In the Section V we define the quotient network, and form the rules for taking quotients. In this section we study our first example case, the networks for which two cell bidirectional ring arise as a quotient. Section VII provides a biological model in which the ideas from the previous sections are used. Section VI is devoted to analyzing the networks for which three-cell bidirectional ring arise as a quotient. In the Section IX we present conclusions and ideas for the future.

II. PRELIMINARIES

A. Boolean functions and dynamical systems

The definitions contained in this subsection come from the classic literature on Boolean functions and networks, see [9], [10], [12], [15]

By *Boolean function* we understand a function $f : \mathbb{F}_2^n \to \mathbb{F}_2$. Let $x_i \in \mathbb{F}_2$, $i = 1, \ldots, n$. The Boolean function can be represented in the form

$$f(x_1, x_2, \ldots, x_n).$$

A *Boolean dynamical system* is a set of n ordered Boolean functions from $\mathbb{F}_2^n$ to $\mathbb{F}_2$. First function is an update function for the first variable, second function is an update function for the second variable, and so on. Thus, a Boolean dynamical system is defined as $F : \mathbb{F}_2^n \to \mathbb{F}_2^n$ where

$$F = (f_1, f_2, \ldots, f_n)$$
$$= F(f_1(x_1, x_2, \ldots, x_n), \ldots, f_n(x_1, x_2, \ldots, x_n))$$

There is some ambiguity in the literature on how multiplication and addition are defined in the Boolean algebra. For example, Francis Robert [10] defines $1 * 1 = 0$, whereas other authors (for example [18]) consider $1 * 1$ to be 1. Except of adding and multiplying variables, we are also allowed to add 1, which is equivalent to negation. Throughout this paper we use the multiplication and addition tables given in Figure 1.

x	y	x+y
1	1	0
1	0	1
0	1	1
0	0	0

x	y	x*y
1	1	1
1	0	0
0	1	0
0	0	0

x	x+1
1	0
0	1

Fig. 1. Multiplication and addition in $\mathbb{F}_2$

Multiplication can be also expressed with the logical operator AND ($\wedge$). If X and Y are Boolean variables, then X AND $Y = 1$ if and only if the value of both X and Y is 1 (the logical value is true). The truth table is then identical to the one for multiplication.

The addition operation is equivalent to XOR ($\veebar$). X XOR Y is true only when either X is true or Y is true, and false when both are true or both are false.

Negation ($\neg$) is equivalent to adding 1 to variable. $X + 1 = 0$ if $X = 1$ and $X + 1 = 1$ if $X = 0$. By adding 1 we flip the value of the variable.

Hence, the alternative formulation with AND, XOR and NEG is

A Boolean dynamical system is a discrete time system. For a system of a size n there are 2^n

x	y	x XOR y
1	1	0
1	0	1
0	1	1
0	0	0

x	y	x AND y
1	1	1
1	0	0
0	1	0
0	0	0

x	NEG x
1	0
0	1

Fig. 2.   Multiplication and addition in $\mathbb{F}_2$

possible states of a system. Throughout this paper we will assume that all variables are updated simultaneously.

While classifying Boolean dynamical systems, we are interested in two phenomena. The first one is the occurrence and the number of steady states (attractors). The *steady state (SS)* of a Boolean system is a state $(x_1, x_2, \ldots, x_n)$ which updates to itself, i.e. $F(x_1, x_2, \ldots, x_n) = (x_1, x_2, \ldots, x_n)$.

The second phenomenon is the number and length of cycles. Let $(x_1, x_2, \ldots, x_n)$ be a starting state. If after $p > 1$ updates the system returns to the starting state, we say that a system has a *cycle* of length $p - 1$.

Unlike for continuous systems, we do not have the tools coming from bifurcation theory. Most methods are based on statistics, algebra, combinatorics and topology, see [9], [12], [13], [14]. We notice that a Boolean dynamical system must either have a steady state or a cycle or both; there is no possibility of oscillations and chaotic behaviors, which occur for continuous networks.

B. Coupled cell networkss: idea

The idea of a coupled cell networks is to look at the dynamical system not as a whole, but rather to look at the dynamics of particular members of the system. In order to do so, we divide the system into separate entities called cells. The cell captures one one or more differential equations. The dynamics of the cell depends upon the cell itself (self-variable(s)) and couplings (variables of other cells). There may be more than one type of coupling since cells may interact with each other in many different ways.

We can easily represent a coupled cell networks as a graph. The vertices of the graph are the cells, and couplings are the edges of the graph. The different types of couplings are shown as different types of edges.

C. Boolean coupled cell networks: formal definitions

Definition 1. By cell *we understand an entity of the n-dimensional Boolean dynamical system together with its update function.*

Definition 2. By coupling *we understand an influence that one cell has on the dynamics of the other cell.*

In this paper, we will consider only *regular networks*. The cells of regular network are all identical and there is only one type of coupling. We assume that every cell has the same number of couplings (this is enforced by the property of all cells being identical). Every cell has only one self-variable.

We assume that if some number of cells couple to cell A, then we can permute the variables of coupling cells and we get the same equation up to permutation of variables. The last statement comes from the assumption that there is only one type of coupling. The statement can be formalized as

$$x_k = f(x_k, \overline{x_{k1}, \ldots, x_{km}})$$

where $x_{k1}, \ldots, x_{km}$ are the variables of coupling cells, and overline indicates that we can permute them. By convention, we write the self-variable in the first position. Here f stands for a function template. Since every cell is governed by the same equation, the template is the same, however, since cells have different couplings (but always the same number of couplings) the functions are not identical.

Thus, the regular Boolean coupled cell networks is represented as

$$x_1 = f(x_1, \overline{x_{11}, \ldots, x_{1m}}),$$
$$x_2 = f(x_2, \overline{x_{21}, \ldots, x_{2m}}),$$
$$\ldots\ldots\ldots\ldots$$
$$x_n = f(x_n, \overline{x_{n1}, \ldots, x_{nm}}).$$

The network shown above has valency m, which means that every cell receives inputs from m other cells.

In this paper we consider only update functions whose formula depends on the variable of self and all the variables of the couplings. For example, in a valency 1 network a function

$$f(x_1, x_2) = x_1 + x_2$$

is a valid function, whereas

$$f(x_1, x_2) = x_2$$

is not a valid function. We call the valid functions *admissible* functions.

In order to fully define Boolean coupled cell networks we need a function template and a graph of connections.

Example 1. *Consider function scheme $F(x, y) = x + x * y$ and a graph of connection (further called architecture graph) shown in Figure 3.*

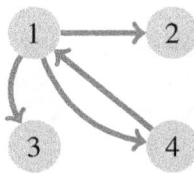

Fig. 3. Example 1: a graph of connections in a regular Boolean CCN

Using the template and the graph, we obtain

$$x_1 = x_1 + x_1 * x_4,$$
$$x_2 = x_2 + x_2 * x_1,$$
$$x_3 = x_3 + x_3 * x_1,$$
$$x_4 = x_4 + x_4 * x_1.$$

Lemma 1. *The number of function templates that could be used in a Boolean CCN with n cells and valency $m < n$ is given as 2^{2m+2}*

Proof: We consider the transition table associated to the Boolean CCN. Every cell in this network is influenced by m other cells. Hence, the template function for this network depends on $m + 1$ variables (variable of self and m variables of couplings). For a given cell, the function is $f(x, y_1, \ldots, y_m)$.

To fully describe a Boolean function on $m + 1$ variables, we need to create a transition table and assign a 0 or a 1 to all possible 2^{m+1} states. This gives 2 choices for every of 2^{m+1} places, which is in total $2^{2^{m+1}}$ possibilities.

As we stated before, couplings are insensitive to permutation. For a given cell, let us set up the variable of self to be 1. Then, we assign 0 or 1 to a state where all couplings are 0s, then 0 or 1 to a state when one coupling is 1 (we emphasize that it does not matter which of the couplings is 1), two couplings are 1s, and so on until we reach the state where all the couplings are 1. In total, we have a choice in 2^{m+1} places.

Next, we set up the variable of self to be 0 and we repeat the same process. We get $2^{m+1} \cdot 2^{m+1}$ possibilities. We have then 2^{2m+2} possible function templates for a network with valency m.

III. PROBLEM STATEMENT

Dynamical systems arising in biology and ecology are often large [16]. Large networks are hard to analyze mathematically, both from discrete and from continuous point of view [15], [17]. Usually in such cases a model reduction technique is applied [13], [17], [18], [19]. The authors of [1] base their model reduction strategy intended for CCNs on cell coloring. They cluster cells with the same color. This clustered network is called a *quotient network*, which is formally defined in Section V. In addition to defining the rules for forming a quotient network, the authors of [1] go further. They look at the quotient network and ask *what are the networks that admits this quotient,* and, *if we know the properties of the quotient, what can be concluded about the original network?*

In the next sections we perform a similar analysis for Boolean CCNs, and we demonstrate analogous results. We give some insights about the influence of the network architecture on the network dynamics.

IV. NETWORK DYNAMICS

A continuous dynamical system is often defined with one or more parameters [20]. We do not have this advantage for Boolean dynamical systems and enforcing the use of parameters is somewhat artificial. This issue is discussed in details in [11] and references therein. Bifurcations are tied to parameters; there are no bifurcations in the Boolean dynamical systems.

The authors of [1] focused on synchrony-breaking (pitchfork) bifurcations that are common in coupled cell networkss and in some cases, quotient is able to predict their existence in the original network.

For Boolean CCNs, instead of looking for bifurcations, we look for steady states and cycles. We show that a small Boolean CCN (with 2, 3 and 4 cells) could not have both cycles and steady states in the same network.

We define a few concepts related to the dynamics of a Boolean CCN.

By *canonical steady state* we understand a state of a system when all the cells are working at the same way. We have 2 such states for a Boolean network: $(0, 0, \ldots, 0)$ and $(1, 1, \ldots, 1)$.

An interesting phenomenon that happens in Boolean coupled cell networkss is that once synchronized, the network could not un-synchronize, because all the cells use the same update function. In all Boolean coupled cell networks we have either canonical steady states or canonical cycle (a cycle when the system alternates between two canonical states).

Canonical Steady States and *Canonical Cycles* are called the *canonical part of the dynamics.*

In addition, Boolean coupled cell networkss often have non-canonical parts, which are steady states and cycles where the system is not synchronized. The appearance of such structures depends on the functions and on the architecture graph.

V. QUOTIENTS NETWORKS

All the results shown in [1] regarding quotients apply directly to Boolean systems, because these results are based on graph theory and combinatorics but not on the network dynamics.

Hence, we just re-state the principles of coloring and taking quotients defined in [1].

By *coloring* we understand the function that assigns a color to every cell (node). Of course one graph could be colored in many different ways.

By *balanced coloring* we understand a coloring for which every cell with color a receives the same number of inputs from the cells with color b, for each b. An example of balanced coloring is shown in Figure 4.

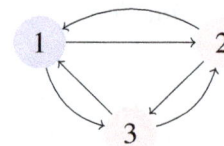

Fig. 4.   An example of balanced coloring in CCN.

A quotient is defined based on coloring. All cells with the same color become one meta-cell. The result is shown in Figure 5.

Fig. 5.   An quotient network for the network from Figure 4 formed based on coloring.

Taking a quotient affects the functions associated with the cells. This means that all the variables of the cells clustered to one meta-cell are replaced by one variable.

In the examples shown in Figure 4 and Figure 5, the original system of three equations

$$x_1 = f(x_1, \overline{x_2, x_3}),$$
$$x_2 = f(x_2, \overline{x_1, x_3}),$$
$$x_3 = f(x_3, \overline{x_1, x_3}),$$

changes to

$$x_1 = f(x_1, \overline{x_2, x_2}),$$
$$x_2 = f(x_2, \overline{x_1, x_2}),$$

VI. NETWORKS THAT ADMIT TWO-CELL BIDIRECTIONAL RING AS A QUOTIENT: CASE STUDY

One of the example cases considered in [1] is a network named two cell bidirectional ring (shown in Figure 6).

Fig. 6.   two cell bidirectional ring

Note. The network presented in Figure 6 should not be confused with the diagram of mutual activation/inhibition that often appears in mathematical biology papers [25]. The function that describes mutual activation/inhibition is a function that assigns the cell the state of its coupling. This is not an admissible function in the context of CCNs, since it does not involve cell's own state variable in the update formula.

A *circuit* in a graph is a path consisting of vertices and edges with the property that we can reach a vertex from itself. We note that networks that admit two-cell bidirectional ring as a quotient have a structure of a bipartite graph with in-degree 1. Such a graph could have only one circuit, and if it had two, it would be disjoint) Hence the graph that admits the two-cell bidirectional ring is a circuit with some attached structure, influenced by the dynamics of the circuit, but not influencing back. We will call all the graphs having this

structure G_{2CBR}. An example of such a structure is shown in Figure 7.

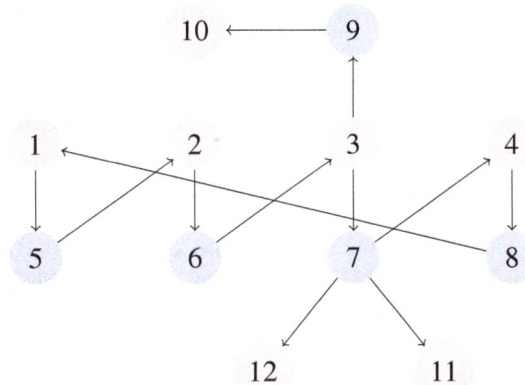

Fig. 7.   An example pf G_{2CBR} graph

Claim. The dynamics of the system built on the G_{2CBR} architecture depends on the dynamics of its single circuit. In particular, if there is a steady state in the circuit, there is a steady state in the system. If there is a cycle of length 2 or more, there is a cycle in the entire system.

Proof: Let us observe that the structure of all G_{2CBR} graphs is a circuit plus some attached structure. We will look at the attached part. The cells belonging to the circuit are influenced only by other cells that belong to the circuit. If a cell belongs to the attached structure, it must receive input from either a cell from the circuit or from another cell that does not belong to the circuit. If the attached structure is non-empty, there is at least one cell in the attached structure that receives input from the cell from the circuit, because the graph is connected.

Assume that the circuit achieves a steady state. Then, all the cells directly influenced by the circuit achieve steady state as well. The same happens with the cells influenced by these cells.

Assume that the circuit achieves a cycle of length greater than 1. This means that the circuit oscillates between two or more states. The inputs received by non-circuit cells are either changing or stay steady. In any case, since the non-circuit part does not influence the circuit part, the entire system could not go to a steady state and there must be a cycle for the entire system.

Lemma 2. *Let G be a graph from the G_{2BCR} family. Let $F(a, b)$ be a two-variable Boolean update function template, where a is the variable of self and b is the variable of a coupling cell.*

The network $X = \{G, F\}$ could either have a non-canonical cycle(s) or non-canonical steady state(s), but never both.

Proof: Let us first use the observation from [1] that the adjacency matrix of such an architecture must have the structure (proved in [1])

$$\begin{bmatrix} C & 0 \\ B_1 & B_2 \end{bmatrix}$$

where C is the matrix of a circuit and B_2 is a lower triangular matrix with 0s on the diagonal. We can imagine such architecture as a circuit of length l with attached non-circuit structure.

We investigate the dynamics of the circuit alone. Based on the previous claim, the dynamics of the non-circuit part strictly depends on the dynamics of the circuit.

Let us assume that there are l cells in the circuit. We will write a state of the circuit as $(s_1, s_2, \ldots, s_l)$ assuming that s_1 sends input to s_2, s_2 to s_3 and so forth and s_l sends input to s_1.

Let us assume that we have a non-canonical steady state in this structure. This means that this steady state of the form $(\ldots, 0, 1, 0, \ldots)$ or $(\ldots, 1, 0, 1, \ldots)$

In both cases we have $F(1, 0) = 1$ and $F(0, 1) = 0$.

Note that we cannot have both $F(0, 0) = 1$ and $F(1, 1) = 0$, since this leads to a function $F(a, b) = b + 1$ (which is not admissible). We could not have $F(1, 1) = 1$ and $F(0, 0) = 0$ because this leads to $F(a, b) = a$, which is not admissible as well.

We have two cases:

- Case 1: $F(0, 0) = 0$ and $F(1, 1) = 0$. We notice that a system driven by such a function could not oscillate. Once changed to 0, a cell could not go back to 1.

- Case 2: $F(0, 0) = 1$ and $F(1, 1) = 1$. We notice that here oscillations are impossible as well. Once a state of a variable is changed to 1, it could not go back to 0.

We conclude that if Boolean CCN from the G_{2BCR} family has a non-canonical steady state, it cannot have a non-canonical cycle.

To prove the converse, let us assume that there is a non-canonical cycle in $\{G, F\}$.

As a part of this cycle we must have a transition between two states of the system as shown below

$$(\ldots, \underbrace{0}_{x_s}, \ldots) \rightarrow (\ldots, \underbrace{1}_{x_s}, \ldots),$$

or

$$(\ldots, \underbrace{1}_{x_s}, \ldots) \rightarrow (\ldots, \underbrace{0}_{x_s}, \ldots),$$

Thus, for the first case we must have $F(0, 0) = 1$ or $F(0, 1) = 1$ and for the second case $F(1, 0) = 0$ or $F(1, 1) = 0$.

- Case 1, $F(0, 0) = 1$. There are 4 possible options (note that either $F(1, 0) = 1$ or $F(1, 1) = 1$ because x_s must eventually return to the original state.

  1)

  $$\begin{aligned} F(0, 1) &= 0 \\ F(1, 0) &= 0 \\ F(1, 1) &= 1 \end{aligned}$$

  In this case $F(a, b) = a + b + 1$. If there exists a non-canonical steady state, we must have for some x_k and x_{k+1}

  $$(\ldots, \underbrace{0}_{x_k}, \underbrace{1}_{x_{k+1}} \ldots) \rightarrow (\ldots, \underbrace{0}_{x_k}, \underbrace{1}_{x_{k+1}} \ldots),$$

  This is, however, impossible because $F(1, 0) = 0$.

  2)

  $$\begin{aligned} F(0, 1) &= 0 \\ F(1, 0) &= 1 \\ F(1, 1) &= 0 \end{aligned}$$

  In this case we have $F(a, b) = 1 + b$ and this is not a valid update function (it does depend only on the coupling variable).

3)

$$F(0,1) = 1$$
$$F(1,0) = 0$$
$$F(1,1) = 1$$

By the same reasoning as in 1) we obtain that this function could not produce a non-canonical steady state.

4)

$$F(0,1) = 1$$
$$F(1,0) = 1$$
$$F(1,1) = 0$$

By the same reasoning as in 1) and in 3), we obtain that this function could not produce a non-canonical steady state.

- Case 1, $F(0,1) = 1$

  Based on similar reasoning as in the previous case, points 1), 3) and 4), the template function with such a property could not produce a non-canonical steady state.

The proof for Case 2 is analogous. Thus, we obtain that if there is a non-canonical cycle in $\{G, X\}$, then there cannot be a non-canonical steady state.

We conclude that non-canonical steady states and non-canonical cycles do not appear together in a Boolean CCN that admits two-cell bidirectional ring as a quotient.

Next, we use the observation that the dynamics of the non-circuit part of the system depends on the circuit. Hence, if there is no oscillation in the circuit, there are no oscillations in the entire system.

Theorem 3. *The following are true:*

1) *If the 2CBR has non-canonical steady states, so does the non-quotient network, from which it arose.*

2) *If the 2CBR has non-canonical cycles, so does the non-quotient network, from which it arose .*

Proof:

There are only $2^{2 \cdot 1+2} = 16$ Boolean coupled cell networks that could be created on a 2CBR architecture. Eight of them yield only canonical dynamics and eight do not. We need to exclude all the networks where we do not have both the influence of self-variable and of the coupling. Eventually we are left with four networks.

These are the systems that have non-canonical cycles:

$$f_1 = x_1 * x_2 + x_1 + 1,$$
$$f_2 = x_2 * x_1 + x_2 + 1,$$

and

$$f_1 = x_1 * x_2 + x_2,$$
$$f_2 = x_2 * x_1 + x_1.$$

These are the systems that have non-canonical steady states.

$$f_1 = x_1 * x_2 + x_2 + 1,$$
$$f_2 = x_2 * x_1 + x_1 + 1,$$

and

$$f_1 = x_1 * x_2 + x_1,$$
$$f_2 = x_2 * x_1 + x_2.$$

Again, we can use the structure that admits 2CBR as a quotient. We know that this structure consists of a circuit and some circuit-dependent cells that do not form a circuit themselves.

Similarly as in the proof of Lemma 2, we can just consider the dynamics of the circuit.

We analyze the above-mentioned four systems separately.

The first system gives $F(0,0) = 1$, $F(1,1) = 1$, $F(0,1) = 1$ and $F(1,0) = 0$. We can assume that the cells influence each other in an order (the first cell influences the second, the second influences the third and so on, the nth cell influences the first cell) and consider any starting state, say

$(1, 0, 0, \ldots, 0)$. From the dynamics we have a sequence of states

$$(1, 0, 0, \ldots, 0)$$
$$(0, 1, 1, \ldots, 1),$$
$$(1, 0, 1, \ldots, 1),$$
$$(1, 1, 0, \ldots, 1),$$
$$\ldots$$
$$(0, 1, 1, \ldots, 1),$$

which is clearly a cycle of length 2 or more. Since the rest of the dynamics is influenced by the dynamics of the cells belonging to the circuit and we have "pulses" of 0s and 1s, we can only end up having a cycle for the entire structure.

By Lemma 2, cycles and steady states do not appear simultaneously and we have a system with additional cycles.

The second case leads to $F(0, 0) = 0$, $F(1, 1) = 0$, $F(0, 1) = 1$, $F(1, 0) = 0$. If we start with $(1, 0, \ldots, 0)$, we obtain

$$(1, 0, 0, \ldots, 0),$$
$$(0, 1, 0, \ldots, 0),$$
$$(0, 0, 1, \ldots, 0),$$
$$\ldots$$
$$(0, 0, 0, \ldots, 1),$$
$$(1, 0, 0, \ldots, 0),$$

which is a cycle of length 2 or more. By the same line of reasoning as for the first case, we get a system with additional cycles.

The third case gives $F(0, 0) = 1$, $F(1, 1) = 1$, $F(0, 1) = 0$ and $F(1, 0) = 1$. Let us use the same argument as for the first case and consider the circuit separately, and assume that it is ordered. Let us take a starting state, say $(1, 0, \ldots, 0)$. Based on F, we obtain a sequence of system states

$$(1, 0, 0, \ldots, 0),$$
$$(1, 0, 1, \ldots, 1),$$
$$(1, 0, 1, \ldots, 1),$$
$$\ldots$$
$$(1, 0, 1, \ldots, 1),$$

and $(1, 0, 1, \ldots, 1)$ is clearly a non-canonical steady state.

Because of the lack of the circuit in the rest of the system architecture, we must have a steady state for the entire system. By the lemma, we must have a system with additional steady states.

The fourth case gives us $F(0, 0) = 0$, $F(1, 1) = 0$, $F(0, 1) = 0$ and $F(1, 0) = 1$. We start with $(1, 0, \ldots, 0)$. We obtain a sequence of system states

$$(1, 0, 0, \ldots, 0),$$
$$(1, 0, 0, \ldots, 0),$$
$$\ldots$$
$$(1, 0, 0, \ldots, 0),$$

and $(1, 0, 0, \ldots, 0)$ is clearly a non-canonical steady state. By similar reasoning as in previous cases and Lemma 2 we obtain that this system must have non-canonical steady states.

VII. BIOLOGICAL EXAMPLE

Most known Boolean models in systems biology are characterized by cooperating species, each of which is governed by a different set of rules [18], [26]. Models based on the idea of identical entities governed by identical sets of rules are quite common in ecology, however these models are usually not Boolean [27].

To illustrate a Boolean network that admits two-cell bidirectional ring as a quotient, we use a simple fish schooling model. A reaction of a fish school to a predator is a well-documented behavior [29]. There exist a couple of theories explaining this phenomenon. One of the theories is a *many eyes* hypothesis [28]. According to this theory, the advantage of swimming in a schools is that the fish can rely on collective vigilance while avoiding predators, and thus spend more time foraging. Once a single fish senses a predator, it sends a *signal* to neighboring fish [30] (for example, changes the direction, and the neighboring fish are able to sense this change rapidly). These neighboring fish send signal further, until the entire school is alert and invokes its defense mechanisms (for example, collectively changes direction). Previously, fish schools have been modeled by both, an ODE model [31], and discrete agent-based model [27].

In very simplistic terms, a fish in a school has two possible states: alert and non-alert. An alert fish spreads the *alert* signal to other fish, causing them to change their state to *alert*. We can assume that a fish might send a signal to multiple fish, but receives a signal only from one other fish (say, closest lateral neighbor). In order for a school to ensure an efficient spread of information, we must have a cycle inside the fish network.

We assume that an alert fish stays alert regardless of the state of its coupling fish. In a school we also have fish who stay "inside" the school and base their safety on the vigilance of the more specialized fish. We note that there might exist some fishes, who are undervigilant and do not inform the surroundings, but are still able to receive the *alert* information.

Hence, in our network the nodes are fish, state 1 means that the fish is alert and state 0 means that it is not. A fish has only one coupling, but might be a coupling to many other fish. A simple illustration of such behavior is shown on the Figure 8.

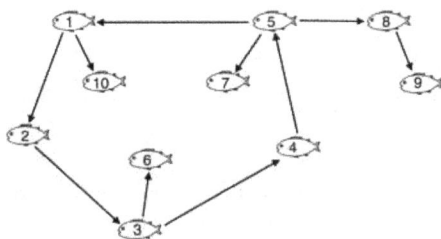

Fig. 8. Simple fish school model.

We note that in such a Boolean network, any state where any of the fish-nodes on the circuit is alert leads to a state where all the fishes are alert. Also a state where none of the fish is alert, is a steady state. We note that such dynamics is governed by a function $F(0,1) = 1$ (non-alert fish changes a state to alert once a coupling is alert), $F(1,0) = 1$, and $F(1,1) = 1$ (alert fish stays alert regardless if its coupling is alert or not)

Time step	1	2	3	4	5	6	7	8	9	10
1	0	1	0	0	0	0	0	0	0	0
2	0	1	1	0	0	0	0	0	0	0
3	0	1	1	1	0	1	0	0	0	0
4	0	1	1	1	1	1	0	0	0	0
5	1	1	1	1	1	1	1	1	0	0
6	1	1	1	1	1	1	1	1	1	1

Fig. 9. Dynamics of fish school model with initial condition: fish 2 senses predator (time step 1).

and $F(0,0) = 0$ (non-alert fish stays non-alert when the coupling is non-alert). In such a network there are canonical steady states $(0,0,\ldots,0)$ and $(1,1,\ldots,1)$, many possible non-canonical steady states, and no cycles. For example, in the network shown in Figure 8, all the states listed in the table below are the non-canonical steady states, and this is not a complete list

(0,0,0,0,0,1,0,0,0,0)
(0,0,0,0,0,0,1,0,0,0)
(0,0,0,0,0,0,0,1,0,0)
(0,0,0,0,0,0,0,0,0,1)
(0,0,0,0,0,1,1,0,0,0)
(0,0,0,0,0,1,0,1,0,0)
(0,0,0,0,0,1,0,0,0,1)
(0,0,0,0,0,0,1,1,0,0)
(0,0,0,0,0,0,1,0,0,1)
(0,0,0,0,0,1,0,1,0,1)
(0,0,0,0,0,1,0,0,1,1)

Fig. 10. Example of non-canonical steady states of the Boolean CCN shown in Figure 8.

The network presented above reduces to a 2CBR with 2 steady states $(0,0)$ and $(1,1)$ and dynamics $(0,1) \rightarrow (1,1)$, $(0,1) \rightarrow (1,1)$, $(0,0) \rightarrow (0,0)$, $(1,1) \rightarrow (1,1)$. The function that governs this system is $F(x1,x2) = (x1+1) * (x2+1) + 1 = \neg(\neg x1 \wedge \neg x2)$.

This shows that regardless of the size of the school, one can interpret its behavior in the same way: if one fish on the circuit is alert (i.e. fish who is specialized in vigilance), all fish become alert, and if none of the fish on the circuit is alert, the school stays non-alert. We note that if the fish that is not on the circuit becomes alert, it does not have an ability to inform the other fish. This can be thought of as either the fish is currently foraging, and the other, more specialized fish in the school would notice the predator anyways, or the fish is too far away from the school, or the fish just received false signal. The quotient network is an simplification of this system where only the dynamics of the cycle is taken into account.

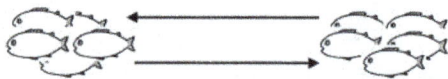

Fig. 11.   Quotient of simple fish school model.

The model presented above is very simple, however, it illustrates the idea well. A behavior where one member of the species informs other members about the danger does occur in plants as well; for example tomato plants are able to spread such information [32].

VIII. NETWORKS THAT ADMIT THREE-CELL BIDIRECTIONAL RING AS A QUOTIENT: CASE STUDY

According to [1], there are two networks with 4 cells that admit three-cell bidirectional ring as a quotient and 12 networks with 5 cells that admit three-cell bidirectional ring as a quotient. The authors of [1] have shown that the dynamics of a three-cell bidirectional ring is a good predictor of the dynamics of a bigger network for both networks with 4 cells and for the 10 out of 12 networks with 5 cells.

The three-cell bidirectional ring is a structure shown in Figure 12.

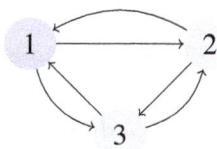

Fig. 12.   three-cell bidirectional ring

The two networks with four cells admitting three-cell bidirectional ring as a quotient are

shown in Figure 13. All twelve five-cell networks

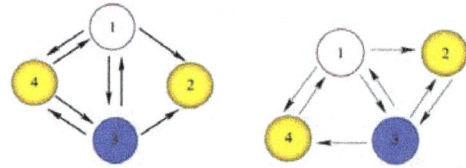

Fig. 13.   Four-cell networks that admit three-cell bidirectional ring as a quotient, taken from [1]

admitting three-cell bidirectional ring as a quotient are shown in Figure 14. The analysis of the net-

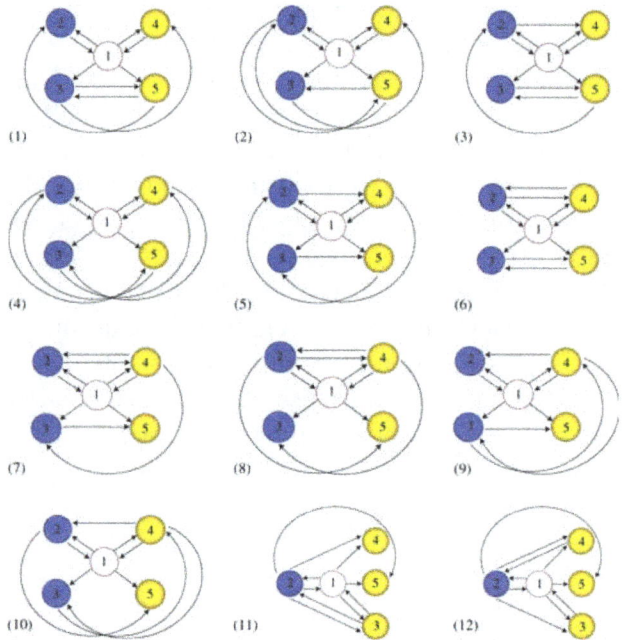

Fig. 14.   Five-cell networks that admit three-cell bidirectional ring as a quotient, taken from [1]

works that admit a particular network as a quotient was based on the topology, not on the dynamics, hence we can use the results from [1].

There are 64 Boolean networks of valency 2 (regular network with two couplings). If we apply those networks to the three-cell bidirectional ring, it turns out that 32 of these networks have only canonical structure, and only 4 have additional cycles and the rest have additional steady states. One of the 4 networks is not admissible since the function template can be written as $F(a, \overline{b, c}) = 1 + a,$

hence only 3 remain. Similarly to the 2CBR case, we cannot have non-canonical steady states and non-canonical cycles in the same network using the three-cell bidirectional ring as an architecture graph.

Theorem 4. *If a Boolean coupled cell networks of valency 2 that admits three-cell bidirectional ring as a quotient is built using any of the three function templates that result in additional cycles for the three-cell bidirectional ring, it could not have non-canonical steady states.*

Proof:

- Network 1: For this network we have

$$
\begin{aligned}
F(1,1,1) &= 0, \\
F(1,1,0) &= 0, \\
F(1,0,1) &= 0, \\
F(1,0,0) &= 0, \\
F(0,1,1) &= 1, \\
F(0,0,1) &= 1, \\
F(0,1,1) &= 1, \\
F(0,0,0) &= 0,
\end{aligned}
$$

We notice that for this network once the cell changes its state to 1, after update it changes to 0 regardless of what is the state of the coupling. Hence the only possible steady state is a canonical state with all 0s.

- Network 2:

$$
\begin{aligned}
F(1,1,1) &= 1, \\
F(1,1,0) &= 0, \\
F(1,0,1) &= 0, \\
F(1,0,0) &= 0, \\
F(0,1,1) &= 1, \\
F(0,0,1) &= 1, \\
F(0,1,1) &= 1, \\
F(0,0,0) &= 1,
\end{aligned}
$$

We use the same line of reasoning and notice that once the cell has a state 0, it must change

the state to 1 regardless what is the state of couplings. The only steady state is such as a system could be a canonical state with all 1s.

- Network 3:

$$
\begin{aligned}
F(1,1,1) &= 1, \\
F(1,1,0) &= 0, \\
F(1,0,1) &= 0, \\
F(1,0,0) &= 0, \\
F(0,1,1) &= 1, \\
F(0,0,1) &= 1, \\
F(0,1,1) &= 1, \\
F(0,0,0) &= 0,
\end{aligned}
$$

Here the situation is not as clear as for the two previous functions.

Assume that we have a non-canonical steady state in this network. Since it is non-canonical, it must have some number of 0s and 1s. If it is $(1,0,0,\ldots,0)$, the first 1 is influenced by 2 0s and we end up in a canonical steady state. The same happens for a state with 2 ones. If we take a state with $(1,1,1,0,0,\ldots,0)$ and assume that first 3 cells influence each other and all the cells with 0s influence each other we conclude that it is a steady state, but in such a case the network is disjoint. We use the same way of reasoning for all the states with 4 or more 1s. We conclude that a network with such an update scheme could not have non-canonical steady states.

The characterization of the networks with non-canonical steady states is much harder. Depending on the architecture, the networks that admit three-cell bidirectional ring may have or may not have both non-canonical steady states and non-canonical cycles.

Claim The following is true for the networks that admit three-cell bidirectional ring as a quotient.

- The dynamics of s three-cell bidirectional ring is a good predictor of a dynamics of the network with 4 cells that has admitted it. If a three-cell bidirectional ring has only non-canonical cycles, so does the bigger network. If a three-cell bidirectional ring has only additional steady states, so does the bigger network.

- For the 4 out of 12 5-cell networks that admit a three-cell bidirectional ring as a quotient the dynamics of a three-cell bidirectional ring is a good predictor of the dynamics of a bigger network. For the rest of the networks, non-canonical cycles may appear, even though the smaller network has only non-canonical steady states.

A contribution towards the proof. As mentioned earlier, there exist 64 function templates for Boolean CCN created based on a three-cell bidirectional ring architecture. 32 of them have non-canonical dynamics. 28 functions out of the 32 have non-canonical steady states. We used a CPP code to test two 4-cell networks and twelve 5-cell networks using each of the 28 functions.

- If a 3CBR quotient has a non-canonical steady state, so does the 4-cell network that has admitted it. If 3CBR has a non-canonical cycle, so does the 4-cell network that admitted it. This result was obtained using exhaustive computer simulation.

- For 4 graphs (8, 9, 11, and 12), if the quotient network has only non-canonical steady states, so does the networks that admitted it. This result was obtained using exhaustive computer simulation.

- If the quotient network has only cycles, so does the network that admitted it. This is true for all 12 of the 5-cell networks. This result was obtained using exhaustive computer simulation.

Figure 15 shows how many (out of 28) functions cause the non-qoutient graph to have both non-canonical cycles and non-canonical steady states.

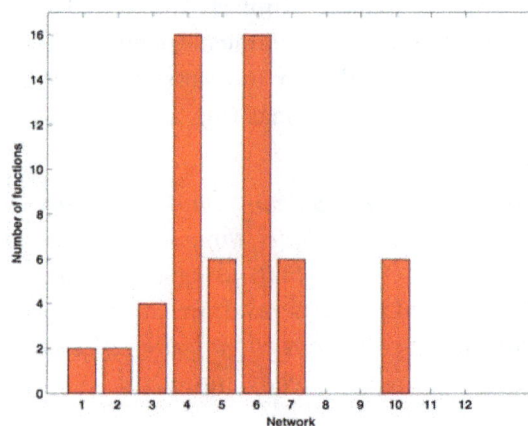

Fig. 15. Number of function templates that have have both non-canonical steady states and non-canonical cycles in a 5-cell non-quotient network.

IX. RELATIONSHIP WITH AGENT-BASED MODELS

Every coupled cell networks can be viewed as an agent-based model. The agent-based model is a model consisting of identical interacting agents. The framework of agent-based models is very general; any object could be considered an agent (network cells, fishes, people susceptible to infection, bugs, trees etc). Every agent is described by the state vector (for example, position, infection, age, alertness) and the state vector is updated according to the same rules.

The model presented in Section VII can be viewed as agent-based model. In this model the fish function as agents. These fish are characterized by a state vector with only one variable: state of alertness (alert or non-alert). In this simple model fish are stationary, which means they do not move throughout the domain. The update rules for the state of a single agent are the same as described in Section VII.

Typically, agent-based models are large. There are not many mathematical methods that allow us to control (predict and change) their behavior [21], [22]. A method that allows a researcher to look at the quotient of the model and predict the behavior of the bigger original model would be a desirable tool. There exist a framework that makes it possible to translate an agent-based model into a system of Boolean equations [3], and thus, at least in some cases, it can be translated to a Boolean CCN.

CONCLUSIONS AND FUTURE WORK

In this paper we show preliminary work regarding Boolean coupled cell networks. We are able to fully characterize the dynamics of the networks that admit two-cell bidirectional ring as a quotient, and we provide a small contribution to characterizing networks that admit three-cell bidirectional ring as a quotient. The main contribution in this paper is showing that CCNs have their Boolean counterpart and that this new setting brings a new perspective on Boolean functions and Boolean networks.

The phenomenon of bifurcations does not arise in Boolean CNNs, yet it does not make investigating their dynamics easy. In fact, predicting dynamics based on the quotient turns out to be a hard problem for the Boolean CCNs. In the future we plan to investigate bigger networks (with more that 5 cells) that bring 3CBR as a quotient, possibly using high performance computing. An efficient algorithm for easy enumeration and generation of such networks is needed as well.

In this paper we do not analyze the dependence between internal symmetries of the network graph and properties of its dynamics. Such analysis is an important part of CNN research [2] and we plan to develop similar techniques and ideas for Boolean CCNs.

Finding a more powerful, biology-related application of Boolean quotient networks is another goal. This would allow us to confirm the importance of the results derived in this paper on a new level.

In this paper we focus on regular Boolean CCNs with one type of coupling. Expanding the research to networks with less regularity and two or more types of coupling is another future goal.

X. ACKNOWLEDGEMENTS

This research was originally inspired by Dr. Reinhard Laubenbacher from the Center for Quantitative Medicine, Uconn Health, University of Connecticut, CT, USA.

The suggestions and comments of two anonymous reviewers on the earlier versions of this manuscript helped to improve the paper and make it more clear and readable.

REFERENCES

[1] D. Aguiar, A.P.S. Dias, M. Golubitsky, M. Leite, *Bifurcations from Regular Quotient Networks: A First Insight.*, Physica D, 238 (2009) 137155, http://dx.doi.org/10.1016/j.physd.2008.10.006

[2] M. Golubitsky and I. Stewart, *The Symmetry Perspective: From Equilibrium to Chaos in Phase Space and Physical Space.*, Birkhauser (2002)

[3] F. Hinkelman, D. Murrugarra, A. Jarrah, R. Laubenbacher, *A Mathematical Framework for Agent Based Models of Complex Biological Networks*, Bulletin of Mathematical Biology, 73(7) (2010), 1583-1602, http://dx.doi.org/10.1007/S11538-010-9582-8

[4] E. Frias-Martinez, G. Williamson, V. Frias-Martinez, *An Agent-Based Model of Epidemic Spread Using Human Mobility and Social Network Information*, Social-Com/PASSAT 2011 Proceedings, Boston, MA, USA, 57-64, http://dx.doi.org/10.1109/PASSAT/SocialCom.2011.142

[5] B. Heath, R. Hill, F. Ciarallo, *A Survey of Agent-Based Modeling Practices (January 1998 to July 2008)*, Journal of Artificial Societies and Social Simulation 12(4) (2009), 1-49, http://jasss.soc.surrey.ac.uk/12/4/9.html

[6] E. Bonabeau, *Agent-Based Modeling: Methods and Techniques For Simulating Human Systems*, PNAS, suppl. 3(99) (2002), 7280-7287, http://dx.doi.org/10.1073/pnas.082080899

[7] S. Maerivoet, B. De Moor, *Cellular Automata Models of Road Traffic*, Physics Reports, 419(1) (2005), 1-64, http://dx.doi.org/10.1016/j.physrep.2005.08.005

[8] V. Grimm, S. F. Railsback *Individual-based Modeling and Ecology*, Princeton Series in Theoretical and Computational Biology (2005)

[9] S. Kaufmann, *Metabolic Stability and Epigenesis in Randomly Constructed Genetic Nets*, Journal of Theoretical Biology, 22(3) (1969), 437-467.

[10] F. Robert, *Discrete Iterations: a Metric Study*, Springer-Verlag (1986)

[11] I. Shmulevich, S. A. Kauffman, *Activities and Sensitivities in Boolean Network Models*, Physics Review Letters, 93(4) (2004), 048701 http://dx.doi.org/10.1103/physrevlett.93.048701

[12] H. Mortveit, C. Reidys, *An Introduction to Sequential Dynamical Systems*, Springer (2008), http://dx.doi.org/10.1007/978-0-387-49879-9

[13] A. Veliz-Cuba, B. Aguilar, F. Hinkelmann, R. Lauben-bacher, *Steady State Analysis of Boolean Molecular Network Models via Model Reduction and Computational Algebra*, BMC Bioinformatics, 15(1) (2014), 1-8, http://dx.doi.org/10.1186/1471-2105-15-221

[14] R. Heckel, S. Schober, M. Bossert, *Harmonic Analysis of Boolean Networks: Determinative Power and Perturbations*, EURASIP Journal of Bioinformatics and System Biology 2013(6), (2013), 1-12, http://dx.doi.org/10.1186/1687-4153-2013-6

[15] R-S. Wang, A. Saadatpour and R. Albert, *Boolean Modeling in Systems Biology: An Overview of Methodology and Applications*, Physical Biology, 9(5) (2012), 055001, http://dx.doi.org/10.1088/1478-3975/9/5/055001

[16] A.T. Adai, S.V. Date, S. Wieland, E. M. Marcotte, *LGL: Creating a Map of Protein Function with an Algorithm for Visualizing Very Large Biological Networks*, Journal of Molecular Biology, 340 (2004), 179190, http://dx.doi.org/10.1016/j.jmb.2004.04.047

[17] A. Shamsul Arefin M. Inostroza-Ponta, L. Mathieson, R. Berretta, P. Moscato, *Clustering Nodes in Large-Scale Biological Networks Using External Memory Algorithms*, Proceedings of ICA3PP, Melbourne, Australia (2011), 375-386, http://dx.doi.org/10.1007/978-3-642-24669-2_36

[18] A. Veliz-Cuba, B. Stigler, *Boolean Models Can Explain Bistability in the lac Operon*, Journal of Computational Biology, 18(6) (2011), 783-794, http://dx.doi.org/10.1089/cmb.2011.0031

[19] A. Veliz-Cuba, B. Aguilar, R. Laubenbacher. *Dimension Reduction of Large Sparse AND-NOT Network Models*. Electronic Notes in Theoretical Computer Science, 316 (2015), 83-95, http://dx.doi.org/10.1016/j.entcs.2015.06.012

[20] J. Guckenheimer, P. Holmes,*Nonlinear Oscillations, Dynamical Systems, and Bifurcations of Vector Fields*, Springer (2008)

[21] M. Oremland, R. Laubenbacher, *Optimization of Agent-Based Models: Scaling Methods and Heuristic Algorithms*, Journal of Artificial Societies and Social Simulation, 17(2) (2014), 6, http://dx.doi.org/10.18564/jasss.2472

[22] M. Oremland, R. Laubenbacher, *Optimal Harvesting of a Predator-Prey Agent-Based Model Using Difference Equations*. Bulletin of Mathematical Biology, 77(3) (2015), 434-459, http://dx.doi.org/10.1007/s11538-014-0060-6

[23] S. Wolfram, *Cellular Automata and Complexity: Collected Papers*, Addison-Wesley (1994)

[24] J. M. Greenberg, S. P. Hastings, *Spatial Patterns for Discrete Models of Diffusion in Excitable Media*, SIAM Journal on Applied Mathematics, 34(3) (1978), 515-523, http://dx.doi.org/10.1137/0134040

[25] D. M. Wittmann, F. Bloechl, D. Truembach, W. Wurst, N. Prakash, F. J. Theis, *Spatial Analysis of Expression Patterns Predicts Genetic Interactions at the Mid-Hindbrain Boundary*, PLOS Computational Biology, 11(5) (2009), 1-16, http://dx.doi.org/10.1371%2Fjournal.pcbi.1000569

[26] J. E. Ferrell Jr., T. Yu-Chen Tsai, Q. Yang, *Modeling the Cell Cycle: Why Do Certain Circuits Oscillate?*, Cell, 144(6) (2011), 874-885, http://dx.doi.org/10.1016/j.cell.2011.03.006

[27] R. Vabø, L Nøtterstad, *An Individual Based Model of Fish School Reactions Predicting Antipredator Behaviour as Observed in Nature*, Fisheries and Oceanography, 6(3) (1997), 155-171, http://dx.doi.org/10.1046/j.1365-2419.1997.00037.x

[28] R.S. Olson, P. B. Haley, F. C. Dyer., C. Adami, *Exploring the Evolution of a Trade-Off Between Vigilance and Foraging in Group-Living Organisms.*, Royal Society Open Science, (2015) 2:150135, http://dx.doi.org/10.1098/rsos.150135

[29] R. Lukeman, Y. X. Li, L. Edelstein-Keshet, *Inferring Individual Rules From Collective Behavior*, Proceedings of the National Academy of Sciences, 107(28) (2010), 12576-12580, http://dx.doi.org/10.1073/pnas.1001763107

[30] F. Gerlotto, S. Bertrand, N. Bez, M. Gutierrez, *Waves of Agitation Inside Anchovy Schools Observed With Multibeam Sonar: a Way to Transmit Information in Response to Predation*, ICES Journal of Marine Science, 63(8) (2006), 1405-1417, https://doi.org/10.1016/j.icesjms.2006.04.023

[31] B. Birnir, *An ODE Model of the Motion of Pelagic Fish*, Journal of Statistical Physics, 128(1) (2007), 535-568, https://doi.org/10.1007/s10955-007-9292-2

[32] Y. Y. Song, R. S. Zeng, J. F. Xu, J. Li, X. Shen, W. G. Yihdego, *Interplant Communication of Tomato Plants through Underground Common Mycorrhizal Networks*, PLOS One, 5(10) (2010), http://dx.doi.org/10.1371/journal.pone.0013324

17

A Particular Solution for a Two-Phase Model with a Sharp Interface

David A. Ekrut, Nicholas G. Cogan
Department of Biological Mathematics
Florida State University
Tallahassee, Florida, United States
ekrut@math.fsu.edu, cogan@math.fsu.edu

Abstract—Two-phase models can be used to describe the dynamics of mixed materials and can be applied to many physical and biological phenomena. For example, these types of models have been used to describe the dynamics of cancer, biofilms, cytoplasm, and hydrogels. Frequently the physical domain separates into a region of mixed material immersed in a region of pure fluid solvent. Previous works have found a perturbation solution to capture the front velocity at the initial time of contact between the polymer network and pure solvent, then approximated the solution to the sharp-interface at other points in time. The primary purpose of this work is to use a symmetry transformation to capture an exact solution to this two-phase problem with a sharp-interface. This solution is useful for a variety of reasons. First, the exact solution replicates the numeric results, but it also captures the dynamics of the volume profile at the boundary between phases for arbitrary time scales. Also, the solution accounts for dispersion of the network further away from the boundary. Further, our findings suggest that an infinite number of exact solutions of various classes exist for the two-phase system, which may give further insights into the behaviors of the general two-phase model.

Keywords-**Multi-phase modeling; Two-phase modelling; Free boundary problems; Gel Dynamics; Analytic solutions; Exact solutions.**

I. INTRODUCTION

Two-phase models are useful for capturing the interactions between fluids and/or viscoelastic material. Each phase is averaged over a control volume, where the volume-averaged phases are incompressible. There is no inertial component to the system, and the phases are immiscible. Each phase is governed by conservation equations. These models have been successful at describing how emergent structures develop though the interactions of the two phases. There are several known applications.

Breward et. al. [1] developed a two-phase model to understand the role of viscosity and drag-friction in avascular tumor growth. An asymptotic solution solved explicitly for the volume fraction revealed that in the absence of viscosity and friction, tumor growth was regulated by oxygen tension. Numerical simulations showed increases in either the drag coefficient or viscosity parameter reduces the speed tumor growth. This leads credence to the notion that the invasiveness of tumor cells is related to the viscosity of the cells. Well-differentiated cells are known to grow more slowly and considered more viscous due to overlapping filopodia. Whereas, poorly differentiated (less viscous) cells repel one another, contributing to the spread of tumors. An extension of this model with an additional phase [2] contrasts the role of the expansive growth (passive response) and foreign body hypotheses (active response) in tumorigenesis. Numerical simulations showed capsule formation could not result from an active response. Another model [3] was used to describe avascular tumor as a two-phase system where tumor spheroids exist in two states, one solid and one liquid. Time independent solutions reveal tumor size increases at an optimal rate of cell proliferation under nutrient-rich stress-free conditions. Simulations also provided a critical region for which a necrotic core forms at the tumor's center.

Several forces are required to balance conservation of momentum. For two-phase models, the viscosity of the phases and interstitial friction must be accounted, but for biofilm morphology, in addition to hydrostatic pressure, osmotic pressure is also needed. One such model [4] describes the role of a network comprised of an extra-cellular polymeric substance (EPS) in structural development in biofilm. Numerical simulations indicate as EPS is produced by bacteria, a rise in osmotic pressure contributes to the expansion of the biofilm region. Two-phase dynamics have also been used to simulate biofilm growth and cell motility [5], [6]. A mobile cell contains polymer network phase comprised of actin filaments, intermediate filaments, and microtubules. This phase is the exoskeleton to a cytoplasmic phase. The network contracts to propel the cell forward. Numerical simulations of these models have shown to contain traveling wave solutions. Another biological model describes to formation of channels in biofilm [7]. Steady-state analysis suggests that there is an optimal range for the pressure gradient to drive the formation of a channel between two flat plates.

When regions occupied by differing materials have free boundaries, numerical methods are useful to track the sharp interface. The location of the interface can be followed explicitly by interface tracking methods [8]. Alternatively, interface capturing can be used to implicitly solve the same equations throughout the domain by capturing the appropriate interface conditions [9].

One such interface capturing method given by Du et. al. [10] has analyzed the behavior of a free boundary problem of a two-phase viscous fluid mixture with a prevalent viscosity in a single phase. The solution found by Du et. al. is perturbation solution of the front velocity at time $t = 0$ for a vanishing solvent phase. This solution was built to explore how the velocity of the interface moves in a consistent manner to develop numerical methods to handle the free boundary problem. The velocity is then tracked numerically for various initial profiles with the interface capturing method developed by the group. In each instance, the numerical solution is compared to the asymptotic solution and found to be accurate.

In part, the purpose of this paper is to explore the accuracy of the perturbation solution given by [10] in comparison to an exact solution, which was found using symmetry analysis, also called Lie's classical method. In each model previously discussed, numerical, perturbation, and semi-analytic methods were used to provide insights into the behaviors of interest. And though these methods have had some successes in assessing two-phase models, few attempts have been made to attain generalized behavior of these systems with exact solutions.

Lie's method produces symmetry transformations which can reduce a system of Partial Differential Equations (PDEs) in one spacial dimension to a system of ODEs. These symmetries are generated by introducing infinitesimal transformations, which leave the original system invariant. For classical symmetry analysis, expansion of this infinitesimal transformation, produces a linear system of PDEs, called determining equations, whose solutions provide the forms for the symmetry transformations. Non-classical methods have also been developed which, in some cases, lead to additional symmetries. The infinitesimal transformations give rise to highly non-linear determining equations and can be difficult or impossible to solve. For this reason, the analysis in the paper only includes the classical method, as it recovers the solution given by [10] that we are seeking. Lie's classical method for producing symmetries has been successful in generating exact solutions for a system of PDEs describing viscous flow through expanding channels [11]. In this work conservation laws and point symmetries provide reductions, some of which lead to exact solutions of the flow in deformable channels. For elliptic, hyperbolic and mixed-type PDEs for Ricci flow, Wang [12] found several solutions, including traveling wave solutions, to hyperbolic geometric flow of Riemann surfaces. The work by Cimpoiasu et. al [13] used Lie symmetries to produce classes of solutions for the 2D nonlinear heat equation. It has also been shown that Lie symmetries generate the similarity solution for a class of (2+1) nonlinear wave equation [14].

In this paper, we generate an exact solution for the two-phase model using a point symmetry. In the first section, we outline a derivation of a two-phase system that represents the simplest version of the model and can be adapted for a variety of physical situations. Next, we briefly discuss how to develop symmetries and find a time translation, scaling symmetry, and a general Galilei time group. In the third section, we use a symmetry transformation to reduce the system of PDEs to an invariant system of ODEs. We make parameter assumptions similar to Du et. al. [10] to recover the exact velocity for their asymptotic solution and compare the exact to the perturbation solution. It is shown that the approximated free boundary solution is a close approximation to the general solution for $t = 0$. In the fourth section, we vary which physical driving forces dominate the two-phase model and generate additional exact solutions to the system. In the final section of this work, we discuss potential uses of exact solutions for the two-phase model and future directions of this work.

II. THE TWO-PHASE MODEL

In this section, we derive the equations to describe a two-phase model as seen in the kinetics of biological gels as described in [5], [10]. Gels swell and deswell due to ionic fluctuations and chemical triggers. An example of this occurs in crawling cells. Myosin converts chemical energy in the form of ATP into mechanical energy by causing actin filament to contract, propelling cells into motion. Neutrophils and macrophages, cells integral to the immune system of humans, respond in this manner. Chemical gradients are left by cells foreign to the immune system, leaving a chemotactic trail for the immunological cells to follow [15].

Like in [10], we assume the viscous terms are prominent forces and inertial terms are negligible. Gels are composed of a polymeric network given by ϕ_1 and a fluid solvent ϕ_2. Both phases are treated as Newtonian fluids that are immiscible. When considering the redistribution of mass within a control volume, the flux of the network is given by $\nabla \cdot (\phi_1 u_1)$, where the network moves with a velocity u_1. A similar argument is made for the solvent to give the following equations to conserve mass.

$$\frac{\partial}{\partial t}(\phi_1) + \frac{\partial}{\partial x}(u\phi_1) = 0, \tag{1}$$

$$\frac{\partial}{\partial t}(\phi_2) + \frac{\partial}{\partial x}(v\phi_2) = 0, \tag{2}$$

where the sum of the volumes saturate to a fixed control volume, $\phi_1 + \phi_2 = 1$.

Several forces act upon the network. The first is the force due to the network stress tensor σ_1, which includes the viscous stress tensor and mass production.

$$\sigma_1 = \hat{\mu}_1(\nabla u_1 + \nabla u_1^T) + \lambda_1 \nabla \cdot u_1,$$

where $\hat{\mu}_1$ is the shear viscosity and λ_1 is the bulk viscosity. In 1-D, this becomes

$$\sigma_1 = \mu_1 \frac{d}{dx} u_1, \tag{3}$$

where $\mu_1 = 2\hat{\mu}_1 + \lambda_1$.

Another force that we include is the frictional force created by interstitial interactions between phases. If both fluids move in unison or if either volume fraction becomes negligible, drag will vanish. With a frictional coefficient given by ξ, this drag force is given by $\xi\phi_1\phi_2(\phi_1 - \phi_2)$. Next, we need to account for both the hydrostatic pressure and osmotic pressure caused by swelling. If P is the total hydrostatic pressure, then the total pressure P acting on the network is given by $\phi_1\nabla P$.

Ionizing chemicals in the solvent can cause the gel to absorb or release the fluid solvent, causing an osmotic pressure gradient $\nabla\psi(\phi_1)$ acting on the network. For this reason, ϕ_1 is considered the active phase. For the form of the osmotic pressure term, we follow Cogan et. al. [5] and the references therein, and assume that $\psi(\phi_1) = k_2\phi_1^2(\phi_1 - \phi_0)$. The constant k_2 accounts for the effects of the ionic environment, polymeric structure, and solvent concentration that contribute to swelling and deswelling. The value of ϕ_0 is a reference volume fraction. This structure allows for osmotic pressure to vanish in the event of $\phi_1 = 0$ or at some reference fraction ϕ_0 that can be determined experimentally for various physical applications.

Assuming constant shear and bulk viscosity, the momentum of these moving fluids can be given by balancing the forces described above.

$$\mu_1\frac{\partial}{\partial x}\left(\phi_1\frac{\partial}{\partial x}u\right) + \phi_1\frac{\partial}{\partial x}P(\phi_1, \phi_2) \quad (4)$$
$$- \frac{\partial}{\partial x}\psi(\phi_1) - \xi\phi_1\phi_2(u - v) = 0$$

Similar arguments can be made to derive the forces of momentum within the solvent. The solvent is a Newtonian fluid with only viscous stresses acting on it. Fluid pressure acts on the solvent, but osmosis does not create pressure on the fluid itself. The fluid is actively absorbed and released by the gel. The final force is the drag or frictional force created by interstitial interactions. Combining these gives the momentum for the solvent.

$$\mu_2\frac{\partial}{\partial x}\left(\phi_2\frac{\partial}{\partial x}v\right) + \phi_2\frac{\partial}{\partial x}P(\phi_1, \phi_2) \quad (5)$$
$$+ \quad \xi\phi_1\phi_2(u - v) = 0,$$

where μ_2 is the viscosity of the solvent. Summing (4) and (5) gives the following equation.

$$\mu_1\frac{\partial}{\partial x}\left(\phi_1\frac{\partial}{\partial x}u\right) + \mu_2\frac{\partial}{\partial x}\left(\phi_2\frac{\partial}{\partial x}v\right)$$
$$+ \quad (\phi_1 + \phi_2)\frac{\partial}{\partial x}P(\phi_1, \phi_2) - \frac{\partial}{\partial x}\psi(\phi_1) = 0.$$

Since $\phi_1 + \phi_2 = 1$, this becomes

$$\mu_1\frac{\partial}{\partial x}\left(\phi_1\frac{\partial}{\partial x}u\right) + \mu_2\frac{\partial}{\partial x}\left(\phi_2\frac{\partial}{\partial x}v\right) \quad (6)$$
$$+ \quad P_x - \frac{\partial}{\partial x}\psi(\phi_1) = 0,$$

where $P_x = \frac{\partial}{\partial x}P(\phi_1, \phi_2)$. Solving for P_x gives

$$P_x = \frac{\partial}{\partial x}\psi(\phi_1) - \mu_1\frac{\partial}{\partial x}\left(\phi_1\frac{\partial}{\partial x}u\right) \quad (7)$$
$$- \quad \mu_2\frac{\partial}{\partial x}\left(\phi_2\frac{\partial}{\partial x}v\right).$$

Next, we substitute $\phi_2 = 1 - \phi_1$ in the equations of mass (1) and (2), and the momentum equation (4) to find the following system for analysis.

$$\frac{\partial}{\partial t}(\phi_1) + \frac{\partial}{\partial x}(u\phi_1) = 0, \quad (8)$$
$$-\frac{\partial}{\partial t}(\phi_1) + \frac{\partial}{\partial x}(v(1 - \phi_1)) = 0, \quad (9)$$
$$\mu_1\frac{\partial}{\partial x}\left(\phi_1\frac{\partial}{\partial x}u\right) - \frac{\partial}{\partial x}\psi(\phi_1) \quad (10)$$
$$+\phi_1 P_x - \xi\phi_1(1 - \phi_1)(u - v) = 0.$$

Together equations (8-10) can be reduced to a system of ODEs using the following transformation.

$$\begin{aligned}u &= f(t - \alpha x),\\ v &= g(t - \alpha x), \quad (11)\\ \phi_1 &= m(t - \alpha x),\end{aligned}$$

where f, g, and m are to be determined and α is an arbitrary constant describing wave speed. Traveling wave solutions have been shown to exist for the two phase system [6]. For this reason, if one were to guess an invariant transformation to reduce this system, the general traveling wave solution (11) may seem like an obvious first choice. But, this specific transformation came from a more general transformation found using symmetry analysis. Before producing the general transformation, a brief explanation of symmetry analysis is given in the following section.

III. SYMMETRY ANALYSIS

In this section, we give a brief explanation of the method for generating the invariant transformations that will be used to generate exact solutions in later sections. For systems of PDEs in 1-D, symmetry transformations reduce the PDEs to a system of ODEs. Derived by Sophus Lie [16], Symmetry Analysis is the mathematical method for finding transformations to a system of PDEs that leaves the set of equations invariant, or unchanged. More recently, there has been substantial literature regarding symmetry methods. For further details, we refer the reader to books by Hydon [17], Bluman and Kumei [18], and Olver [19].

The following coordinate change is called the infinitesimal transformations. These can be thought of as a local perturbation on the original coordinate system.

$$
\begin{aligned}
\bar{\phi}_1 &= \phi_1 + \Phi_1(t,x,u,v)\epsilon + O(\epsilon^2), \\
\bar{t} &= t + T(t,x,u,v)\epsilon + O(\epsilon^2), \\
\bar{x} &= x + X(t,x,u,v)\epsilon + O(\epsilon^2), \\
\bar{u} &= u + U(t,x,u,v)\epsilon + O(\epsilon^2), \quad (12) \\
\bar{v} &= v + V(t,x,u,v)\epsilon + O(\epsilon^2),
\end{aligned}
$$

where Φ_1, T, X, U, and V are called the infinitesimals. In general, one seeks to find invariance of a system of differential equations of the form

$$F_i(t,x,u,v,\phi_1,u_t,v_t,\phi_{1t},u_x,v_x,\phi_{1x},...) = 0, \quad (13)$$

with $i = 1, 2, \ldots, n$, where u, v, ϕ_1 are functions of t, x. In the specific case of our two-phase model, the system F_i is given by the equations (8-10). Under (12), a set of differential equations is produced for the infinitesimals T, X, U, and V. These differential equations are called the determining equations because they determine the form for the infinitesimals. Solving these determining equations produced by (12) provides invariant transformations for the differential equations given by (13).

The following is called the invariant surface condition, so called because it leaves the solution surface invariant under the change of coordinates.

$$
\begin{aligned}
Tu_t + Xu_x &= U, & (14) \\
Tv_t + Xv_x &= V, & (15) \\
T\phi_{1t} + X\phi_{1x} &= \Phi_1. & (16)
\end{aligned}
$$

When the infinitesimals are solved in conjunction with the invariant surface condition given by (14-16), the solutions u, v, and ϕ_1 provide a transformation which reduces the original PDE (13) to an ODE. In other words, by using Lie's method to find an infinitesimal change of coordinates, a two variable PDE can be reduced to an equation of a single variable to become an ordinary differential equation (ODE). Taking the physical nature of the problem into account, these reductions can lead to exact solutions to the PDE.

Applying the transformation given by (12) on (8-10) yields a large system of linear PDEs.

The determining equations are solved interactively to give the forms of the infinitesimals.

$$
\begin{aligned}
\Phi_1 &= 0, \\
T &= \alpha, \\
X &= \delta x + \Gamma(t), \quad (17) \\
U &= \delta u + \frac{d}{dt}\Gamma(t), \\
V &= \delta v + \frac{d}{dt}\Gamma(t).
\end{aligned}
$$

Due to the size of the equations, details of the determining equations are omitted. For more details on an example, see the details given in the Appendix (A).

In order for the PDEs given by the determining equations to be satisfied, two cases arise. Either $\delta = 0$ or $\delta \neq 0$. If $\delta \neq 0$, then the friction coefficient ξ given in the momentum equations vanishes. The transformation given by α is a time translation, δ is a scaling symmetry, and $\Gamma(t)$ is a general time dependant Galilei group, as used in fluid mechanics [20]. These symmetries can be used to find invariant reductions in the original system. Notice that for $\delta = 0$ and $\Gamma(t) = 1$ in (17) and solving for u, v, and ϕ_1 in (14-16) gives the transformation

$$u = \frac{1}{\alpha} + \hat{f}(t - \alpha x),$$

$$v = \frac{1}{\alpha} + \hat{g}(t - \alpha x),$$

$$\phi_1 = m(t - \alpha x),$$

Letting $\hat{f} = f(t - \alpha x) - \frac{1}{\alpha}$ and $\hat{g} = g(t - \alpha x) - \frac{1}{\alpha}$ gives the transformation (11).

It should be noted that for our purposes, we are only interested in pursuing a classical symmetry analysis to recover the solution presented by Du et. al. [10]. It is possible that more solutions will arise from other methods as well. Non-classical symmetries arise in many cases. In the work performed by Arrigo et. al. [21], a nonclassical symmetry is emitted by a class of Burgers' system. The Steinbergs symmetry method has provided exact solutions and reductions to the Calogero-Bogoyavlenskii-Schiff equation [22]. The Gardner method can generate an infinite hierarchy of symmetries, as was shown with the KdV equations, Camassa-Holm, and sine-Gordon equations [23]. Non-classical symmetries have also been generated for the fourth-order thin film equation using non-classical methods [24].

Further analysis could include any of these methods, as well as a classification of parameters which has the potential to produce more symmetries. The purpose of this work is not an exhaustive search for symmetries, but an introduction to using symmetry methods to recover a more general solution to the two-phase problem described above and partially recovered by Du et. al. [10].

IV. RECOVERING THE EXACT SOLUTION FOR A FREE BOUNDARY PROBLEM

As discussed in [10], since the viscosity of the solvent is of a much higher magnitude than that of the fluid, we assume the solvent viscosity μ_2 is zero. Since, $\phi_1 + \phi_2 = 1$, we have $\phi_2 = 1 - \phi_1$. Now, we replace ϕ_2 in the equations of momentum (4-5) and find

$$\mu_1 \frac{\partial}{\partial x} \left(\phi_1 \frac{\partial}{\partial x} u \right) + \phi_1 P_x - \frac{\partial}{\partial x} \psi(\phi_1) \quad (18)$$
$$-\xi \phi_1 (1 - \phi_1)(u - v) = 0,$$

$$(1 - \phi_1) P_x + \xi \phi_1 (1 - \phi_1)(u - v) = 0, \quad (19)$$

where u, v, and ϕ_1 are all functions of t, x as previously discussed and P_x is the pressure gradient. Next, we solve (19) for P_x to find

$$P_x = -\xi \phi_1 (u - v). \quad (20)$$

We see the mass equations (1-2) have now become

$$\frac{\partial}{\partial t} (\phi_1) + \frac{\partial}{\partial x} (u \phi_1) = 0,$$

$$-\frac{\partial}{\partial t} (\phi_1) + \frac{\partial}{\partial x} (v(1 - \phi_1)) = 0,$$

Summing these two equations of mass gives

$$\frac{\partial}{\partial x} (u \phi_1 + v(1 - \phi_1)) = 0.$$

Imposing the average velocity is zero, we have

$$u \phi_1 + v(1 - \phi_1) = 0,$$

which gives

$$v = -\frac{\phi_1}{1 - \phi_1} u. \quad (21)$$

To match the form of equations given by Du et. al. [10], we let the osmotic swelling term take the form $\psi(\phi_1) = \phi_1 \Psi(\phi_1)$. This, together with

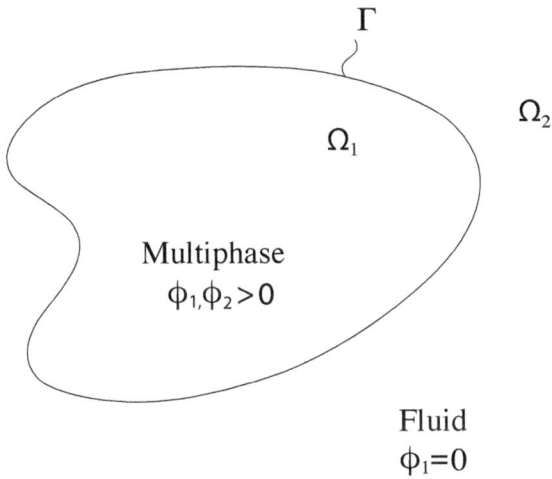

Fig. 1. This shows the region Ω_2 of pure solvent ($\phi_1 = 0$) separated at the boundary Γ from the region Ω_1 containing the mixture of both phases.

(20-21), reduces the equation (18) to the following equation.

$$(\mu_1\phi_1(u)_x)_x - (\phi_1\Psi(\phi_1))_x - \frac{\xi\phi_1}{1-\phi_1}u = 0. \quad (22)$$

As in Figure (1), we assume the mixture occupies the interior region (Ω_1), while pure solvent occupies the external region (Ω_2). As the gel-mixture swells/deswells, the interface between the regions (Γ) moves. To specify the motion Du et. al. impose standard jump conditions:

$$[\mu_1\phi_1(u)_x - \phi_1\Psi(\phi_1)] = 0$$
$$[u] = 0.$$

The solution found by Du et. al. [10] approximates the front velocity for the free boundary problem at time $t = 0$. The solution for a piecewise constant profile is given by,

$$\phi^1 = \begin{cases} \phi_- & \text{if } x < 0 \\ \phi_+ & \text{if } x > 0 \end{cases},$$

and the following can be derived

$$u = \begin{cases} Ce^{\beta_- x} & \text{if } x < 0 \\ Ce^{-\beta_+ x} & \text{if } x > 0 \end{cases}, \quad (23)$$

where

$$\beta_\pm = \sqrt{\frac{\xi}{\mu_1(1-\phi_\pm^1)}},$$

and

$$C = \frac{-\phi_+\Psi(\phi_+) + \phi_-\Psi(\phi_-)}{\mu_1(\phi_+\beta_+ + \phi_-\beta_-)}.$$

The solution (23) was derived by assuming $\phi_1+ \to 0$ at $t = 0$. In biological gels, regions of gel separate from regions of pure solvent. So, it is reasonable to assume that the network phase vanishes in this region of pure solvent. To make a graph of the solution given by (23), we assign the following initial profile.

$$\phi^1 = \begin{cases} \phi_- = \frac{1}{6} & \text{if } x < 0 \\ \phi_+ = 0 & \text{if } x > 0 \end{cases}. \quad (24)$$

The parameters used to generate the graphs are taken from [10], but are repeated in (I) for convenience. The graph *Figure (2)* represents the velocity front for a swelling gel in contact with a fluid solvent. This perturbation solution is an approximation for the velocity front at $t = 0$. However, there exists an exact solution to this system that captures this behavior for all values of ϕ_1 at any point in time.

For the infinitesimals given by (17), let $\delta = 0$ and $\Gamma(t) = 1$. Solving the invariant surface condition for u, v, and ϕ_1 will lead to (11) in terms of the variable $r = t - \alpha x$. As with the case found with solving for (23), we assume the viscosity of the second phase is negligible in comparison to that of the first phase, letting $\mu_2 = 0$. To make the analysis easier, we allow only for swelling in the active phase, making $\phi_0 = 0$. Applying (11) reduces (8-10) to a single ODE.

$$\mu_1\alpha^4(\alpha f - 1)^2 f'^2 - \mu_1\alpha^3(\alpha f - 1)^3 f'' \quad (25)$$
$$+3k_2\gamma^2\alpha^3 f' + \xi(\alpha f - 1)^4 = 0,$$

where

$$g = \frac{1}{\alpha}, \quad (26)$$

$$m = \frac{\gamma}{\alpha f - 1}, \quad (27)$$

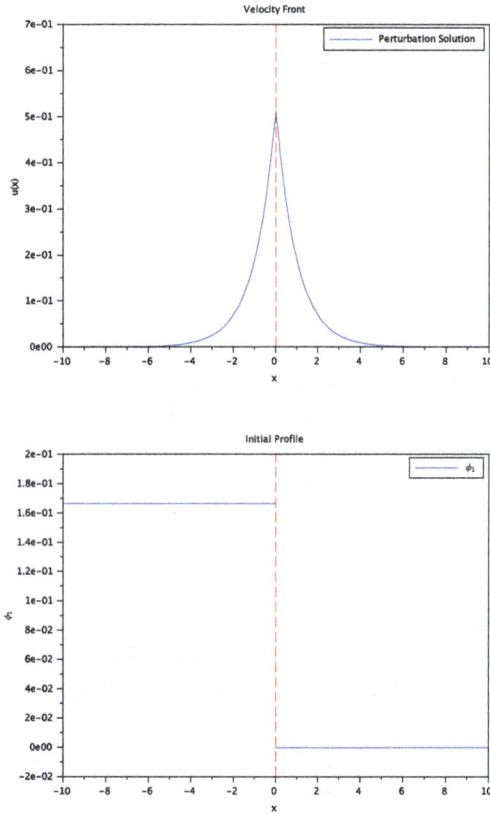

Fig. 2. This is the perturbation solution given by (23) at time $t = 0$. This shows the velocity for a region of vanishing network (top) in the region $x > 0$ for an initial profile (24) (bottom). This would represent expectations of a velocity front for a swelling gel to contact a fluid solvent.

It should be noted that we could have just as easily solved (8-10) for f and left m to be determined. We are choosing to leave f general to assess the behavior of the velocity, because we wish to show the exact solution approximated by Du et. al. [10] can be recovered.

Multiple solutions exist for (25). First, we attempt to recover an exponential solution similar in form to (23). If we assume the viscosity of the network phase ϕ_1 has a greater impact on the system than viscosity and interstitial friction, as was assumed by Du et al [10]. we can divide by μ_1. This gives the following equation from (25).

$$\alpha^4(\alpha f - 1)^2 f'^2 - \alpha^3(\alpha f - 1)^3 f'' \qquad (28)$$
$$+3\frac{k_2}{\mu_1}\gamma^2\alpha^3 f' + \frac{\xi}{\mu_1}(\alpha f - 1)^4 = 0.$$

For μ_1 of a much larger magnitude than k_2 and ξ, this becomes the following:

$$\alpha(\alpha f - 1)^2 f'^2 - (\alpha f - 1)^3 f'' = 0, \qquad (29)$$

whose solution is

$$f(r) = \frac{e^{\alpha(\kappa r + \lambda)}}{\alpha} + \frac{1}{\alpha}. \qquad (30)$$

This makes the analytic solution for the original system (1-2) and (4-5) to be

$$\phi_1 = \frac{\hat{\gamma}}{\alpha f - 1} = \gamma e^{-\alpha(\kappa(t - \alpha x) + \lambda)},$$
$$\phi_2 = 1 - \gamma e^{-\alpha(\kappa(t - \alpha x) + \lambda)},$$
$$u = \frac{e^{\alpha(\kappa(t - \alpha x) + \lambda)}}{\alpha} + \frac{1}{\alpha},$$
$$v = \frac{1}{\alpha},$$

with $\mu_1 = 0$, $k_2 = 0$, $\xi = 0$.

The parameters of this solution can be matched to the parameters of the solution given by (23). We can see that if $k = -\frac{\beta}{\alpha^2}$ and $\lambda = \frac{1}{\alpha}\ln(\alpha C - 1)$, then the solution found above becomes:

$$\phi_1 = \frac{\gamma}{C}e^{\frac{\beta}{\alpha}t - \beta x},$$
$$\phi_2 = 1 - \frac{\gamma}{C}e^{\frac{\beta}{\alpha}t - \beta x},$$
$$u = Ce^{-\frac{\beta}{\alpha}t + \beta x} + \frac{1}{\alpha}, \qquad (31)$$
$$v = \frac{1}{\alpha},$$

The parameter α remaining in the velocity of (31) gives flexibility on scaling time and adjusting the orientation of the velocity. Notice, as $\alpha \to \infty$, this solution is the same as (23). The velocity becomes identical, and the volume fraction becomes constant, as in the perturbation solution provided by (23). So, in essence, we have recovered the time function that was missed by the perturbation method used to find (23). Next, we match the numerical results of (23) for the parameters given by (I). To do this, we set $t = 0$ and separate the solution for the velocity as follows.

$$u = \begin{cases} Ce^{\beta x} + \frac{1}{\alpha}, & \text{if } x \leq 0 \\ Ce^{-\beta x} + \frac{1}{\alpha} & \text{if } x > 0 \end{cases}, \quad (32)$$

In *Figure (3)*, we can see the solutions given by (23) and (32) super-imposed on the same graph for parameter values given by (I). It is clear that the perturbation solution is a close approximation for the exact solution for large values of α. As we would expect from inspection of the solution (32), smaller values of α will adjust the exact solution away from the perturbation solution. The largest impact α has on the system is in regards to the time scale and solvent velocity. Large values of α require larger time steps for movement in the system, while decreasing the solvent velocity.

β	μ_1	ξ	α
1	0.0108	0.018	1000
10	0.0037	0.616	10000
100	0.000338	5.64	100000

TABLE I
THE PARAMETERS GIVEN IN THE ROW BEGINNING WITH $\beta = 1$ GENERATES THE RESULTS IN (3) TOP. THE NEXT ROW FOR $\beta = 10$ GIVES (3) MIDDLE WITH THE FINAL ROW GENERATING (3) BOTTOM WITH $\beta = 100$.

There are several benefits of finding the exact solution, instead of using numerical methods. First, numerical results have a difficult time capturing the behavior at the region of contact between the phases, while the analytical solution easily gives interface behaviour without computationally expensive coding, as can be seen in 4. Here we can see the region of network at $t = 0$ moving uniformly away from the initial contact region $x = 0$. Smaller values of β fail to capture the sharp interface. But as β increases to $\beta = 100$, we see the interface remains sharp as time increases. This is expected, as these results coincide with the numerical simulations found in [10] by a moving mesh.

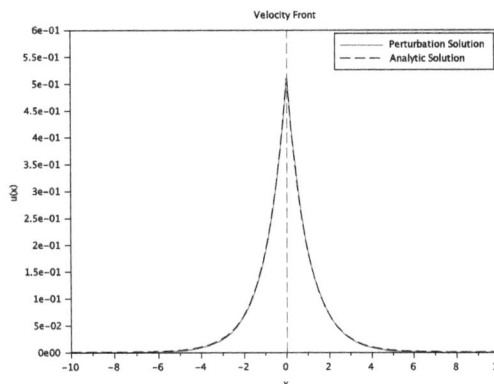

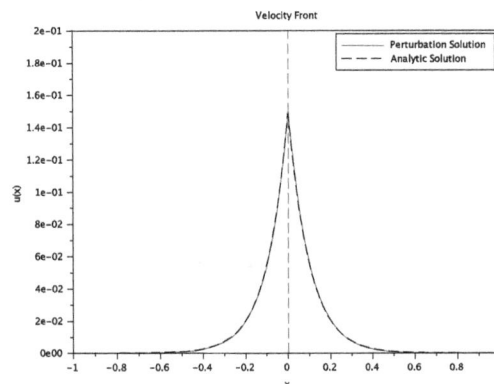

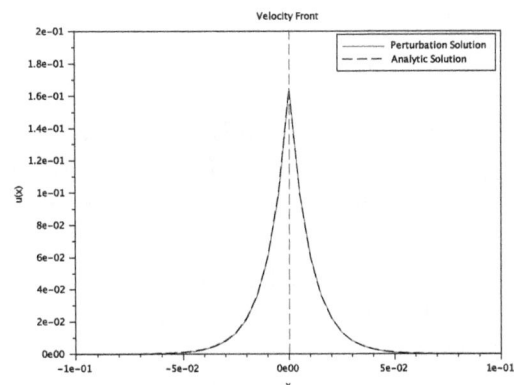

Fig. 3. These are the perturbation solutions given by (23) at time $t = 0$ graphed with the solution given by 32 with $\beta = 1$ and the corresponding values for μ_1 and ξ described in (I) given by the top, $\beta = 10$ middle, and $\beta = 100$ on bottom. The perturbation solutions are a close approximations for the exact solutions near the region of separation. We can see that the shape of each solution is preserved for each set of parameters, though the scale is modified.

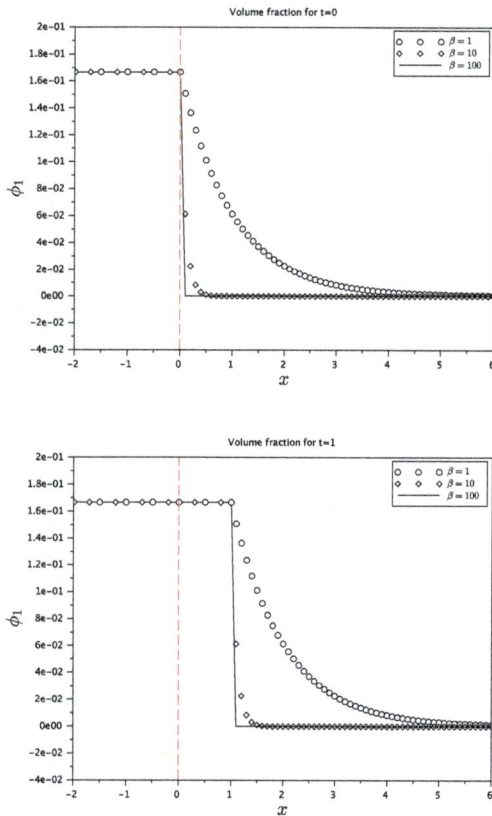

Fig. 4. For network (ϕ_1) profiles for $t = 0$ (top) and $t = 1000$ (bottom). As β increases, the exact solution becomes more accurate at capturing the expected behavior at the sharp interface.

As we have seen, the analytic solution recovers the perturbation solution as well as the numerical results given by Du et. al. [10]. However, this is just a single solution to the nonlinear equation given by (25). It is possible that the other solutions are extraneous, but more likely, additional solutions describe other physical or biological phenomenon yet to be determined. Further exploration will be required to fit these solutions, but we look at the others here.

A. Other Solutions to (25)

Being a non-linear system, the solution to (25) is not unique. Even though the transformation given by (11) will clearly give traveling wave

solutions, the structure of the traveling wave for each solution can vary widely as can be seen with the next two examples.

If the viscosity of the network is negligible $\mu_1 = 0$, the following solution to (25) is given by

$$f(r) = -\frac{k_2^{\frac{1}{3}} \gamma^{\frac{2}{3}}}{\xi^{\frac{1}{3}} \alpha^{\frac{1}{3}} (r + \delta)^{\frac{1}{3}}} + \frac{1}{\alpha}, \qquad (33)$$

where $r = t - \alpha x$.

The structure of this solution is different from (30) in several ways. When plotting at a single moment in time $t = 0$, it looks like a pulse as seen by the first curve in figure (5). When animated (30) can be seen as a traveling wave solution, given by the black curves which moves in the positive t direction.

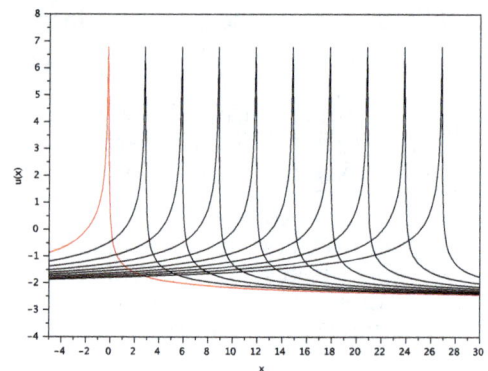

Fig. 5. The solution given by (33) plotted at $t = (0, 10)$. The first curve is at $t = 0$. As seen by the black curves, the velocity front travels like a wave as time increases.

Alternatively, if the osmotic pressure has less of an impact than viscosity and friction, then with $k_2 = 0$ as seen in the absence of ionizing agents for gels, we find the following solution

$$f(r) = \frac{e^{\alpha(-\kappa r + \lambda) + \frac{\xi}{2\mu_1} r^2}}{\alpha} + \frac{1}{\alpha}. \qquad (34)$$

Like (30) this solution is exponential, but as seen in (6) the quadratic term gives an unbounded traveling wave. The velocity at $t = 0$ is given by the first curve. As time increases, the front velocity travels as a wave moving to the right. This does not seem to have any physical analogue since the velocities are unbounded.

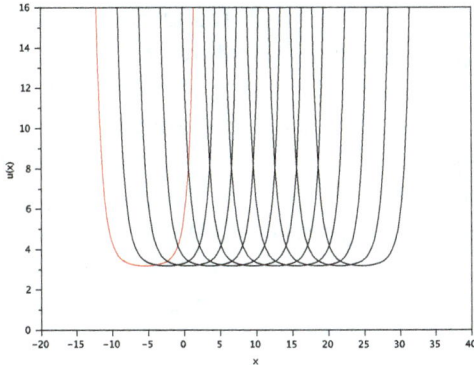

Fig. 6. The solution given by (34) plotted at $t = [0, 10]$. The first curve is at $t = 0$. The velocity front shifts to the right as time increases, moving as a traveling wave.

V. ADDITIONAL SOLUTIONS TO THE TWO-PHASE MODEL

This section also provides theoretical solutions, which may or may not have physical relevance. We explore them here to account for the multitude of solutions that are emitted by the system (8-10).

There are other two-phase models from physics that might have solutions contained here. For example, one such model describes granular flows where air is considered a non-viscous (nondense) phase with the rocks, debris, and other materials considered as a second highly viscous (dense) phase [25], [26]. Within these works, numerical simulations describe the flow behaviors air has on granular flow. The results suggest that drag has more than a negligible effect on the flow of granular materials of finite mass.

The traveling wave solutions provided by (11) are given by a simple choice for $\Gamma(t)$ in (17). Here, we explore different choices for the transformation and follow the reduction of the PDEs to ODEs. Then, we derive solutions to the ODEs by considering various changes in the physical nature of the problem. By adjusting which physical parameters

are the dominating driving force in the problem, we can generate different solutions, which may prove useful in exploring the nature of physical and biological phenomenon.

First, we let $\delta = 0$ and $\Gamma = \dfrac{1}{t}$ in (17). Solved with (14-16) will give the following transformation.

$$u = \frac{1}{\alpha t} + f\left(x - \frac{\ln(t)}{\alpha}\right),$$

$$v = \frac{1}{\alpha t} + g\left(x - \frac{\ln(t)}{\alpha}\right),$$

$$\phi_1 = m\left(x - \frac{\ln(t)}{\alpha}\right).$$

Neglecting solvent viscosity, $\mu_2 = 0$, reduces the original system (8-10) to the following ODE,

$$\mu_1 f^2(\alpha - f)f'^2 - \mu_1 f^3 f'' \tag{35}$$
$$-3k_2\alpha(2\phi_0 f - 3\alpha)(\alpha - f)f' - \xi f^5 = 0.$$

with

$$m = \frac{\alpha}{f},$$

$$g = \frac{\alpha f}{\alpha - f}.$$

Again, if we assume the dominating force is the viscosity and set k_2 and ξ_2 to zero, this can be solved to give

$$f = \kappa e^{\lambda\left(x - \frac{\ln(t)}{\alpha}\right)} = \frac{\kappa}{\sqrt[\alpha]{t^\lambda}}e^{\lambda x}. \tag{36}$$

The complete solution to (1-2) and (4-5) becomes

$$u = \frac{1}{\alpha t} + \frac{\kappa}{\sqrt[\alpha]{t^\lambda}}e^{\lambda x},$$

$$v = \frac{1}{\alpha t} + \frac{\alpha}{\alpha - \frac{\kappa}{\sqrt[\alpha]{t^\lambda}}e^{\lambda x}}\frac{\kappa}{\sqrt[\alpha]{t^\lambda}}e^{\lambda x},$$

$$\phi_1 = \frac{\kappa}{\sqrt[\alpha]{t^\lambda}}e^{\lambda x},$$

$$\phi_2 = 1 - \frac{\kappa}{\sqrt[\alpha]{t^\lambda}}e^{\lambda x}.$$

If we assume friction and pressure dominate and let $\mu_1 = 0$, the ODE yields no real solution without further assumptions on the constants of integration.

This may imply that viscosity is required for non-constant solutions.

Another choice for (17) is to let δ remain arbitrary and to consider $\xi = 0$, a requirement for invariance to be satisfied. Again we consider cases where the viscosity of the solvent is negligible, $\mu_2 = 0$. Choosing $\Gamma = 0$ we find the following transformation

$$
\begin{aligned}
u &= f(xe^{-\frac{\delta}{\alpha}t})e^{\frac{\delta}{\alpha}t}, \\
v &= g(xe^{-\frac{\delta}{\alpha}t})e^{\frac{\delta}{\alpha}t}, \\
\phi_1 &= m(xe^{-\frac{\delta}{\alpha}t}),
\end{aligned}
$$

which reduces (8-10) to the following three ODEs.

$$
\begin{aligned}
(\alpha f - r\delta)m' + \alpha m f' &= 0, \\
-(\alpha g - r\delta)m' + \alpha(1 - m)g' &= 0, \quad (37) \\
(1 - m)(\mu_1 f' - k_2 m(2\phi_0 - 3m))m' & \\
+ \mu_1 m(1 - m)f'' &= 0,
\end{aligned}
$$

with $r = xe^{-\frac{\delta}{\alpha}t}$. If both osmotic pressure and viscosity are negligible such that $k_2 = 0$ and $\mu_1 = 0$, then the following solution satisfies (37).

$$
\begin{aligned}
f &= \gamma e^{-\lambda r} + \delta \frac{\lambda r - 1}{\lambda \alpha}, \\
g &= \frac{(\lambda r - 1)\delta e^{\lambda r}}{\lambda \alpha(\delta e^{\lambda r} - 1)} + \frac{\kappa}{\lambda \alpha(\delta e^{\lambda r} - 1)}, \\
m &= \delta e^{\lambda r}.
\end{aligned}
$$

The complete solution to (1-2) and (4-5), is

$$
u = \gamma e^{-\lambda xe^{-\frac{\delta}{\alpha}t}} + \delta \frac{\lambda xe^{-\frac{\delta}{\alpha}t} - 1}{\lambda \alpha}e^{\frac{\delta}{\alpha}t},
$$

$$
\begin{aligned}
v &= \frac{(\lambda xe^{-\frac{\delta}{\alpha}t} - 1)\delta e^{\lambda xe^{-\frac{\delta}{\alpha}t}}}{\lambda \alpha(\delta e^{\lambda xe^{-\frac{\delta}{\alpha}t}} - 1)} \\
&\quad + \frac{\kappa}{\lambda \alpha(\delta e^{\lambda xe^{-\frac{\delta}{\alpha}t}} - 1)} \quad (38)
\end{aligned}
$$

$$
\phi_1 = \delta e^{\lambda xe^{-\frac{\delta}{\alpha}t}}, \quad (39)
$$

$$
\phi_2 = 1 - \phi_1.
$$

It should be noted that if either viscosity is the dominating force with $k_2 = 0$, or if osmotic pressure is the dominating force with $\mu_2 = 0$, then m, f, and g are constant. This implies that friction is required for non-constant solutions. This is different from before, where we found viscosity to be the driving force for the model.

In summary, we have found that each solution requires a dominating force to generate non-constant solutions. This gives flexibility in assessing the two-phase model and suggests that exact solutions may exist for many differing physical phenomenon of interest. For example, it is possible that the solution given by (39) can be matched to results consistent to granular flow, since friction as a necessary component for granular flow [25], [26].

VI. Discussion

In this work, we found an exact solution which accurately replicates the results from a previously found numerical results. It has been shown that for $\alpha \to \infty$, the analytic solution found here is exactly the perturbation solution found by Du et. al. [10]. The exact solution has the benefit of time dependence, which is useful for assessing behavior of the two-phase system without the implementation of numerical methods. Additionally, we showed that many traveling wave solutions arise from the two-phase problem. Due to the time dependent general Galilei group, we have an unlimited number of choices to adjust the speed of the wave through time. These solutions also require specific dominating forces to attain. It is possible that such solutions only arise in specific physical circumstances. Though some of these solutions may be extraneous, further investigation is warranted to determine their uses.

Although asymptotic and numerical methods yield useful information concerning the behavior of multi-phase systems, these methods require substantial efforts. Exact solutions have the benefit of being computationally inexpensive to simulate, and with Lie symmetries, are relatively simple to generate.

There are several directions for future analysis that arise from this work. First, exploring the behavior of the additional solutions may give further insights into the nature of dominating forces in the two phase system. This may give insight into specific physical phenomenon in which these additional solutions may be esoterically relevant. Also, additional symmetries may exist, which could be found using non-classical methods. Solutions arising from non-classical methods would then need to be assessed to determine relevant matching physical or biological behavior. Additionally, biofilms typically include growth terms to account for the production of new network. It is possible that symmetry solutions can capture this behavior as well.

APPENDIX

Deriving the infinitesimals for the two-phase model generates a large system of linear PDEs. For this reason, we have provided details of the process by way of an example in this appendix. For further details, see [27].

Consider the following nonlinear first order PDE

$$u_t = u_x^2 \tag{40}$$

Under the transformation

$$
\begin{aligned}
\bar{t} &= t + \epsilon T(t, x, u) + O(\epsilon^2), \\
\bar{x} &= x + \epsilon X(t, x, u) + O(\epsilon^2), \\
\bar{u} &= u + \epsilon U(t, x, u) + O(\epsilon^2),
\end{aligned}
$$

to order ϵ^2 (40) becomes

$$
\begin{aligned}
U_t &+ u_t U_u - u_t \left(T_t + u_t T_u \right) - u_x \left(X_t + u_t X_u \right) \\
&- 2u_x (U_x + u_x U_u - u_t \left(T_x + u_x T_u \right) \\
&- u_x \left(X_x + u_x X_u \right)) = 0.
\end{aligned}
$$

Using the original equation (40) to eliminate u_t and grouping coefficients of u_x, we have

$$
\begin{aligned}
U_t &+ u_x^2 U_u - u_x^2 \left(T_t + u_x^2 T_u \right) \\
&- u_x \left(X_t + u_x^2 X_u \right) - 2u_x (U_x + u_x U_u \\
&- u_x^2 \left(T_x + u_x T_u \right) - u_x \left(X_x + u_x X_u \right)) \\
&= U_t - \left(X_x + 2U_x \right) u_x + \left(2X_x - T_t - U_u \right) u_x^2 \\
&= \left(X_u + 2T_x \right) u_x^3 + T_u u_x^4 \\
&= 0.
\end{aligned}
$$

Invariance requires the coefficients of u_x to be zero, providing us with the following system.

$$
\begin{aligned}
U(t, x, u)_t &= 0, \\
X(t, x, u)_x + 2U(t, x, u)_x &= 0, \\
2X(t, x, u)_x - T(t, x, u)_t - U(t, x, u)_u &= 0, \\
X_u + 2T_x &= 0, \\
T_u &= 0.
\end{aligned}
$$

These are called the determining equations, because they determine the forms of the infinitesimals. These are linear PDEs, which are easily solved with standard techniques of integration. So, we have the following form for the infinitesimals.

$$
\begin{aligned}
T(t, x, u) &= c_1 + c_2 t + c_3 x + c_4 t^2 + c_5 tx + c_6 x^2, \\
X(t, x, u) &= c_7 + c_8 + c_9 x + c_4 tx + \frac{1}{2} k_5 x^2 \\
&\quad - \left(2k_3 + 2k_5 t - 4k_6 x \right) u, \\
U(t, x, u) &= k_{10} - \frac{1}{2} k_8 x - \frac{1}{4} k_4 x^2 \\
&\quad + \left(2k_9 - k_2 \right) u + k_5 xu - 4k_6 u^2,
\end{aligned}
$$

where c_i and k_i are arbitrary constants of integration. Together with the invariant surface condition given by $Tu_t + Xu_x = U$ we can find a transformation to reduce (40) to an ODE. The form for the transformation will vary depending on choices for the constants c_i and k_i.

REFERENCES

[1] C. J. Breward, H. M. Byrne, C. E. Lewis The role of cell-cell interactions in a two-phase model for avascular tumor growth J. Math. Biol, 45 (2002) 125-152

[2] S. R. Lubkin, T Jackson Multiphase mechanics of capsule formation in tumors J. Biomech. Eng.124 (2002) 237-243 http://dx.doi.org/10.1115/1.1427925

[3] H. M. Byrne, L. Preziosi Modelling solid tumor growth using the theory of mixtures Math. Med. Biol. 20 (2003) 341-366 http://dx.doi.org/10.1093/imammb/20.4.341

[4] N. G. Cogan, J. P. Keener The role of the biofilm matrix in structural development Math. Med. Biol. 21 (2005) 147166

[5] N. G. Cogan, R. D. Guy Multiphase flow models of biogels from crawling cells to bacterial biofilms HSFP Journal4 (2009) 11-25

[6]   J. M. Oliver, L. S. Kimpton, J. P. Whiteley, S. L. Waters, J. R. King Multiple travelling-wave solutions in a minimal model for cell motility *Math. Med. Biol.*30 (2013) 241272

[7]   N. G. Cogan, J. P. Keener Channel formation in gels *SIAM J. Appl. Math.* 65 (2005) 1839-1854 http://dx.doi.org/10.1137/040605515

[8]   R. D. Guy, A. L. Fogelson, J. P. Keener Fibrin Gel Formation in a Shear Flow *Math. Med. and Bio.* 24 (2007) 111-130

[9]   W. J. Boettinger, J. A. Warren, C. Beckermann, A. Karma Phase-field simulation of solidification *Annu. Rev. Mater. Res.* 32 (2002) 163-194 http://dx.doi.org/10.1146/annurev.matsci.32.101901.155803

[10] J. Du, R. D. Guy, A. L. Fogelson, G. B. Wright, J. P. Keener An interface-capturing regularization method for solving the equations for two-fluid mixtures *Commun. Comput. Phys.* 14 (2013) 1-25

[11] S. Asghar, M. Mushtaq, A. H. Kara Exact solutions using symmetry methods and conservation laws for the viscous flow through expanding-contracting channels *App. Math. Model.*32 (2008) 29362940 http://dx.doi.org/10.1016/j.apm.2007.10.006

[12] J. Wang Symmetries and solutions to geometrical flows *Sci China Math* 56 (2013) 16891704 http://dx.doi.org/10.1007/s11425-013-4635-8

[13] R. Cimpoiasu, R. Constantinescu Lie symmetries and invariants for a 2D nonlinear heat equation *Nonlinear Analysis* 68 (2007) 22612268 http://dx.doi.org/10.1016/j.na.2007.01.053

[14] M. Nadjafikhah, R. Bakhshandeh-Chamazkoti, A. Mahdipour-Shirayeh A symmetry classification for a class of (2+1)-nonlinear wave equation *Nonlinear Analysis* 71 (2009) 5164-5169 http://dx.doi.org/10.1016/j.na.2009.03.087

[15] S. Nagy, B. L. Ricca, M. F. Norstram, D. S. Courson, C. M. Brawley, P. A. Smithback , R. S. Rock A myosin motor that selects bundled actin for motility *Proc. Natl. Acad. Sci.*105 (2008) 96169620 http://dx.doi.org/10.1073/pnas.0802592105

[16] S. Lie Klassifikation und Integration von gewohnlichen Differentialgleichen zwischen x, y die eine Gruppe von Transformationen gestatten *Math. Ann.* 32 (1888) 213281http://dx.doi.org/10.1007/BF01444068

[17] [10.1017/CBO9780511623967] P. E. Hydon *Symmetry Methods for Differential Equations: A Beginner's Guide* 6th edition, New York : Cambridge University Press (2000)

[18] [10.1007/978-1-4757-4307-4] G. Bluman, S. Kumei *Symmetries and Differential Equations* Springer-Verlag, New York Inc. (1989)

[19] [10.1007/978-1-4684-0274-2] P. J. Olver *Applications of Lie Groups to Differential Equations* Springer-Verlag, New York Inc. (1993)

[20] T. Kambe Gauge principle and variational formulation for flows of an ideal fluid *Acta Mechanica Sinica*19 (2003) 437-452

[21] D. Arrigo, D. Ekrut, J. Fliss, L. Le Nonclassical symmetries of a class of Burgers systems *J. Math. Anal. Appl.* 371 (2010) 813-820 http://dx.doi.org/10.1016/j.jmaa.2010.06.026

[22] G. M. Moatimid, R. M. El-Shiekh, A. A. A. H. Al-Nowehy Exact solutions for CalogeroBogoyavlenskiiSchiff equation using symmetry method *Appl. Math. and Comput.*220 (2013) 455-462 http://dx.doi.org/10.1016/j.amc.2013.06.034

[23] M. K. Srivastava, Y. Wang, X. G. Zhang, D.M.C. Nicholson, H. P. Cheng The Gardner method for symmetries *Physical Review* (2013)

[24] K. Charalambous, C. Sophocleous Symmetry properties for a generalised thin film equation *J Eng Math* 46 (2013) 1-16

[25] L. T. Sheng, Y. C. Tai, C. Y. Kuo, S. S. Hsiau A two-phase model for dry density-varying granular flows *Advanced Powder Technology* 24 (2013) 132-142 http://dx.doi.org/10.1016/j.apt.2012.04.001

[26] S. P. Pudasaini A general two-phase debris flow model *J. Geophys. Res.* 117 (2011) 1-28

[27] D. Arrigo, Symmetry Analysis of Differential Equations: An Introduction, *Wiley* (2015) 73-84

18

Mathematical Analysis of a Size Structured Tree-Grass Competition Model for Savanna Ecosystems

Valaire Yatat[1], Yves Dumont[2], Jean Jules Tewa[1], Pierre Couteron[3] and Samuel Bowong[1]
[1]UMMISCO, LIRIMA project team GRIMCAPE, Yaounde, Cameroon
Email: yatatvalaire@yahoo.fr; tewajules@gmail.com; sbowong@gmail.com
[2]CIRAD, Umr AMAP, Montpellier, France
[3]IRD, Umr AMAP, Montpellier, France
Email: yves.dumont@cirad.fr; pierre.couteron@ird.fr

Abstract—**Several continuous-time tree-grass competition models have been developed to study conditions of long-lasting coexistence of trees and grass in savanna ecosystems according to environmental parameters such as climate or fire regime. In those models, fire intensity is a fixed parameter while the relationship between woody plant size and fire-sensitivity is not systematically considered. In this paper, we propose a mathematical model for the tree-grass interaction that takes into account both fire intensity and size-dependent sensitivity. The fire intensity is modeled by an increasing function of grass biomass and fire return time is a function of climate. We carry out a qualitative analysis that highlights ecological thresholds that summarize the dynamics of the system. Finally, we develop a non-standard numerical scheme and present some simulations to illustrate our analytical results.**

Keywords-**Asymmetric competition, Savanna, fire, continuous-time modelling, qualitative analysis, Non-standard numerical scheme.**

I. INTRODUCTION

Savannas are tropical ecosystems characterized by the durable co-occurrence of trees and grasses (Scholes 2003, Sankaran et al. 2005) that have been the focus of researches since many years. Savanna-like vegetations cover extensive areas, especially in Africa and understanding savannas history and dynamics is important both to understand the contribution of those areas to biosphere-climate interactions and to sustainably manage the natural resources provided by savanna ecosystems. At biome scale, vegetation cover is known to display complex interactions with climate that often feature delays and feed-backs. For instance any shift from savanna to forest vegetation not only means increase in vegetation biomass and carbon sequestration but also may translate into changes in the regional patterns of rainfall (Scheffer et al. 2003, Bond et al. 2005). In the face of the ongoing global change, it is therefore important to understand how climate along with local factors drive the dynamics of savannas ecosystems. In many temperate and humid tropical biomes, forest vegetation in known to recover quickly from disturbances and woody species are expected to take over herbaceous species. Yet in the dry tropics, it is well-known that grassy and woody species may coexist over decades although their relative proportion may show strong variations (Scholes 2003, Sankaran et al. 2005, 2008).

Savanna-like ecosystems are diverse and explanations found in the literature about the long-lasting coexistence of woody and grassy vegetation components therefore relate to diverse factors and processes depending on the location and the ecological context. Several studies have pointed towards the role of stable ecological factors in shaping the tree to grass ratio along large-scale gradients of rainfall or soil fertility (Sankaran et al. 2005, 2008). Other studies have rather emphasized the reaction of vegetation to recurrent disturbances such as herbivory or fire (Langevelde et al. 2003, D'Odorico et al. 2006, Sankaran et al. 2008, Smit et al. 2010, Favier et al. 2012 and references therein). Those two points of view are not mutually-exclusive since both environmental control and disturbances may co-occur in a given area, although their relative importance generally varies among ecosystems. Bond et al. (2003) proposed the name of climate-dependent for ecosystems that are highly dependent on climatic conditions (rainfall, soil moisture) and fire-dependent or herbivore-dependent for ecosystems which evolution are strongly dependent on fires or herbivores. In a synthesis gathering data from 854 sites across Africa, Sankaran et al. (2005) showed that the maximal observed woody cover appears as water-controlled in arid to semi-arid sites since it directly increase with mean annual precipitation (MAP) while it shows no obvious dependence on rainfall in wetter locations, say above c. 650 mm MAP where it is probably controlled by disturbance regimes. Above this threshold, fire, grazing and browsing are therefore required to prevent tree canopy closure and allow the coexistence of trees and grasses.

Several models using a system of ordinary differential equations (ODES) have been proposed to depict and understand the dynamics of woody and herbaceous components in savanna-like vegetation. A first attempt (Walker et al. 1981) was orientated towards semiarid savannas and analyzed the effect of herbivory and drought on the balance between woody and herbaceous biomass. This model refers to ecosystems immune to fire due to insufficient annual rainfall. Indeed, fires in savanna-like ecosystems mostly rely on herbaceous biomass that has dried up during the dry season. As long as rainfall is sufficient, fire can thus indirectly increase the inhibition of grass on tree establishment in a way far more pervasive than the direct competition between grass tufts and woody seedlings.

More recently, several attempts have been made (see Accatino et al. 2010, De Michele et al. 2011 and references therein) to model the dynamics of fire-prone savannas on the basis of the initial framework of Tilman (1994) that used coupled ODES to model the competitive interactions between two kinds of plants. On analogous grounds, Langevelde et al. (2003) have developed a model taking into account fires, browsers, grazers and Walter's (1971) hypothesis of niche separation by rooting zone depth. Models relying on stochastic differential equations have also been used (Baudena et al. 2010). Notably, Accatino et al. (2010) and De Michele et al. (2011) focused on the domain of stability of tree-grass coexistence with respect to influencing "biophysical" variables (climate, herbivory). However, fire was considered as a forcing factor independent of climate and vegetation, while woody cover was treated as a single variable with no distinction between seedling/saplings which are highly fire sensitive and mature trees which are largely immune to fire damages. The way in which the fundamental, indirect retroaction of grass onto tree dynamics is modeled is therefore to be questioned. In the present paper we therefore a model that differs in this respect.

Thus, to take into account the role of fire in

savanna dynamics, we consider a tree-grass compartmental model with one compartment for grass and two for trees, namely fire-sensitive individuals (like seedlings, saplings, shrubs) and non-sensitive mature trees. Based on field observations and experiments reported by Scholes and Archer (1997) and by Scholes (2003), we develop a system of three coupled non-linear ordinary differential equations (ODES), one equation per vegetation compartment that describes savanna dynamics. In addition, we model fire intensity (i.e. impact on sensitive woody plants) as an increasing function of grass biomass. Compared to existing models, our model aims to properly acknowledge two major phenomena, namely the fire-mediated negative feedback of grasses onto sensitive trees and the negative feed-back of grown-up, fire insensitive trees on grasses. We therefore explicitly model the asymmetric nature of tree-grass competitive interactions in fire-prone savannas.

After some theoretical results of the continuous fire model, though which we highlighted some ecological thresholds that summarize savanna dynamics and some interesting bistability, we present an appropriate non-standard numerical scheme (see Anguelov et al. 2012, 2013, 2014 and Dumont et al. 2010, 2012) for the model considered and we end with numerical simulations. We show that the fire frequency and the competition parameters are bifurcation parameters which allow the continuous fire model of asymmetric tree-grass competition to converge to different steady states.

II. THE CONTINUOUS FIRE MODEL OF ASYMMETRIC TREE-GRASS COMPETITION (COFAC)

As we have mentioned before, we consider the class of sensitive tree biomass (T_S), the class of non-sensitive tree biomass (T_{NS}) and the class of grass biomass (G). We model the fire intensity by an increasing function of grass biomass $w(G)$. To built up our model, we consider the following assumptions.

1) The grass vs. sensitive-tree competition has a negative feedback on sensitive tree dynamics.

2) The grass vs. non sensitive-tree competition has a negative feedback on grass dynamics.
3) After an average time expressed in years, the sensitive tree biomass becomes non sensitive to fire.
4) Fire only impacts grass and sensitive Tree.

We also consider the following parameters.

- There exists a carrying capacity K_T for tree biomass (in tons per hectare, $t.ha^{-1}$).
- There exists a carrying capacity K_G for grass biomass (in tons per hectare, $t.ha^{-1}$).
- Sensitive tree biomass is made up from non sensitive tree biomass with the rate γ_{NS} (in yr^{-1}) and from existing sensitive tree biomass with the rate γ_S (in yr^{-1}).
- Sensitive tree biomass has a natural death rate μ_S (in yr^{-1}).
- Non sensitive tree biomass has a natural death rate μ_{NS} (in yr^{-1}).
- f is the fire frequency (in yr^{-1}).
- Grass biomass has a natural death rate μ_G (in yr^{-1}).
- $\dfrac{1}{\omega_S}$ is the average time, expressed in year, that a sensitive tree takes to become non sensitive to fire.
- $\dfrac{1}{\omega_S + \mu_S}$ is the average time that a tree spends in the sensitive tree class without competition and fires.
- σ_G is the competition rate, for light or/and nutrients, between sensitive tree and grass (in $ha.t^{-1}.yr^{-1}$).
- σ_{NS} is the competition rate, for light or/and nutrients, between non sensitive tree and grass (in $ha.t^{-1}.yr^{-1}$).
- η_S is the proportion of sensitive tree biomass that is consumed by fire.
- η_G is the proportion of grass biomass that is consumed by fire.

Remark 1. *Competition parameters σ_G and σ_{NS} are asymmetric, indeed σ_G inhibits sensitive tree (T_S) growth and there is no reciprocal inhibition; likewise, σ_{NS} inhibits grass (G) growth.*

Based on these ecological premises, and taking

into account the effect of fire as a forcing continuous in time, which is the classical approach, we propose a model for the savanna vegetation dynamics through a system of three interrelate non-linear equations.

The COFAC is given by

$$
\begin{cases}
\dfrac{dT_S}{dt} &= (\gamma_S T_S + \gamma_{NS} T_{NS})\left(1 - \dfrac{T_S + T_{NS}}{K_T}\right) \\
&\quad - T_S(\mu_S + \omega_S + \sigma_G G + f\eta_S w(G)), \\[2mm]
\dfrac{dT_{NS}}{dt} &= \omega_S T_S - \mu_{NS} T_{NS}, \\[2mm]
\dfrac{dG}{dt} &= \gamma_G\left(1 - \dfrac{G}{K_G}\right)G - (\sigma_{NS} T_{NS} + f\eta_G + \mu_G)G,
\end{cases}
$$
$$(1)$$

with

$$
T_S(0) = T_{S_0} > 0, \; T_{NS}(0) = T_{NS_0} \geq 0 \;, \; G(0) = G0 > 0.
$$
$$(2)$$

For this continuous fire model, the fire intensity function w is chosen as a sigmoidal function of grass biomass because we want first to investigate the ecological consequences of the non-linear response of fire intensity to grass biomass, while nearly all published models using differential equations so far assumed a linear response. Non linearity is justified since whenever grass biomass is low fires are virtually absent while fire impact increases rapidly with grass biomass before reaching saturation. Thus,

$$
w(G) = \frac{G^2}{G^2 + g_0^2},
$$
$$(3)$$

where $G_0 = g_0^2$ is the value of grass biomass at which fire intensity reaches its half saturation (g_0 in tons per hectare, $t.ha^{-1}$).

The feasible region for system (1) is the set Ω defined by
$\Omega = \{(T_S, T_{NS}, G) \in \mathbf{R}_+^3 \mid 0 \leq T_S + T_{NS} \leq K_T, \; 0 \leq G \leq K_G\}$.

III. MATHEMATICAL ANALYSIS

A. *Existence of equilibria, ecological thresholds and stability analysis*

We set
$$
\mathcal{R}_1^0 = \frac{\gamma_S \mu_{NS} + \gamma_{NS}\omega_S}{\mu_{NS}(\mu_S + \omega_S)} \quad and \quad \mathcal{R}_2^0 = \frac{\gamma_G}{f\eta_G + \mu_G}.
$$

1) *Existence of equilibria:*

Setting the right hand-side of system (1) to zero, straightforward computations lead to the following proposition

Proposition 1. *System* (1) *has four kinds of equilibria*

- *The desert equilibrium point $E_0 = (0,0,0)$ which always exists.*
- *The forest equilibrium point $E_T = (\overline{T}_S; \overline{T}_{NS}; 0)$, with*

$$
\overline{T}_S = \frac{K_T \mu_{NS}}{\omega_S + \mu_{NS}}\left(1 - \frac{1}{\mathcal{R}_1^0}\right) \quad and
$$

$$
\overline{T}_{NS} = \frac{K_T \omega_S}{\omega_S + \mu_{NS}}\left(1 - \frac{1}{\mathcal{R}_1^0}\right)
$$

which is ecologically meaningful whenever $\mathcal{R}_1^0 > 1$.

- *The point $E_G = (0,0,\overline{G})$, with*

$$
\overline{G} = K_G\left(1 - \frac{1}{\mathcal{R}_2^0}\right),
$$

is ecologically meaningful when $\mathcal{R}_2^0 > 1$
The point E_G when it exists is the grassland equilibrium.

- *The savanna equilibrium point $E_{TG} = (T_S^*, T_{NS}^*, G^*)$, with T_S^*, T_{NS}^* and G^* given in Appendix A, has an ecological significance whenever*

$$
\mathcal{R}_1^0 > 1, \; \mathcal{R}_2^0 > 1 \; and \; 0 < G^* < K_G\left(1 - \frac{1}{\mathcal{R}_2^0}\right).
$$

Remark 2. *The number of savanna equilibria depends on the form of the function w.*

- *If $w(G) = G$, then the COFAC has at most one savanna equilibrium.*
- *If $w(G) = \dfrac{G}{G + G_0}$ (the Holling type II function), then the COFAC has at most two savanna equilibria.*

- If $w(G) = \dfrac{G^2}{G^2 + G_0}$ (the Holling type III function), then the COFAC has at most three savanna equilibria.

2) Ecological thresholds interpretation:

The qualitative behaviors of the COFAC depend on the following thresholds

$$\mathcal{R}_1^0, \; \mathcal{R}_2^0,$$
$$\mathcal{R}_1^{\overline{G}} = \frac{\gamma_S \mu_{NS} + \gamma_{NS} \omega_S}{\mu_{NS}(\mu_S + \omega_S + \sigma_G \overline{G} + f\eta_S w(\overline{G}))},$$

$$\mathcal{R}_2^{\overline{T}_{NS}} = \frac{\gamma_G}{f\eta_G + \mu_G + \sigma_{NS}\overline{T}_{NS}},$$

where

- $\mathcal{R}_1^0$ is the sum of the average amount of biomass produced by a sensitive/young plant, without fires and competition with grass, and the average amount of biomass produced by a mature plant multiplied by the proportion of young plants which reach the mature stage.

- $\mathcal{R}_1^{\overline{G}}$ is the sum of the average amount of biomass produced by a sensitive/young plant, in presence of fires and competition with grass, and the average amount of biomass produced by a mature plant multiplied by the proportion of young plants which reach the mature stage.

- $\mathcal{R}_2^0$ is the average amount of biomass produced per unit of grass biomass during its whole lifespan in presence of fires and and free from competition with non-sensitive trees.

- $\mathcal{R}_2^{\overline{T}_{NS}}$ is the average biomass produced per unit of grass biomass during its whole lifespan in presence of fires and experiencing competition from non-sensitive trees.

Remark 3. *The following relations hold*

$$\mathcal{R}_1^{\overline{G}} < \mathcal{R}_1^0, \; \mathcal{R}_2^{\overline{T}_{NS}} < \mathcal{R}_2^0.$$

3) Stability analysis:

Let

$$\mathcal{R} = \mathcal{R}(G^*) = \frac{\gamma_G (\mu_{NS} + \omega_S)(\gamma_S \mu_{NS} + \gamma_{NS}\omega_S)}{K_G K_T \mu_{NS}\omega_S \sigma_{NS}(\sigma_G + f\eta_S w'(G^\star))}.$$

We have the following result:

Theorem 1. *If $\mathcal{R}_1^0 < 1$ and $\mathcal{R}_2^0 < 1$, then the desert equilibrium E_0 is globally asymptotically stable.*

Proof: See Appendix B.

Theorem 2. *If $\mathcal{R}_1^0 > 1$, then the forest equilibrium E_T exists.*

- *If $\mathcal{R}_2^{\overline{T}_{NS}} < 1$, then the forest equilibrium E_T is locally asymptotically stable.*
- *If $\mathcal{R}_2^0 < 1$, then the forest equilibrium E_T is globally asymptotically stable.*
- *If $\mathcal{R}_2^0 > 1$, $\mathcal{R}_2^{\overline{T}_{NS}} < 1$, $\mathcal{R}_1^{\overline{G}} > 1$ and $\mathcal{R} < 1$, then the forest equilibrium E_T is globally asymptotically stable.*

Proof: See Appendix C.

Furthermore, using the same approach as in the proof of Theorem 2, we derive the following results

Theorem 3. *Suppose $\mathcal{R}_2^0 > 1$ so that the grassland equilibrium E_G exists.*

- *If $\mathcal{R}_1^{\overline{G}} < 1$, then the grassland equilibrium E_G is locally asymptotically stable.*
- *If $\mathcal{R}_1^0 < 1$, then the grassland equilibrium E_G is globally asymptotically stable.*
- *If $\mathcal{R}_1^0 > 1$, $\mathcal{R}_2^{\overline{T}_{NS}} > 1$, $\mathcal{R}_1^{\overline{G}} < 1$ and $\mathcal{R} < 1$, then the grassland equilibrium E_G is globally asymptotically stable.*

Theorem 4. *Suppose that $\mathcal{R}_1^0 > 1$, $\mathcal{R}_2^0 > 1$ and $\mathcal{R} > 1$. We have the following three cases:*

- *The savanna equilibrium E_{TG} is locally asymptotically stable (LAS) when it is unique.*
- *When there exists two savanna equilibria, one is LAS and the other is unstable.*
- *When there exists three savanna equilibria, two are LAS and one is unstable. Thus System (1) will converges to one of the two stable savanna equilibria depending on initial conditions.*

Proof: See Appendix D.

4) Summary table of the qualitative analysis:

The qualitative behavior of system (1) is summarized in the following Table in which we present only the most realistic, from an ecological point of view, case i.e $\mathcal{R}_1^0 > 1$ and $\mathcal{R}_2^0 > 1$.

TABLE I

SUMMARY TABLE OF THE QUALITATIVE ANALYSIS OF SYSTEM (1)

Thresholds					E_0	E_T	E_G	E_{TG}
$\mathcal{R}_1^0$	$\mathcal{R}_2^0$	$\mathcal{R}_1^{\bar{G}}$	$\mathcal{R}_2^{\bar{T}_{NS}} > 1$	$\mathcal{R} > 1$	U	U	U	L*
$>$	$>$	$>$	$\mathcal{R}_2^{\bar{T}_{NS}} < 1$	$\mathcal{R} > 1$	U	L	U	L
1	1	1		$\mathcal{R} < 1$	U	G	U	U
		$\mathcal{R}_1^{\bar{G}}$	$\mathcal{R}_2^{\bar{T}_{NS}} > 1$	$\mathcal{R} > 1$	U	U	L	L
		$<$		$\mathcal{R} < 1$	U	U	G	U
		1	$\mathcal{R}_2^{\bar{T}_{NS}} < 1$	$\mathcal{R} > 1$	U	L	L	L
				$\mathcal{R} < 1$	U	L	L	U

In Table I, the notations U, L and G stand for unstable, locally asymptotically stable, globally asymptotically stable, respectively, while the notation $L^\star$ means that we have the global stability if there are no periodic solutions.

Remark 4. *From an ecological point of view, Lignes 1 to 7 of Table I are interesting because in these cases, one ton of grass biomass will produce during it lifespan at least one ton of grass biomass ($\mathcal{R}_2^0 > 1$) and simultaneously, one ton of tree biomass (sensitive and non sensitive) will produce during it lifespan at least one ton of tree biomass ($\mathcal{R}_1^0 > 1$). Moreover, it is also in these cases that we have the most interesting situations of savanna dynamics, namely **bistability** cases (Lines 2, 4, 7 in Table 1) and a **tristability** case (Line 6 in Table 1).*

IV. NUMERICAL SIMULATIONS

Compartmental models are usually solved using standard numerical methods, for example, Euler or Runge Kutta methods included in software package such as Scilab [18] and Matlab [19]. Unfortunately, these methods can sometimes present spurious behaviors which are not in adequacy with the continuous system properties that they aim to approximate i.e, lead to negative solutions, exhibit numerical instabilities, or even converge to the wrong equilibrium for certain values of the time discretization or the model parameters (see Anguelov et al. 2012, Dumont et al. 2010 for further investigations). For instance, we provided in Appendix E some numerical simulations done with Runge Kutta schemes to illustrate some of its spurious behaviors. In this section, following Anguelov et al. 2012, 2013, 2014 and Dumont et al. 2010, 2012, we perform numerical simulations using an implicit nonstandard algorithm to illustrate and validate analytical results obtained in the previous sections.

A. A nonstandard scheme for the COFAC

System (1) is discretized as follows:

$$\frac{T_{NS}^{k+1} - T_{NS}^k}{\phi(h)} = \omega_S T_S^{k+1} - \mu_{NS} T_{NS}^{k+1},$$

$$\frac{G^{k+1} - G^k}{\phi(h)} = \gamma_G \left(1 - \frac{G^k}{K_G}\right) G^{k+1} - \sigma_{NS} T_{NS}^k G^{k+1}$$
$$- (\mu_G + f \eta_G) G^{k+1},$$

$$\frac{T_S^{k+1} - T_S^k}{\phi(h)} = (\gamma_S - (\mu_S + \omega_S)) T_S^{k+1} + \gamma_{NS} T_{NS}^{k+1}$$

$$- \frac{\gamma_S}{K_T} T_S^{k+1}(T_S^k + T_{NS}^k) - \frac{\gamma_{NS}}{K_T} T_{NS}^k T_{NS}^{k+1}$$

$$- \left(\frac{\gamma_{NS}}{K_T} T_{NS}^k + (\sigma_G G^k + f \eta_S w(G^k))\right) T_S^{k+1},$$
$$\tag{4}$$

where the denominator function ϕ is such that $\phi(h) = h + O(h^2)$, $\forall h > 0$. Systems (1) $-$ (2) can be written in the following matrix form:

$$\begin{cases} \dfrac{dX}{dt} &= \mathcal{A}(X)X, \\ X(0) &= X_0, \end{cases} \tag{5}$$

where $X = (T_{NS}, G, T_S) \in \mathbb{R}_+^3$ and $\mathcal{A}(X) = (\mathcal{A}_{ij})_{1 \le i, j \le 3}$ with $\mathcal{A}_{11} = -\mu_{NS}$, $\mathcal{A}_{12} = 0$, $\mathcal{A}_{13} = \omega_S$, $\mathcal{A}_{21} = 0$, $\mathcal{A}_{22}(X) = \gamma_G \left(1 - \frac{G}{K_G}\right) - (\sigma_{NS} T_{NS} + f\eta_G + \mu_G)$, $\mathcal{A}_{23} = 0$, $\mathcal{A}_{31}(X) = \gamma_{NS} \left(1 - \frac{T_S + T_{NS}}{K_T}\right)$, $\mathcal{A}_{32} = 0$, $\mathcal{A}_{33}(X) = \gamma_S \left(1 - \frac{T_S + T_{NS}}{K_T}\right) - (\mu_S + \omega_S + \sigma_G G + f\eta_S w(G))$.

Using (5), the numerical scheme (4) can be rewritten as follows:

$$\mathcal{B}(X^k) X^{k+1} = X^k,$$

where

$$\mathcal{B}(X^k) = (Id_3 - \phi(h)\mathcal{A}(X^k)). \qquad (6)$$

Thus $\mathcal{B}(X^k) =$

$$\begin{pmatrix} 1 + \phi(h)\mu_{NS} & 0 & -\omega_S \phi(h) \\ 0 & 1 - \phi(h)\mathcal{A}_{22}^k & 0 \\ -\phi(h)\mathcal{A}_{31}^k & 0 & 1 - \phi(h)\mathcal{A}_{33}^k \end{pmatrix}$$

It suffices now to choose $\phi(h)$ such that the matrix $\mathcal{B}(X^k))$ is an M-matrix for all $h > 0$, which implies that $\mathcal{B}^{-1}(X^k)$ is a nonnegative matrix, for all $h > 0$. In particular, choosing ϕ such that

$$\begin{aligned} 1 - \phi(h)(\gamma_G - (\mu_G + f\eta_G)) &\ge 0 \\ 1 - \phi(h)(\gamma_S - (\mu_S + \omega_S)) &\ge 0, \end{aligned} \qquad (7)$$

lead to positive diagonal terms and nonpositive off diagonal terms. We need to show that $\mathcal{B}(X^k)$ is invertible. Obviously $1 - \phi(h)\mathcal{A}_{22}^k$ is a positive eigenvalue. Let us define N, a submatrix of matrix $\mathcal{B}(X^k)$, as follows

$$N = \begin{pmatrix} 1 + \phi(h)\mu_{NS} & -\omega_S \phi(h) \\ -\phi(h)\mathcal{A}_{31}^k & 1 - \phi(h)\mathcal{A}_{33}^k \end{pmatrix}.$$

We already have $trace(N) > 0$. Then, a direct computation shows that $det(N) > 0$ if $\phi(h)$ is choosen such that

$$1 - \phi(h)\left(\gamma_S + \frac{\gamma_{NS}\omega_S}{\mu_{NS}} - (\mu_S + \omega_S)\right) \ge 0.$$

Thus we have $\alpha(N) > 0$, i.e. the eigenvalues have positive real parts, which implies that $\mathcal{B}(X^k)$ is invertible. Finally, choosing

$$\phi(h) = \frac{1 - e^{-Qh}}{Q}, \qquad (8)$$

with

$$Q \ge \max\left(\gamma_G - (\mu_G + f\eta_G), \gamma_S - (\mu_S + \omega_S) + \frac{\gamma_{NS}\omega_S}{\mu_{NS}}\right), \qquad (9)$$

matrix $\mathcal{B}(X^k)$ is an M-matrix. Furthermore, assuming $X^k \ge 0$, we deduce

$$X^{k+1} = \mathcal{B}^{-1}(X^k)X^k \ge 0.$$

Lemma 1. *Using the expression of ϕ defined in (8), the numerical scheme (4) is positively stable (i.e for $X^k \ge 0$, we obtain $X^{k+1} \ge 0$).*

An equilibrium X_e of the continuous model (1) verifies $\mathcal{A}(X_e)X_e = 0$. Multiplying the above expression by $\phi(h)$ and summing with X_e yields

$$(Id_3 - \phi(h)\mathcal{A}(X_e))X_e = X_e,$$

Thus, we deduce that the numerical scheme (4) and the continuous model (1) have the same equilibria which are (assumed to be) hyperbolic.

The dynamics of model (1) can be captured by any number Q satisfying

$$Q \ge \max\left\{\frac{|\lambda|^2}{2|Re(\lambda)|}\right\}, \qquad (10)$$

where $\lambda \in sp(J)$ with $J_{ij} = \frac{\partial A_i}{\partial X_j}$. We also have the following result:

Lemma 2. *If $\phi(h)$ is chosen as in Eqs. (8), (9) and (10), then the numerical scheme (4) is elementary stable (i.e local stability properties of equilibria are preserved).*

The proof of Lemma 2 follows the proof of Theorem 2 in Dumont et al., 2010.

B. NUMERICAL SIMULATIONS AND BIFURCATION PARAMETERS

In literature we found the following parameters values

TABLE II
PARAMETERS VALUES FOUND IN LITERATURE

Parameters	values	References
f	0-1	Langevelde et al. 2003
	0-2	Accatino et al. 2010
γ_G	$0.4^{(1)} - 4.6^{(2)}$	[1] Penning de Vries 1982
		[2] Menaut et al. 1979
$\gamma_S + \gamma_{NS}$	0.456-7.2	Breman et al. 1995
$\mu_S + \mu_{NS}$	0.03-0.3	Accatino et al. 2010
	0.4	Langevelde et al. 2003
μ_G	0.9	Langevelde et al. 2003
η_G	$0.1^{(a)}$-$1^{(b)}$	[a] Van de Vijver 1999
		[b] Accatino et al. 2010
η_S	0.02-0.6	Accatino et al. 2010
ω_S	0.05-0.2	Walkeling et al. 2011

We now provide some numerical simulations to illustrate the theoretical results and for discussions.

1) Some monostability and bistability situations:

- Monostability.
  We choose

γ_S	γ_{NS}	γ_G	η_S	η_G
1	2	3.1	0.5	0.5
μ_S	μ_{NS}	μ_G	σ_{NS}	σ_G
0.1	0.3	0.3	0.3	0.05
G_0	ω_S	K_T	K_G	f
2	0.05	50	12	0.5 yr^{-1}
$\mathcal{R}_1^0$	$\mathcal{R}_2^0$	$\mathcal{R}_1^{\overline{G}}$	$\mathcal{R}_2^{\overline{T}NS}$	$\mathcal{R}$
8.8889	5.6364	1.5006	1.2644	2.9424

With the chosen parameters, the savanna equilibrium is stable, i.e. sensitive trees, non-sensitive trees and grasses coexist. Figure 1 presents the 3D plot of the trajectories of system (1). It illustrates that the savanna equilibrium point is stable. Figure 1 also illustrates the monostability situation presented in Ligne 1 of Table I.

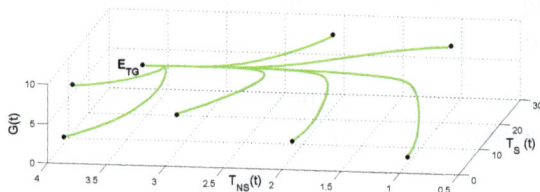

Fig. 1. 3D plot of the trajectories of system (1) showing that the savanna equilibrium point E_{TG} point is stable. The red bullets represent different initial conditions.

- Bistability

– Bistability involving forest and grassland equilibria. The state trajectories of the model will converge to a state depending of initial quantity.
  We choose

γ_S	γ_{NS}	γ_G	μ_S	μ_{NS}	μ_G	f
0.4	2	2.1	0.1	0.3	0.3	0.5 yr^{-1}
$\eta_S =0.5$,		$\eta_G = 0.5$,		$K_T =50$,		$K_G = 12$
$\mathcal{R}_1^0$	$\mathcal{R}_2^0$	$\mathcal{R}_1^{\overline{G}}$	$\mathcal{R}_2^{\overline{T}NS}$	$\mathcal{R}$	σ_G	σ_{NS}
4.8889	3.8182	0.5731	0.9315	0.2958	0.1	0.3

The 3D plot of the trajectories of system (1) is depicted in Figure 2. It clearly appears that the forest and grassland equilibria are stables. Figure 2 illustrates the bistability situation presented in Ligne 7 of Table I.

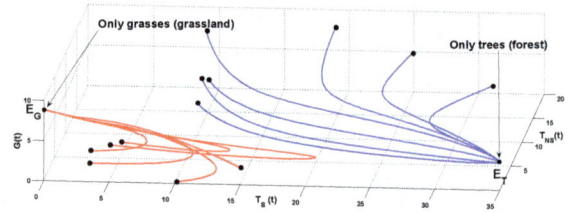

Fig. 2. 3D plot of the trajectories of system (1) showing that the forest (E_T) and grassland (E_G) equilibria are stable. The green bullets represent different initial conditions.

– Bistability involving forest and savanna equilibria. The state trajectories of the model will converge to a state depending on initial quantity.
  We choose

γ_S	γ_{NS}	γ_G	μ_S	μ_{NS}	μ_G	f
0.6	2	2.1	0.1	0.3	0.3	0.5 yr^{-1}
$\eta_S =0.5$,		$\eta_G = 0.5$,		$K_T =50$,		$K_G = 12$
$\mathcal{R}_1^0$	$\mathcal{R}_2^0$	$\mathcal{R}_1^{\overline{G}}$	$\mathcal{R}_2^{\overline{T}NS}$	$\mathcal{R}$	σ_G	σ_{NS}
6.2222	3.8182	1.1156	0.8942	1.2985	0.05	0.3

For these parameters, there exist two savanna equilibria but only one is stable as shown in Figure 3. Figure 3 also illustrates the bistability situation presented in Ligne 2 of Table I.

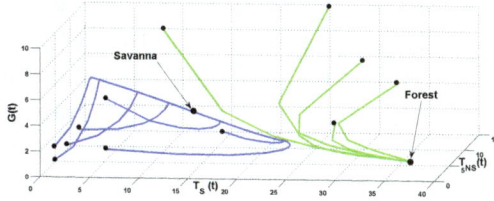

Fig. 3.   3D plot of the trajectories of system (1) showing that the forest (E_T) and savanna (E_{TG}) equilibria are stable. The green and red bullets represent different initial conditions.

Remark 5. *Note also that we didn't observed periodic behaviors in the previous simulations, considering the set of parameters presented in Table 2, while their existence cannot be completely ruled out by the analytical analysis.*

C. Some bifurcation parameters

In this section we emphasize on some bifurcation parameters of system (1) which are such that the COFAC can converge to different steady state depending on the variation of these parameters.

- The grass vs. sensitive-tree competition parameter σ_G is a bifurcation parameter. Figure 4 presents how the system (1) changes from the savanna state to the grassland state as a function of the grass vs. sensitive-tree competition parameter σ_G.
  We choose

γ_S	γ_{NS}	γ_G	μ_S	μ_{NS}	μ_G	η_S	σ_{NS}	f
0.4	1	4	0.1	0.3	0.1	0.5	0.3	0.2 yr^{-1}
$\eta_G = 0.5$, K_T=45, $K_G = 10$								

For these parameters values, system (1) undergoes a transcritical bifurcation. Indeed, we move from ligne 1 to ligne 5 of Table I. From left to right, ($\mathcal{R}_1^0 = 3.7778$, $\mathcal{R}_2^0 = 20$, $\mathcal{R}_1^{\overline{G}} = 2.2865$, $\mathcal{R}_2^{\overline{T}_{NS}} = 2.4721$, $\mathcal{R} = 99.3058$) $\rightarrow$ ($\mathcal{R}_1^0 = 3.7778$, $\mathcal{R}_2^0 = 20$, $\mathcal{R}_1^{\overline{G}} = 1.2943$, $\mathcal{R}_2^{\overline{T}_{NS}} = 2.4721$, $\mathcal{R} = 5.6803$) $\rightarrow$ ($\mathcal{R}_1^0 = 3.7778$, $\mathcal{R}_2^0 = 20$, $\mathcal{R}_1^{\overline{G}} = 0.9026$, $\mathcal{R}_2^{\overline{T}_{NS}} = 2.4721$). For the last case, the savanna equilibrium E_{TG} is undefined.

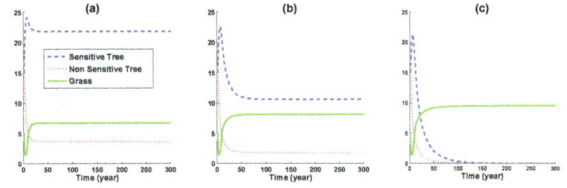

Fig. 4.   From savanna to grassland as a function of the grass vs. sensitive-tree competition parameter σ_G. From left to right, the fire period $\tau = \frac{1}{f}$ is fixed, while the grass vs. sensitive-tree competition parameter σ_G increases. In (a) ($\tau = 5$, $\sigma_G = 0$), in (b) ($\tau = 5$, $\sigma_G = 0.02$) and in (c) ($\tau = 5$, $\sigma_G = 0.04$)

- The fire period parameter $\tau = \frac{1}{f}$ is a bifurcation parameter. Figure 5 presents a shift of the convergence of system (1) from the forest state to the grassland state as a function of the fire period τ.
  We choose

TABLE III
PARAMETERS VALUES FOR FIGURES 5 AND 6

γ_S	γ_{NS}	γ_G	μ_S	μ_{NS}	μ_G	η_S	σ_{NS}	f
0.4	2	2.1	0.1	0.3	0.3	0.5	0.3	yr^{-1}
$\eta_G = 0.5$, K_T=50, $K_G = 12$								

For these parameters values, system (1) undergoes a forward bifurcation. Indeed, we move from ligne 3 to ligne 7 of Table I. From left to right, ($\mathcal{R}_1^0 = 4.8889$, $\mathcal{R}_2^0 = 1.0678$, $\mathcal{R}_1^{\overline{G}} = 1.3025$, $\mathcal{R}_2^{\overline{T}_{NS}} = 0.5720$) $\rightarrow$ ($\mathcal{R}_1^0 = 4.8889$, $\mathcal{R}_2^0 = 1.3548$, $\mathcal{R}_1^{\overline{G}} = 0.5446$, $\mathcal{R}_2^{\overline{T}_{NS}} = 0.6453$, $\mathcal{R} = 0.0983$) $\rightarrow$ ($\mathcal{R}_1^0 = 4.8889$, $\mathcal{R}_2^0 = 1.6154$, $\mathcal{R}_1^{\overline{G}} = 0.5679$, $\mathcal{R}_2^{\overline{T}_{NS}} = 0.6989$, $\mathcal{R} = 0.1201$). For the first case, the savanna equilibrium E_{TG} is undefined.

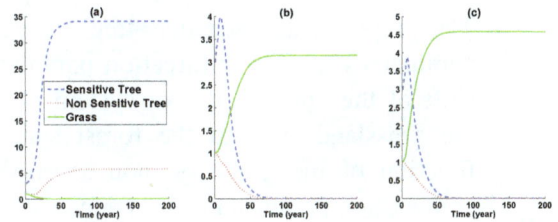

Fig. 5.   From forest to grassland as a function of the fire period τ. From left to right, the fire period τ increased, while the sensitive tree-grass competition parameter σ_G is fixed. In (a) ($\tau = 0.3$, $\sigma_G = 0.05$), in (b) ($\tau = 0.4$, $\sigma_G = 0.05$) and in (c) ($\tau = 0.5$, $\sigma_G = 0.05$)

Suppose now that fire period is fixed and the grass vs. sensitive-tree competition parameter σ_G varies. Figure 6 illustrates a shift of the convergence of system (1) from the forest state to the grassland state through a savanna state as a function of the grass vs. sensitive-tree competition parameter σ_G.

For the parameters values in Table III, system (1) exhibits to bifurcation phenomena: a pitchfork bifurcation and a transcritical bifurcation. We move from ligne 3 (figure 6 (a), -(b)) to ligne 7 (figure 6 (e), -(f)) through ligne 2 (figure 6 (c), -(d)) of Table I. Indeed, in figure 6 (a), -(b) E_{TG} is undefined, E_G is unstable, E_T is stable. In in figure 6 (c), -(d), E_G remains unstable but we have bistability between E_{TG} and E_T: it is a case of pitchfork bifurcation. In figure 6 (d), -(f), E_{TG} becomes unstable and we have a bistability between E_T and E_G: it is a case of transcritical bifurcation. Values of $\mathcal{R}_1^0$, $\mathcal{R}_2^0$, $\mathcal{R}_1^{\overline{G}}$, $\mathcal{R}_2^{\overline{T}_{NS}}$ and $\mathcal{R}$ are given in Appendix F.

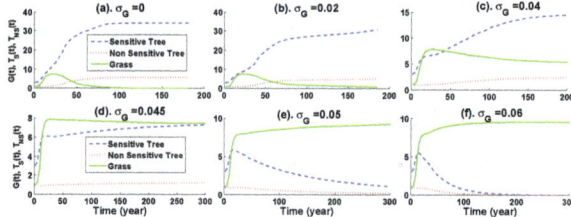

Fig. 6. From forest to grassland, with a transition through a savanna state, as a function of the sensitive tree-grass competition parameter. From left to right, the fire period τ is fixed at 4, while the grass vs. sensitive-tree competition parameter σ_G increased.

- The grass vs. non sensitive-tree competition parameter σ_{NS} is a bifurcation parameter. A shift of the convergence of system (1) from the grassland state to the forest state as a function of the grass vs. non sensitive-tree competition parameter σ_{NS} is depicted in Fig. 7.

We choose

γ_S	γ_{NS}	γ_G	μ_S	μ_{NS}	μ_G	σ_G	η_S	f
0.4	1	4	0.1	0.3	0.1	0.05	0.5	yr^{-1}
$\eta_G = 0.5$,		$K_T = 45$,		$K_G = 10$				

For these parameters values, system (1) exhibits a pitchfork bifurcation. We move from ligne 5 (figure 7 (a), -(b)) to ligne 7 (figure 7 (c), -(d)) of Table I. Indeed, in figure 7 (a), -(b) E_{TG} is undefined, E_T is unstable, E_G is stable. In figure 7 (c), -(d), E_{TG} exits but it is unstable and we have bistability between E_G and E_T: it is a case of pitchfork bifurcation. Values of $\mathcal{R}_1^0$, $\mathcal{R}_2^0$, $\mathcal{R}_1^{\overline{G}}$, $\mathcal{R}_2^{\overline{T}_{NS}}$ and $\mathcal{R}$ are given in Appendix G.

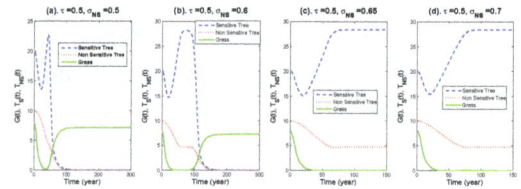

Fig. 7. From grassland to forest as a function of the grass vs. non sensitive-tree competition parameter σ_{NS}. From left to right, the fire period τ is fixed, while the grass vs. non sensitive-tree competition parameter σ_{NS} increases.

V. CONCLUSION AND DISCUSSION

In this work, we present and analyze a new mathematical model to study the interaction of tree and grass that explicitly makes fire intensity dependent on the grass biomass and distinguishes two levels of fire sensitivity within the woody biomass (implicitly relating to plant size and bark thickness). Fire was considered as a time-continuous forcing as in several existing models (Langevelde et al. 2003, Accatino et al. 2010, De Michele et al. 2011 and reference therein) with a constant frequency of fire return that can be interpreted as mainly expressing an external forcing to the tree-grass system from climate and human practices. What is novel in our model is that fire impact on tree biomass is modeled as a non-linear function w of the grass biomass. Using a non-linear function is to our knowledge only found in Staver et al. 2011. But this latter model made peculiar assumptions and does not predict grassland and forest as possible equilibria (only desert and savanna). The advantage of a non-linear function is that it can account for the absence of fire at low biomass. As a consequence and although keeping the same modeling paradigm as in Langevelde et al. 2003, Accatino et al. 2010, De Michele et al. 2011, we reached different results and predictions.

Distinguishing fire sensitive vs. fire insensitive woody biomass lead to three variables expressing fractions of the above ground phytomass, namely grass and both fire-sensitive and -insensitive woody vegetation. It featured three coupled, non-linear ordinary differential equations.As several existing models (Baudena et al. 2010, Staver et al. 2011), our model acknowledges two major phenomena that regulate savanna dynamics, namely the fire-mediated negative feedback of grasses onto sensitive trees and the negative feedback of grown-up, fire insensitive trees on grasses. We therefore explicitly model the asymmetric nature of tree-grass competitive interactions in fire-prone savannas.

The analytical study of the model reveals three possible equilibria excluding tree-grass coexistence (desert, grassland, forest) along with equilibria for which woody and grassy components show durable coexistence (i.e. savanna vegetation). The number of such equilibrium points depends on the function used to model the increase of fire intensity with grass biomass(see Remark 2); for our model, we can have at most three savanna equilibria. We identified four ecologically meaningful thresholds that defined in parameter space regions of monostability, bistability as in Accatino et al. 2010, De Michele et al. 2011 and tristability with respect to the equilibria. Tristability of equilibria may mean that shifts from one stable state to another may often be less spectacular that hypothesized from previous models and that scenarios of vegetation changes may be more complex.

The model features some parameters that have been analytically identified as liable to trigger bifurcations (i.e., the state variables of the model converges to different steady states), notably parameters σ_{NS} and σ_G of asymmetric competition that embody the depressing influence of fire insensitive trees on grasses and of grasses on sensitive woody biomass respectively. Since tree-grass asymmetric competition is largely mediated by fire, this finding of the role of those two parameters is not intuitive and is the result of the modeling effort and of the analytical analysis. Since such parameters that quantify direct interactions between woody and grassy components appear crucial to understand the tree-grass dynamics in savanna ecosystems and for enhanced parameter assessment, they could be the focus of straightforward field experiments that would not request manipulating fire regime. Another bifurcation parameter is the fire frequency, f, (or fire period parameter $\tau = \frac{1}{f}$) which has been assumed to be an external forcing parameter that integrates both climatic and human influences. Frequent fires preclude tree-grass coexistence and turn savannas into grasslands. In the wettest situations, or under subequatorial climates, very high fire frequencies (above one fire per year) seem to be needed to prevent the progression of forests over savannas (unpublished data of experiments carry out at La Lopé National Park in Gabon).

However, it is questionable to model fire as a continuous forcing that regularly removes fractions of fire sensitive biomass. Indeed, several months can past between two successive fires, such that fire may be considered as an instantaneous perturbation of the savanna ecosystem. Several recent papers have proposed to model fires as stochastic events while keeping the continuous-time differential equation framework (Beckage et al. 2011) or using time discrete matrix models (Accatino & De Michele 2013). But in all those examples, fire characteristics remain mainly a linear function of grass biomass. Another framework that we will explore in a forthcoming work in order to acknowledge the discrete nature of fire events is based on system of impulsive differential equations (Lakshmikantham et al. 1989, Bainov and Simeonov 1993).

REFERENCES

[1] F. Accatino, C. De Michele, Humid savanna-forest dynamics: a matrix model with vegetation-fire interactions and seasonality. Eco. Mod. 265, pp. 170-179, 2013.

[2] F. Accatino, C. De Michele, R. Vezzoli, D. Donzelli, R. Scholes, "Tree-grass co-existence in savanna: interactions of rain and fire." J. Theor. Biol. 267, pp. 235-242, 2010. http://dx.doi.org/10.1016/j.jtbi.2010.08.012

[3] R. Anguelov, Y. Dumont, J.M-S. Lubuma and M. Shillor, Comparison of some standard and nonstandard numerical methods for the MSEIR epidemiological model, In: T. Simos, G. Psihoyios, Ch Tsitouras (eds), Proceedings of the International Conference of Numerical Analysis and Applied Mathematics, Crete, Greece, 18-22 September 2009, American Institute of Physics Conference Proceedings-AIP 1168, Volume 2, 2009, pp. 1209-1212. http://dx.doi.org/10.1063/1.3241285

[4] R. Anguelov, Y. Dumont, and J.M.-S. Lubuma, On nonstandard finite difference schemes in biosciences. AIP Conf. Proc. 1487, pp. 212-223, 2012. http://dx.doi.org/10.1063/1.4758961

[5] R. Anguelov, Y. Dumont, J.M.-S. Lubuma, and E. Mureithi, "Stability Analysis and Dynamics Preserving Non-Standar Finite Difference Schemes for Malaria Model", Mathematical Population Studies, 20 (2), pp. 101-122, 2013. http://dx.doi.org/10.1080/08898480.2013.777240

[6] R. Anguelov, Y. Dumont, J.M.-S. Lubuma, and M. Shillor, Dynamically consistent nonstandard finite difference schemes for epidemiological Models. Journal of Computational and Applied Mathematics, 255, pp. 161-182, 2014. http://dx.doi.org/10.1016/j.cam.2013.04.042

[7] D.D. Bainov, and P.S. Simeonov, Impulsive Differential Equations: Periodic Solutions and Applications. longman, England, 1993.

[8] M. Baudena, F. D'Andrea, A. Provenzale, An idealized model for tree-grass coexistence in savannas: the role of life stage structure and fire disturbances. J. Ecol. 98, pp. 74-80, 2010. http://dx.doi.org/10.1111/j.1365-2745.2009.01588.x

[9] B. Beckage, L.J. Gross, W.J. Platt, Grass feedbacks on fire stabilize savannas. Eco. Mod. 222, pp. 2227-2233, 2011. http://dx.doi.org/10.1016/j.ecolmodel.2011.01.015

[10] W.J. Bond, G.F. Midgley, F.I. Woodward, What controls SouthAfrican vegetation-climate or fire? S. Afr. J. Bot. 69, pp. 79-91, 2003.

[11] W.J. Bond, F.I. Woodward, G.F. Midgley, The global distribution of ecosystems in a world without fire. New Phytologist, 165, pp. 525-538, 2005. http://dx.doi.org/10.1111/j.1469-8137.2004.01252.x

[12] H. Breman and J.J. Kessler. Woody plants in agroecosystems of semi-arid regions. With an emphasis on the Sahelian countries. Advanced series in Agricultural 23, Springer-Verlag, Berlin. 1995.

[13] C. De Michele, F. Accatino, R. Vezzoli, R.J. Scholes, Savanna domain in the herbivores-fire parameter space exploiting a tree-grass-soil water dynamic model. J. Theor. Biol. 289, pp. 74-82, 2011.

http://dx.doi.org/10.1016/j.jtbi.2011.08.014

[14] Y. Dumont, J.C. Russell, V. Lecomte and M. Le Corre, Conservation of endangered endemic seabirds within a multi-predatot context:The Barau's petrel in Reunion island. Natural Ressource Modelling, 23, pp. 381-436, 2010. http://dx.doi.org/10.1111/j.1939-7445.2010.00068.x

[15] Y. Dumont, J.M. Tchuenche, Mathematical Studies on the Sterile Insect Technique for the Chikungunya Disease and Aedes albopictus. Journal of mathematical Biology, 65 (5), pp. 809-854, 2012. http://dx.doi.org/10.1007/s00285-011-0477-6

[16] P. D'Odorico, F. Laio, and L.A. Ridolfi, probabilistic analysis of fire-induced tree-grass coexistence in savannas. The American Naturalist, 167, pp. E79-E87, 2006.

[17] C. Favier, J. Aleman, L. Bremond, M.A. Dubois, V. Freycon, and J.M. Yangakola, Abrupt shifts in African savanna tree cover along a climatic gradient. Global Ecology and Biogeography, 21, pp. 787-797, 2012. http://dx.doi.org/10.1111/j.1466-8238.2011.00725.x

[18] http://www.scilab.org

[19] http://www.mathworks.com

[20] V. Lakshmikantham, D.D. Bainov and P.S. Simeonov, Theory of Impulsive Differential Equations, World Scientific, Singapore, 1989.

[21] V.F. Langevelde, C. van de Vijver, L. Kumar, J. van de Koppel, N. de Ridder, J. van Andel, et al., Effects of fire and herbivory on the stability of savanna ecosystems. Ecology, 84 (2), pp. 337-350, 2003.

[22] M.Y. Li and L. Wang. A criterion for stability of Matrices. J. Math. Ana. App. 225, pp. 249-264, 1998.

[23] J.C. Menaut and J. Csar. Structure and primary productivity of Lamto savannas, Ivory Coast. Ecology. 60, pp. 1197-1210, 1979.

[24] F.W.T. Penning de Vries and M.A. Djiteye. La productivité des paturages sahéliens. Une étude des sols, des végétations et de l'exploitation de cette ressource naturelle. PUDOC, Wageningen, 1982.

[25] M. Sankaran, J. Ratnam, and N. Hanan. Woody cover in African savannas: the role of resources, fire and herbivory. Global Ecology and Biogeography, 17, pp. 236-245, 2008. http://dx.doi.org/10.1111/j.1466-8238.2007.00360.x

[26] M. Sankaran, N.P. Hanan, R.J. Scholes, J. Ratnam, D.J. Augustine, B.S. Cade, J. Gignoux, S.I. Higgins, X. LeRoux, F. Ludwig, J. Ardo, F. Banyikwa, A. Bronn, G. Bucini, K.K. Caylor, M.B. Coughenour, A. Diouf, W. Ekaya, C.J. Feral, E.C. February, P.G.H. Frost, P. Hiernaux, H. Hrabar, K.L. Metzger, H.H.T. Prins, S. Ringrose, W. Sea, J. Tews, J. Worden, N. Zambatis. Determinants of woody coverin African savannas. Nature, 438, pp. 846-849, 2005. http://dx.doi.org/10.1038/nature04070

[27] M. Scheffer, and Stephen R. Carpenter. Catastrophic regime shifts in ecosystems: linking theory to observation. TRENDS in Ecology and Evolution. Vol.18 No.12, pp. 648-656, December 2003.

http://dx.doi.org/10.1016/j.tree.2003.09.002

[28] R.J. Scholes, Convex relationships in ecosystems containing mixtures of trees and grass. Environnemental and Ressource Economics, 26, pp. 559-574, 2003.

[29] R.J. Scholes, S.R. Archer, Tree-Grass interactions in savannas. Annu. Rev. Ecol. Syst. 28, pp. 517-544, 1997.

[30] I.P.J. Smit, G. Asner, N. Govender, T. Kennedy-Bowdoin, D. KNAPP, and J. Jacobson, Effects of fire on woody vegetation structure in African savanna. Ecological Applications, 20 (7), pp. 1865-1875, 2010.

[31] A.C. Staver, S. Archibald, S. Levin, Tree cover in sub-Saharan Africa: Rainfall and fire constrain forest and savanna as alternative stable states. Ecology. 92(5), pp. 1063-1072, 2011.

[32] D. Tilman, Competition and biodiversity in spatially structured habitats. Ecology, 75, pp. 2-16, 1994.

[33] C.A. Van de Vijver, Foley and H. Olff, Changes in the woody component of an East African savanna during 25 years. Journal of Tropical Ecology 15, pp. 545-564, 1999.

[34] J.L. Walkeling, A.C. Staver and W.J. Bond, Simply the best: the transition of savanna saplings to trees. Oikos, 120, pp. 1448-1451, 2011.

[35] B. Walker, D. Ludwig, C.S. Holling, and R.M. Peterman, Stability of semi-arid savanna grazing systems. Journal of Ecology, 69, pp. 473-498, 1981.

[36] H. Walter, Ecology of tropical and subtropical vegetation. Oliver and Boyd, Edinburgh, UK., 1971.

APPENDIX A: EXPRESSIONS OF T_S^*, T_{NS}^* AND G^*

After straightforward but long computation, we show that

$$T_{NS}^* = \frac{\gamma_G}{\sigma_{NS}}\left(1 - \frac{1}{\mathcal{R}_2^0} - \frac{G^*}{K_G}\right),$$

$$T_S^* = \frac{\mu_{NS}}{\omega_S}T_{NS}^*,$$

$$\frac{1}{K_T}\left(1 + \frac{\omega_S}{\mu_{NS}}\right)T_S^* = 1 - \frac{1}{\mathcal{R}_1^0} - \frac{\mu_{NS}(\sigma_G G^* + f\eta_S w(G^*))}{\gamma_S\mu_{NS} + \gamma_{NS}\omega_S},$$

where G^* is solution of

$$w(G) = AG + B = F(G), \qquad (11)$$

with
$$A = \frac{1}{f\eta_S}\left(\frac{\gamma_G(\omega_S + \mu_{NS})(\gamma_S\mu_{NS} + \gamma_{NS}\omega_S)}{K_G K_T \sigma_{NS}\mu_{NS}\omega_S} - \sigma_G\right),$$

$B =$

$$\frac{(\gamma_S\mu_{NS} + \gamma_{NS}\omega_S)\left[1 - \frac{1}{\mathcal{R}_1^0} - \frac{\gamma_G(\omega_S+\mu_{NS})}{K_T\sigma_{NS}\omega_S}\left(1 - \frac{1}{\mathcal{R}_2^0}\right)\right]}{f\eta_S\mu_{NS}}.$$

We summarize the problem of existence of solutions of equation (11) in the following Table

TABLE IV
EXISTENCE OF SOLUTIONS OF EQUATION (11)

A	B	Number of solutions
> 0	> 0	0 or 2 solutions
	< 0	1 or 3 solutions
< 0	> 0	1 solution
	< 0	0 solution

Note that solutions G^* of (11) that give rise to savanna equilibria must satisfy $0 < G^* < K_G\left(1 - \frac{1}{\mathcal{R}_2^0}\right)$.

APPENDIX B: PROOF OF THEOREM 1

let $\mathcal{R}_0^0 = \dfrac{\gamma_S}{\mu_S + \omega_S + \mu_{NS}}$. In a matrical writing, System (1) reads as

$$\frac{dX}{dt} = \mathcal{A}(X)X < \mathcal{A}^{\max}(X)X, \qquad (12)$$

with $X = (T_S, T_{NS}, G) \in \mathbb{R}_+^3$, $\mathcal{A}(X) = (\mathcal{A}_{ij})_{1\leq i,j\leq 3}$ with $\mathcal{A}_{11} = \gamma_S\left(1 - \frac{T_S+T_{NS}}{K_T}\right) - (\mu_S + \omega_S + \sigma_G G + f\eta_S w(G))$, $\mathcal{A}_{12} = \gamma_{NS}\left(1 - \frac{T_S+T_{NS}}{K_T}\right)$, $\mathcal{A}_{13} = 0$, $\mathcal{A}_{21} = \omega_S$, $\mathcal{A}_{22} = -\mu_{NS}$, $\mathcal{A}_{23} = 0$, $\mathcal{A}_{31} = 0$, $\mathcal{A}_{32} = 0$, $\mathcal{A}_{33} = \gamma_G\left(1 - \frac{G}{K_G}\right) - (\sigma_{NS}T_{NS} + f\eta_G + \mu_G)$.
and

$$\mathcal{A}^{\max}(X) = \begin{pmatrix} \gamma_S - \mu_S - \omega_S & \gamma_{NS} & 0 \\ \omega_S & -\mu_{NS} & 0 \\ 0 & 0 & \gamma_G\left(1 - \frac{1}{\mathcal{R}_2^0}\right) \end{pmatrix} = \begin{pmatrix} A & B \\ C & D \end{pmatrix},$$

with $A = \begin{pmatrix} \gamma_S - \mu_S - \omega_S & \gamma_{NS} \\ \omega_S & -\mu_{NS} \end{pmatrix}$, $B = \begin{pmatrix} 0 \\ 0 \end{pmatrix}$, $C = \begin{pmatrix} 0 & 0 \end{pmatrix}$, and $D = \gamma_G\left(1 - \frac{1}{\mathcal{R}_2^0}\right)$.

Matrix $\mathcal{A}^{\max}(X)$ is a Metzler matrix (i.e all its off-diagonal terms are nonnegative) and $\alpha(\mathcal{A}^{\max}(X)) \leq 0$ if $\alpha(A) \leq 0$ and $\alpha(D) \leq 0$ where α denotes the stability modulus. Moreover, for matrix D, $\alpha(D) \leq 0$ if

$$\mathcal{R}_2^0 < 1. \qquad (13)$$

For matrix A, $\alpha(A) \leq 0$ if $trace(A) < 0$ and $det(A) > 0$.

$$trace(A) = \gamma_S - \mu_S - \omega_S - \mu_{NS}$$
$$= \gamma_S\left(1 - \frac{1}{\mathcal{R}_0^0}\right). \quad (14)$$

$$det(A) = \mu_{NS}(\mu_S + \omega_S) - (\mu_{NS}\gamma_S + \omega_S\gamma_{NS})$$
$$= \mu_{NS}(\mu_S + \omega_S)(1 - \mathcal{R}_1^0). \quad (15)$$

Furthermore,

$$\mathcal{R}_0^0 < \mathcal{R}_1^0 \quad (16)$$

Thus, from relations (13), (14), (15) and (16) we deduce that the desert equilibrium $(0; 0; 0)$ is globally asymptotically stable whenever $\mathcal{R}_1^0 < 1$ and $\mathcal{R}_2^0 < 1$.

APPENDIX C: PROOF OF THEOREM 2

If $\mathcal{R}_1^0 > 1$, then the forest equilibrium E_T exists.

- Using the Jacobian matrix of system (1) at E_T, one can prove that E_T is locally asymptotically stable if $\mathcal{R}_2^{\bar{T}_{NS}} < 1$.
- The solution G of system (1) verify

$$\frac{dG}{dt} \leq (\gamma_G - (f\eta_G + \mu_G))G,$$
$$\leq \gamma_G\left(1 - \frac{1}{\mathcal{R}_2^0}\right). \quad (17)$$

So, if $\mathcal{R}_2^0 < 1$, then

$$\lim_{t \to +\infty} G(t) = 0. \quad (18)$$

Moreover, the solutions T_S and T_{NS} of system (1) admit as a limit system, the system:

$$\begin{cases} \dfrac{dT_S}{dt} = (\gamma_S T_S + \gamma_{NS}T_{NS})\left(1 - \dfrac{T_S + T_{NS}}{K_T}\right) \\ \qquad\quad -T_S(\mu_S + \omega_S) = F_1(T_S, T_{NS}), \\ \dfrac{dT_{NS}}{dt} = \omega_S T_S - \mu_{NS}T_{NS} = F_2(T_S; T_{NS}). \end{cases} \quad (19)$$

Now, let $h(T_S, T_{NS}) = T_S^{-1}$. Then, one has
$$\frac{\partial F_1 h}{\partial T_S} + \frac{\partial F_2 h}{\partial T_{NS}} = -\gamma_{NS}\frac{T_{NS}}{T_S^2}\left(1 - \frac{T_S + T_{NS}}{K_T}\right)$$
$$-\frac{1}{K_T T_S}(\gamma_S T_S + \gamma_{NS}T_{NS}) - \mu_{NS}T_S^{-1}.$$
Furthermore, we have

$$\frac{\partial F_1 h}{\partial T_S} + \frac{\partial F_2 h}{\partial T_{NS}} < 0 \text{ in } \Omega_2^\circ \text{ where } \Omega_2 = \{(T_S, T_{NS}) \in \mathbf{R}_+^2 \mid 0 \leq T_S + T_{NS} \leq K_T\}$, and by the Bendixson-Dulac theorem, we deduce that system (19) don't admits a periodic solution in Ω_2.

Moreover, the equilibrium $(\bar{T}_S, \bar{T}_{NS})$ exists if $\mathcal{R}_1^0 > 1$ and using the Jacobian matrix of system (19), we deduce that $(\bar{T}_S, \bar{T}_{NS})$ is locally asymptotically stable and then, globally asymptotically stable since there is no periodic solution. Thus, if $\mathcal{R}_2^0 < 1$, then one has

$$\lim_{t \to +\infty}(T_S, T_{NS}, G)(t) = E_T.$$

- Suppose that
$\mathcal{R}_1^0 = \dfrac{\gamma_S\mu_{NS} + \gamma_{NS}\omega_S}{\mu_{NS}(\mu_S + \omega_S)} > 1$ and $\mathcal{R}_2^0 = \dfrac{\gamma_G}{f\eta_G + \mu_G} > 1$,
then equilibria $(\bar{T}, \bar{T}_{NS}, 0)$, $(0, 0, \bar{G})$ and $(T_S^\star, T_{NS}^\star, G^\star)$ are defined.
The Jacobian matrix of system (1) at an arbitrarily equilibrium point is

$$J = \begin{pmatrix} J_{11} & J_{12} & J_{13} \\ J_{21} & J_{22} & 0 \\ 0 & J_{32} & J_{33} \end{pmatrix},$$

where
$$\begin{aligned} J_{11} &= \gamma_S\left(1 - \frac{X+Y}{K_T}\right) - \frac{1}{K_T}(\gamma_S X + \gamma_{NS}Y) \\ &\quad -\mu_S - \omega_S - \sigma_G Z - f\eta_S w(Z), \\ J_{12} &= \gamma_{NS}\left(1 - \frac{X+Y}{K_T}\right) - \frac{1}{K_T}(\gamma_S X + \gamma_{NS}Y), \\ J_{13} &= -\sigma_G X - Xf\eta_S w'(Z), \\ J_{21} &= \omega_S, \\ J_{22} &= -\mu_{NS}, \\ J_{32} &= -\sigma_{NS}Z, \\ J_{33} &= \gamma_G - 2\frac{\gamma_G}{K_G}Z - \sigma_{NS}Y - f\eta_G - \mu_G. \end{aligned}$$

The second additive compound matrix of J is

$$J^{[2]} = \begin{pmatrix} J_{11} + J_{22} & 0 & -J_{13} \\ J_{32} & J_{11} + J_{33} & J_{12} \\ 0 & J_{21} & J_{22} + J_{33} \end{pmatrix}. \quad (20)$$

From the Jacobian matrix of system (1), the equilibria $(0, 0, \bar{G})$ and $(T_S^\star, T_{NS}^\star, G^\star)$ are unstable if $\mathcal{R}_2^{\bar{G}} > 1$ and $\mathcal{R} < 1$.

In the sequel, we suppose that $\mathcal{R}_2^{\bar{T}_{NS}} < 1$ to process with the discussion.

The second additive compound matrix (20) at the equilibrium $(\bar{T}_S, \bar{T}_{NS}, 0)$ is

$$J^{[2]}(\bar{T}_S, \bar{T}_{NS}, 0) =$$
$$\begin{pmatrix} J_{11} + J_{22} & 0 & \sigma\bar{T}_S \\ 0 & J_{11} + J_{33} & J_{12} \\ 0 & J_{21} & J_{22} + J_{33} \end{pmatrix}_{(\bar{T}_S, \bar{T}_{NS}, 0)}.$$

Let

$$B = \begin{pmatrix} J_{11} + J_{33} & J_{12} \\ J_{21} & J_{22} + J_{33} \end{pmatrix}.$$

Then, a simple calculation gives

$$(J_{11} + J_{22})_{(\bar{T}_S, \bar{T}_{NS}, 0)} = \gamma_S \left(1 - \frac{\bar{T}_S + \bar{T}_{NS}}{K_T}\right) - \frac{1}{K_T}(\gamma_S\bar{T}_S + \gamma_{NS}\bar{T}_{NS}) - \mu_S - \omega_S - \mu_{NS}.$$

Using the relations

$$-\mu_S - \omega_S = -\left(\gamma_S + \gamma_{NS}\frac{\omega_S}{\mu_{NS}}\right)\left(1 - \frac{\bar{T}_S + \bar{T}_{NS}}{K_T}\right)$$

and $\left(1 - \frac{\bar{T}_S + \bar{T}_{NS}}{K_T}\right) > 0$, we have

$$(J_{11} + J_{22})_{(\bar{T}_S, \bar{T}_{NS}, 0)} = -\frac{1}{K_T}\bar{T}_S\left(\gamma_S + \gamma_{NS}\frac{\omega_S}{\mu_{NS}}\right)$$
$$-\gamma_{NS}\frac{\omega_S}{\mu_{NS}}\left(1 - \frac{\bar{T}_S + \bar{T}_{NS}}{K_T}\right) - \mu_{NS} < 0.$$

Since $J_{11} + J_{22} < 0$ and $\mathcal{R}_2^{\bar{T}_{NS}} < 1$, one has

$$tr(B) = (J_{11} + J_{22} + 2J_{33})_{(\bar{T}_S, \bar{T}_{NS}, 0)},$$
$$= J_{11} + J_{22} + 2\gamma_G\left(1 - \frac{1}{\mathcal{R}_2^{\bar{T}_{NS}}}\right) < 0.$$

Also, if $\mathcal{R}_2^{\bar{T}_{NS}} < 1$, one has

$$J_{11}J_{33} = \left(-\frac{\gamma_G}{K_T}\bar{T}_S\left(\gamma_S + \gamma_{NS}\frac{\omega_S}{\mu_{NS}}\right)\right.$$
$$\left. -\gamma_G\gamma_{NS}\frac{\omega_S}{\mu_{NS}}\left(1 - \frac{\bar{T}_S + \bar{T}_{NS}}{K_T}\right)\right)$$
$$\times\left(1 - \frac{1}{\mathcal{R}_2^{\bar{T}_{NS}}}\right) > 0,$$

and

$$J_{33}(J_{22} + J_{33}) =$$
$$\gamma_G\left(1 - \frac{1}{\mathcal{R}_2^{\bar{T}_{NS}}}\right)\left(-\mu_{NS} + \gamma_G\left(1 - \frac{1}{\mathcal{R}_2^{\bar{T}_{NS}}}\right)\right) > 0.$$

With this in mind, we have

$$det(B) = (J_{11} + J_{33})(J_{22} + J_{33}) - J_{12}J_{21},$$
$$= J_{11}J_{22} - J_{21}J_{12} + J_{11}J_{33}$$
$$+ J_{33}(J_{22} + J_{33}),$$
$$= \frac{\mu_{NS}}{K_T}\bar{T}_S\left(\gamma_S + \gamma_{NS}\frac{\omega_S}{\mu_{NS}}\right)$$
$$+ \frac{\omega_S}{K_T}(\gamma_S\bar{T}_S + \gamma_{NS}\bar{T}_{NS})$$
$$+ J_{11}J_{33} + J_{33}(J_{22} + J_{33}) > 0.$$

Thus, if $\mathcal{R}_2^{\bar{T}_{NS}} < 1$, one has $(J_{11} + J_{22})_{(\bar{T}_S, \bar{T}_{NS}, 0)} < 0$, $tr(B) < 0$ and $det(B) > 0$. This implies that $s(J^{[2]}(\bar{T}_S, \bar{T}_{NS}, 0)) < 0$ where s denotes the stability modulus. Following Theorem 3.3 in Li and Wang 1998, we can deduce that there is no hopf bifurcation points for $J(\bar{T}_S, \bar{T}_{NS}, 0)$. Since $\mathcal{R}_2^{\bar{T}_{NS}} < 1$, the equilibrium point $(\bar{T}_S, \bar{T}_{NS}, 0)$ is locally asymptotically stable and one can conclude that this equilibrium point is globally asymptotically stable if $\mathcal{R}_2^{\bar{T}_{NS}} < 1, \mathcal{R}_2^{\bar{G}} > 1$ and $\mathcal{R} < 1$. This completes the proof.

APPENDIX D: PROOF OF THEOREM 4

Suppose that the savanna equilibrium E_{TG} exists. The Jacobian matrix of system (1) at E_{TG} is

$$J = \begin{pmatrix} J_{11} & J_{12} & J_{13} \\ J_{21} & J_{22} & 0 \\ 0 & J_{32} & J_{33} \end{pmatrix},$$

where

$$J_{11} = \gamma_S\left(1 - \frac{T_S + T_{NS}}{K_T}\right) - \frac{1}{K_T}(\gamma_S T_S + \gamma_{NS} T_{NS})$$
$$-\mu_S - \omega_S - \sigma_G G - f\eta_S w(G),$$
$$J_{12} = \gamma_{NS}\left(1 - \frac{T_S + T_{NS}}{K_T}\right) - \frac{1}{K_T}(\gamma_S T_S + \gamma_{NS} T_{NS}),$$
$$J_{13} = -\sigma_G T_S - T_S f\eta_S w'(G),$$
$$J_{21} = \omega_S,$$
$$J_{22} = -\mu_{NS},$$
$$J_{32} = -\sigma_{NS} G,$$
$$J_{33} = -\frac{\gamma_G}{K_G}G.$$

Let

$$A_1 = -J_{11}J_{22}J_{33},$$
$$A_2 = J_{21}J_{12}J_{33},$$
$$A_3 = -J_{21}J_{32}J_{13},$$
$$C_1 = -J_{11} - J_{22} - J_{33},$$
$$C_2 = A_1 + A_2 + A_3,$$
$$C_3 = J_{11}J_{33} + J_{11}J_{22} - J_{21}J_{12} + J_{22}J_{33} - \frac{C_2}{C_1}.$$

Note that, by the Routh-Hurwitz theorem, the savanna equilibrium E_{TG} is locally asymptotically stable if

$C_1 > 0$, $C_2 > 0$ and $C_3 > 0$.

Moreover, components of the savanna equilibrium E_{TG} satisfy

$$\begin{cases} T_{NS} = \frac{\omega_S}{\mu_{NS}} T_S, \\[2mm] -\mu_S - \omega_S - \sigma_G G - f\eta_S w(G) = -\left(\gamma_S + \gamma_{NS} \frac{\omega_S}{\mu_{NS}}\right) \times \\[1mm] \hspace{4cm} \left(1 - \frac{T_S + T_{NS}}{K_T}\right), \end{cases}$$

thus,

$$\begin{aligned} C_1 &= -J_{11} - J_{22} - J_{33}, \\ &= \frac{1}{K_T}(\gamma_S T_S + \gamma_{NS} T_{NS}) + \gamma_{NS} \frac{\omega_S}{\mu_{NS}}\left(1 - \frac{T_S + T_{NS}}{K_T}\right) \\ &\quad + \mu_{NS} + \frac{\gamma_G}{K_G} G, \\ &> 0. \end{aligned}$$

$$\begin{aligned} C_3 &= J_{11}J_{33} + J_{11}J_{22} - J_{21}J_{12} + J_{22}J_{33} - \frac{C_2}{C_1}, \\[2mm] &= \frac{1}{K_T}(\mu_{NS} + \omega_S)(\gamma_S T_S + \gamma_{NS} T_{NS}) \\[2mm] &\quad + \frac{\gamma_G G}{K_G}\left(\frac{1}{K_T}(\gamma_S T_S + \gamma_{NS} T_{NS})\right. \\ &\quad \left. + \gamma_{NS}\frac{\omega_S}{\mu_{NS}}\left(1 - \frac{T_S + T_{NS}}{K_T}\right) + \mu_{NS}\right) - \frac{C_2}{C_1}, \\[2mm] &= \frac{1}{K_T}(\mu_{NS} + \omega_S)(\gamma_S T_S + \gamma_{NS} T_{NS}) \\[2mm] &\quad - \frac{\frac{\gamma_G G}{K_G K_T}(\mu_{NS} + \omega_S)(\gamma_S T_S + \gamma_{NS} T_{NS})}{C_1} \\ &\quad + \frac{\omega_S \sigma_{NS} G(\sigma_G T_S f\eta_S w'(G)T_S)}{C_1} \\ &\quad + \frac{\gamma_G G}{K_G}\left(\frac{1}{K_T}(\gamma_S T_S + \gamma_{NS} T_{NS})\right. \\ &\quad \left. + \gamma_{NS}\frac{\omega_S}{\mu_{NS}}\left(1 - \frac{T_S + T_{NS}}{K_T}\right) + \mu_{NS}\right), \\[2mm] &= \frac{1}{K_T}(\mu_{NS} + \omega_S)(\gamma_S T_S + \gamma_{NS} T_{NS})\left(1 - \frac{\frac{\gamma_G G}{K_G}}{C_1}\right) \\[2mm] &\quad + \frac{\omega_S \sigma_{NS} G(\sigma_G T_S f\eta_S w'(G)T_S)}{C_1} \\ &\quad + \frac{\gamma_G G}{K_G}\left(\frac{1}{K_T}(\gamma_S T_S + \gamma_{NS} T_{NS})\right. \\ &\quad \left. + \gamma_{NS}\frac{\omega_S}{\mu_{NS}}\left(1 - \frac{T_S + T_{NS}}{K_T}\right) + \mu_{NS}\right), \\[2mm] &= \frac{1}{K_T C_1}(\mu_{NS} + \omega_S)(\gamma_S T_S + \gamma_{NS} T_{NS}) \times \\ &\quad \left(\frac{1}{K_T}(\gamma_S T_S + \gamma_{NS} T_{NS})\right. \\ &\quad \left. + \gamma_{NS}\frac{\omega_S}{\mu_{NS}}\left(1 - \frac{T_S + T_{NS}}{K_T}\right) + \mu_{NS}\right) \\[2mm] &\quad + \frac{\omega_S \sigma_{NS} G(\sigma_G T_S f\eta_S w'(G)T_S)}{C_1} + \frac{\gamma_G G}{K_G}\left(\frac{1}{K_T}(\gamma_S T_S\right. \\ &\quad \left. + \gamma_{NS} T_{NS}) + \gamma_{NS}\frac{\omega_S}{\mu_{NS}}\left(1 - \frac{T_S + T_{NS}}{K_T}\right) + \mu_{NS}\right), \\ C_3 &> 0. \end{aligned}$$

$$\begin{aligned} C_2 &= A_1 + A_2 + A_3, \\ &= \frac{\gamma_G G}{K_G K_T}(\mu_{NS} + \omega_S)(\gamma_S T_S + \gamma_{NS} T_{NS}) \\ &\quad - \omega_S \sigma_{NS} G(\sigma_G T_S + f\eta_S w'(G)), \\ &= \omega_S \sigma_{NS} G T_S \left(\frac{\gamma_G}{K_G K_T \omega_S \sigma_{NS}}(\mu_{NS} + \omega_S) \times \right. \\ &\quad \left. \left(\gamma_S + \frac{\gamma_{NS}\omega_S}{\mu_{NS}}\right) - \sigma_G - f\eta_S w'(G)\right), \\ &= \omega_S \sigma_{NS} G T_S (\sigma_G + f\eta_S w'(G))(\mathcal{R} - 1). \end{aligned}$$

Thus, $C_2 > 0$ if and only if $\mathcal{R} > 1$.

Finally, we deduce that the savanna equilibrium E_{TG}, when it is unique, is locally asymptotically stable if $\mathcal{R} = \mathcal{R}(G) > 1$. The first part of Theorem 4 holds.

One should note that $C_2 > 0$ means that the slope of w (the sigmoidal function) is less than the slope of F where F is given by (11).

Furthermore, by using relation (11) we deduce part 2 and part 3 of Theorem 4 graphically as follow

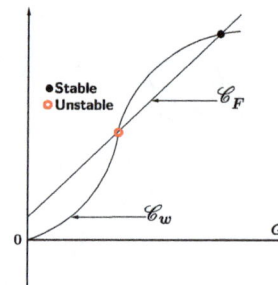

Fig. 8. There exist two savanna equilibria but one is stable and the other is unstable.

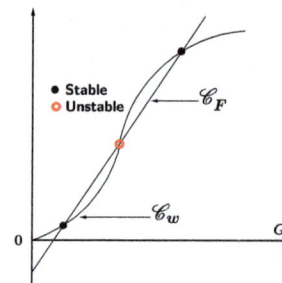

Fig. 9. There exist three savanna equilibria two are stable and one is unstable. Thus system (1) will converge to one of the two stable equilibria depending on initial conditions.

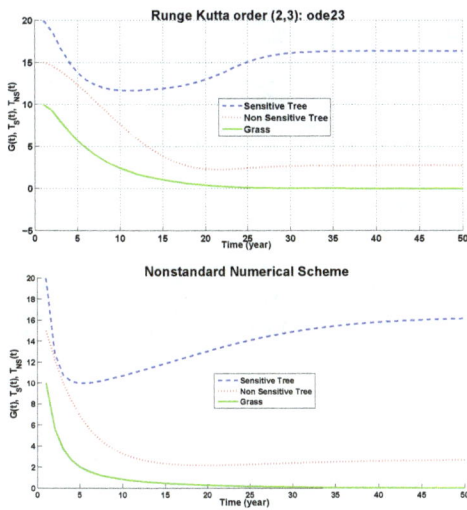

Fig. 10.   For these figure, $\mu_G = 0.4$. The ODE's routine which is a standard numerical algorithms shows a spurious negative solutions.

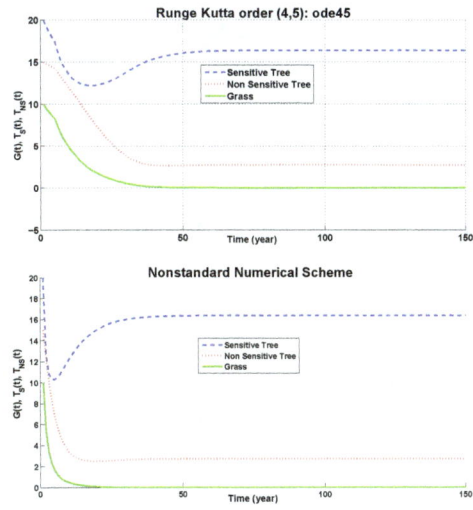

Fig. 11.   For these figure, $\mu_G = 0.5$. The ODE's routine which is a standard numerical algorithms shows again a spurious negative solutions.

APPENDIX E: SPURIOUS BEHAVIORS OF RUNGE KUTTA METHODS TO APPROXIMATE SOLUTIONS OF SYSTEM (1)

For the following figures we choose

γ_S	γ_{NS}	γ_G	η_S	η_G
0.1	2	0.6	0.5	0.5

μ_S	μ_{NS}	μ_G	σ_{NS}	σ_G
0.3	0.3		0.02	0.05

G_0	ω_S	K_T	K_G	f
2	0.05	50	12	0.5

Other examples of spurious solutions given by standard methods are also given in (Anguelov et al. 2009).

APPENDIX F: VALUES OF $\mathcal{R}_1^0$, $\mathcal{R}_2^0$, $\mathcal{R}_1^{\overline{G}}$, $\mathcal{R}_2^{\overline{T}_{NS}}$ AND $\mathcal{R}$ IN FIGURE 6

- Figure 6 (a):

$\mathcal{R}_1^0$	$\mathcal{R}_2^0$	$\mathcal{R}_1^{\overline{G}}$	$\mathcal{R}_2^{\overline{T}_{NS}}$
4.8889	4.9412	2.6928	0.9863

- Figure 6 (b):

$\mathcal{R}_1^0$	$\mathcal{R}_2^0$	$\mathcal{R}_1^{\overline{G}}$	$\mathcal{R}_2^{\overline{T}_{NS}}$
4.8889	4.9412	1.5813	0.9861

- Figure 6 (c):

$\mathcal{R}_1^0$	$\mathcal{R}_2^0$	$\mathcal{R}_1^{\overline{G}}$	$\mathcal{R}_2^{\overline{T}_{NS}}$	$\mathcal{R}$
4.8889	4.9412	1.1193	0.9861	1.3984

- Figure 6 (d):

$\mathcal{R}_1^0$	$\mathcal{R}_2^0$	$\mathcal{R}_1^{\overline{G}}$	$\mathcal{R}_2^{\overline{T}_{NS}}$	$\mathcal{R}$
4.8889	4.9412	1.0431	0.9861	1.2980

- Figure 6 (e):

$\mathcal{R}_1^0$	$\mathcal{R}_2^0$	$\mathcal{R}_1^{\overline{G}}$	$\mathcal{R}_2^{\overline{T}_{NS}}$	$\mathcal{R}$
4.8889	4.9412	0.9766	0.9861	0.5936

- Figure 6 (f):

$\mathcal{R}_1^0$	$\mathcal{R}_2^0$	$\mathcal{R}_1^{\overline{G}}$	$\mathcal{R}_2^{\overline{T}_{NS}}$	$\mathcal{R}$
4.8889	4.9412	0.8662	0.9861	0.5746

APPENDIX G: VALUES OF $\mathcal{R}_1^0$, $\mathcal{R}_2^0$, $\mathcal{R}_1^{\overline{G}}$, $\mathcal{R}_2^{\overline{T}_{NS}}$ AND $\mathcal{R}$ IN FIGURE 7

- Figure 7 (a):

$\mathcal{R}_1^0$	$\mathcal{R}_2^0$	$\mathcal{R}_1^{\overline{G}}$	$\mathcal{R}_2^{\overline{T}_{NS}}$
3.7778	3.6364	0.3840	1.1549

- Figure 7 (b):

$\mathcal{R}_1^0$	$\mathcal{R}_2^0$	$\mathcal{R}_1^{\overline{G}}$	$\mathcal{R}_2^{\overline{T}_{NS}}$
3.7778	3.6364	0.3840	1.0162

- Figure 7 (c):

$\mathcal{R}_1^0$	$\mathcal{R}_2^0$	$\mathcal{R}_1^{\overline{G}}$	$\mathcal{R}_2^{\overline{T}_{NS}}$	$\mathcal{R}$
3.7778	3.6364	0.3840	0.9587	0.2066

- Figure 7 (d):

$\mathcal{R}_1^0$	$\mathcal{R}_2^0$	$\mathcal{R}_1^{\overline{G}}$	$\mathcal{R}_2^{\overline{T}_{NS}}$	$\mathcal{R}$
3.7778	3.6364	0.3840	0.9073	0.1453

PERMISSIONS

All chapters in this book were first published in BIOMATH, by Biomath Forum; hereby published with permission under the Creative Commons Attribution License or equivalent. Every chapter published in this book has been scrutinized by our experts. Their significance has been extensively debated. The topics covered herein carry significant findings which will fuel the growth of the discipline. They may even be implemented as practical applications or may be referred to as a beginning point for another development.

The contributors of this book come from diverse backgrounds, making this book a truly international effort. This book will bring forth new frontiers with its revolutionizing research information and detailed analysis of the nascent developments around the world.

We would like to thank all the contributing authors for lending their expertise to make the book truly unique. They have played a crucial role in the development of this book. Without their invaluable contributions this book wouldn't have been possible. They have made vital efforts to compile up to date information on the varied aspects of this subject to make this book a valuable addition to the collection of many professionals and students.

This book was conceptualized with the vision of imparting up-to-date information and advanced data in this field. To ensure the same, a matchless editorial board was set up. Every individual on the board went through rigorous rounds of assessment to prove their worth. After which they invested a large part of their time researching and compiling the most relevant data for our readers.

The editorial board has been involved in producing this book since its inception. They have spent rigorous hours researching and exploring the diverse topics which have resulted in the successful publishing of this book. They have passed on their knowledge of decades through this book. To expedite this challenging task, the publisher supported the team at every step. A small team of assistant editors was also appointed to further simplify the editing procedure and attain best results for the readers.

Apart from the editorial board, the designing team has also invested a significant amount of their time in understanding the subject and creating the most relevant covers. They scrutinized every image to scout for the most suitable representation of the subject and create an appropriate cover for the book.

The publishing team has been an ardent support to the editorial, designing and production team. Their endless efforts to recruit the best for this project, has resulted in the accomplishment of this book. They are a veteran in the field of academics and their pool of knowledge is as vast as their experience in printing. Their expertise and guidance has proved useful at every step. Their uncompromising quality standards have made this book an exceptional effort. Their encouragement from time to time has been an inspiration for everyone.

The publisher and the editorial board hope that this book will prove to be a valuable piece of knowledge for researchers, students, practitioners and scholars across the globe.

LIST OF CONTRIBUTORS

Olaposi Idowu Omotuyi and Hiroshi Ueda
Department of Pharmacology and Therapeutic Innovation University Graduate School of Biomedical Sciences, 852-8521 Nagasaki, Japan

Edward T. Dougherty
Department of Mathematics Rowan University Glassboro, NJ, USA

Anton Iliev
Faculty of Mathematics and Informatics Paisii Hilendarski University of Plovdiv, Plovdiv, Bulgaria
Institute of Mathematics and Informatics Bulgarian Academy of Sciences, Sofia, Bulgaria

Nikolay Kyurkchiev and Svetoslav Markov
Institute of Mathematics and Informatics Bulgarian Academy of Sciences, Sofia, Bulgaria

José Renato Campos
Area of Sciences Federal Institute of Education, Science and Technology of São Paulo, Votuporanga, SP, Brazil

Edvaldo Assunção
Department of Electrical Engineering UNESP - Univ Estadual Paulista, Ilha Solteira, SP, Brazil

Geraldo Nunes Silva
Department of Applied Mathematics UNESP - Univ Estadual Paulista, São José do Rio Preto, SP, Brazil

Weldon Alexander Lodwick
Department of Mathematical and Statistical Sciences University of Colorado, Denver, Colorado, USA

Peter Kumberger
Center for Modeling and Simulation in the Biosciences BioQuant-Center, Heidelberg University, Germany

Christina Kuttler
Zentrum Mathematik Technische Universität München, Germany

Peter Czuppon
Mathematisches Institut: Abt. f. Math. Stochastik Universität Freiburg, Germany

Burkhard A. Hense
Institute of Computational Biology Helmholtz-Zentrum München, Germany

David H. Margarit and Lilia Romanelli
Instituto de Ciencias, Universidad Nacional de General Sarmiento, Buenos Aires, Argentina
Consejo Nacional de Investigaciones Científicas y Técnicas (CONICET), Buenos Aires, Argentina

Burcu Gürbüz and Mehmet Sezer
Department of Mathematics, Manisa Celal Bayar University, Manisa, Turkey

Nicholas A. Battista
Dept. of Mathematics and Statistics, The College of New Jersey Ewing Township, NJ, USA

Laura A. Miller
Dept. of Biology, Dept. of Mathematics, University of North Carolina at Chapel Hill Chapel Hill, NC, USA

Julia E. Samson
Dept. of Biology

Shilpa Khatri
Applied Mathematics Unit, School of Natural Sciences University of California Merced, Merced, CA, USA

Emilie Peynaud
CIRAD, UMR AMAP, Yaoundé, Cameroun
AMAP, University of Montpellier, CIRAD, CNRS, INRA, IRD, Montpellier, France
University of Yaoundé 1, National Advanced School of Engineering, Yaoundé, Cameroon

Bradley J. Roth
Department of Physics, Oakland University Rochester, MI, USA

Nikolaos Sfakianakis
Institute of Applied Mathematics, Heidelberg University Im Neuenheimerfeld 205, 69120, Heidelberg, Germany

Diane Peurichard
Laboratoire Jacques-Louis Lions, INRIA, Sorbonne University Place Jussieu 4, Paris, France

Aaron Brunk
Institute of Mathematics, Johannes Gutenberg University Staudingerweg 9, 55128, Mainz, Germany

Christian Schmeiser
Faculty of Mathematics, University of Vienna Oskar-Morgenstern-Platz 1, 1090, Vienna, Austria

Adejimi Adesola Adeniji, Igor Fedotov and Michael Y. Shatalov
Department of Mathematics and Statistics, Tshwane University of Technology

Abdoulaye Diouf and Diène Ngom
Département de Mathmatiques, UMI 2019-IRD & UMMISCO-UGB, Université Assane Seck de Ziguinchor, B.P. 523 Ziguinchor, Sénégal

Baba Issa Camara, Vincent Felten, Jean-François Masfaraud and Jean-François Férard
Université de Lorraine - CNRS UMR 7360, Laboratoire Interdisciplinaire des Environnements Continentaux, Campus Bridoux - 8 Rue du Général Delestraint, 57070 Metz, France

Héla Toumi
Université de Lorraine - CNRS UMR 7360, Laboratoire Interdisciplinaire des Environnements Continentaux, Campus Bridoux - 8 Rue du Général Delestraint, 57070 Metz, France
Laboratoire de Bio-surveillance de l'Environnement (LBE), Université de Carthage, Faculté des Sciences de Bizerte, 7021 Zarzouna, Bizerte, Tunisie

N. F. Britton
Department of Mathematical Sciences and Centre for Mathematical Biology University of Bath, Bath BA2 7AY, UK

M. Almeida Neto
Departmento de Ecologia, Universidade Federal de Goiás, 74001-970 Goiânia-GO, Brazil

Gilberto Corso
Departamento de Biofísica e Farmacologia, Centro de Biociências, Universidade Federal do Rio Grande do Norte, 59072-970 Natal-RN, Brazil

Katarzyna (Kasia) Świrydowicz
Department of Mathematics Virginia Polytechnic Institute Blacksburg, VA, USA

David A. Ekrut and Nicholas G. Cogan
Department of Biological Mathematics Florida State University Tallahassee, Florida, United States

Valaire Yatat, Jean Jules Tewa and Samuel Bowong
UMMISCO, LIRIMA project team GRIMCAPE, Yaounde, Cameroon

Yves Dumont
CIRAD, Umr AMAP, Montpellier, France

Pierre Couteron
IRD, Umr AMAP, Montpellier, France

Index